HOWARD FLOREY

The making of a great scientist

Howard Florey

Howard Florey

THE MAKING OF
A GREAT SCIENTIST

———

BY

GWYN MACFARLANE

OXFORD UNIVERSITY PRESS

1979

Oxford University Press, Walton Street, Oxford OX2 6DP

OXFORD LONDON GLASGOW
NEW YORK TORONTO MELBOURNE WELLINGTON
KUALA LUMPUR SINGAPORE JAKARTA HONG KONG TOKYO
DELHI BOMBAY CALCUTTA MADRAS KARACHI
IBADAN NAIROBI DAR ES SALAAM CAPE TOWN

ISBN 0 19 858161 0

British Library Cataloguing in Publication Data

Macfarlane, Robert Gwyn
    Howard Florey.
    1. Florey, Howard, *Baron Florey*
    2. Pathologists—Great Britain—Biography
    616.07′092′4   R489.F/   78–41115

ISBN 0–19–858161–0

Printed in Great Britain
by Butler & Tanner Ltd.,
Frome and London

One owes consideration to the living; to the dead one owes only the truth.

VOLTAIRE (Lettres sur Œdipe)

PREFACE

I first met Florey in 1938 when I was a junior pathologist at the British Postgraduate Medical School, London, and he came there to give a lecture. How many spellbinding speakers have I heard in the course of my life, and how little do I remember of what they had to say! Florey had none of the calculated rhetoric that makes the lecturer rather than his subject memorable, but he described his research with a simple clarity that was enthralling. Even after forty years I can recall much of what he told us and the excitement of the new ideas that he conveyed so quietly. It was a style, I was to learn, that expressed his mode of work—straightforward, logical exploration with none of the emotional misdirection of pet theories or preconceived ideas, but always with that underlying, infectious excitement of true discovery.

Two years later I met him again under rather different circumstances. I was then working at the Wellcome Physiological Research Laboratories at Beckenham, and Florey came there to try to persuade the Director to take up the large-scale production of a mould that most of us had never heard of, but which Florey seemed sure would prove to be of paramount importance. He went away disappointed. Although our laboratories were equipped for such work and, in normal times, the results of his animal experiments might have decided the issue in his favour, he could not compete with the demands of war combined with the disruption caused by the blitz. In 1941, when I went to Oxford as a hospital pathologist, I saw for myself Florey's reaction to this and other rebuffs from the British pharmaceutical industry. He had refused to accept defeat, and had created a penicillin factory in his own university department. A few months later I saw, too, the vindication of this courageous venture when the first effective trial of penicillin in human patients was carried out with convincing success at the Radcliffe Infirmary, Oxford. Finally, while in the army in 1944, I was able to see the almost miraculous effect of penicillin in the treatment of battle casualties during the campaigns in Normandy and north-west Europe.

During the next twenty years I came to know Florey fairly well, and to recognize his stature as a scientist and one of the great medical benefactors of all time. Though never a member of his staff, I taught regularly in his department, attended its meetings, and contributed to the book on general pathology that he edited. I discussed my own research problems with him and soon learned that his advice was always worth taking. I found him affable, considerate, and, in his peculiarly astringent way, amusing. I worked with Dr. Ethel Florey, his first wife, in the Oxford blood transfusion service, and during her early clinical studies of the antibiotics; and Florey's sister, Dr. Hilda Gardner of Melbourne, worked for a time in my laboratory. I felt an almost personal pride, therefore, in the lengthening list of honours that Florey received, and in the achievements of his department. The climax came in 1960, when he was elected President of the Royal Society. This happened to coincide with my own membership of the Council of the Society so that I was able to see the way in which he set about the practical reforms that finally brought the Society to its magnificent new quarters in Carlton House Terrace and, more important, into the mainstream of applied science in our modern technological world.

Florey died in 1968, and a few years later I began to consider the possibility of writing a book about him. There were two main reasons. First, it would be a tribute to a man I had liked and much admired, and to his leadership of a revolution in medicine. Secondly, I had a general interest in tracing the development of scientific ideas and the qualities of character and ability that are needed to make any scientist a great one. Florey, probably the most effective medical scientist since Lister, would, I felt, be an admirable subject for such a study. I had first-hand knowledge of much of his career and most of his lines of research. Moreover, I had known almost everyone who had played a significant part in his professional life, with one important exception—to my regret I had never met Sir Charles Sherrington.

I discussed this tentative idea with Lady (Margaret) Florey, and found her sympathetic and encouraging. She gave me access to all the letters, papers, documents, notebooks, and diaries that, with great thoroughness and devotion, she had

collected and arranged to form the Florey archives at the
Royal Society. A first inspection of this mass of material was
enough to show that a full-scale biography was, for me, out
of the question. It would be a far larger work than the one
I had in mind, and even so would have to be incomplete
because, during the latter part of his life, Florey had been concerned with official matters that were still confidential. I decided, therefore, on a study of his development as a scientist
taken in detail up to the climax of the work on penicillin.
This would be followed by an epilogue briefly outlining his
subsequent career, and preceded by an historical introduction
sketching the growth of knowledge that had created the field
in which Florey was to work.

Even this limited project would entail three or four years
of research and travel, and before beginning it I explored the
prospects of eventual publication. The results were utterly
discouraging. It was made only too plain to me that a biography
by an unknown author had no attractions whatever for commercial publishers unless the subject's name—famous or in-
famous—was already a household word. I had supposed that
Florey's name would be known to everyone, but I learned to
my dismay that this was not the case. On the other hand, it
appeared that everyone did know of Fleming, for reasons that
I was later to discover, and one London publisher even suggested that it was his biography that I might write with profit.
Nothing could have made me the more determined to produce
a book on Florey, whatever the financial difficulties.

That these difficulties were overcome is due to the influence
of three good friends. Sir George Pickering approached the
Leverhulme Trust Fund on my behalf, and I was awarded
a three-year grant that covered the cost of research, the production of the draft manuscript, and part of the expenses of
a visit to Australia. In order to underwrite the cost of publication, Professor Gustav Born obtained a donation from Messrs.
Merck, Sharp, and Dohme of New Jersey, and Professor Fritz
Koller's support resulted in a most generous guarantee from
Messrs. Hoffmann-La Roche of Basel. I wish here to record
my gratitude for this financial assistance, which has made the
production of the book possible.

The list of people who have helped me with the book itself

is a long one. My first and best thanks go to Lady Florey, who has provided me with so much of the material that appears in this book and has approved its publication. Above all, she has taken a constructive interest in every stage of the writing, corrected mistakes, and made many helpful suggestions and legitimate criticisms. Though I have made most of the detailed alterations she proposed, Lady Florey and I could not always agree on matters of opinion, emphasis, or interpretation, and I take full responsibility for the views expressed in this book and also, of course, for the errors that will inevitably remain undetected until it is too late to correct them.

I wish to thank other members of Florey's family who have been helpful. His son and daughter, Dr. Charles Florey and Mrs. J. McMichael, have given me interviews and information in letters and read the manuscript. Florey's cousin, Mrs. Hilda Kinton, has written of their grandfather, Walter Florey, and his family in London. To Dr. Joan Gardner, Florey's niece in Melbourne, I owe many thanks. I had several interviews with her during my visit there, and she produced Florey family papers, documents, and photographs, from which she had allowed me to select material for publication. I also wish to record my gratitude to the late Mrs. Emmeline Brebner of Adelaide, who told me much about her sister Ethel, her family background, and her marriage to Howard Florey.

In Adelaide I received every help from Mrs. Mollie Bowen who, since she had known Howard Florey from their childhood, must be his oldest friend. Mrs. Bowen, full of life and vigour, gave me vivid accounts of the Florey family and their life in Adelaide during the early years of this century, and took me to see what remains of the Florey houses, and the places and institutions that featured in Howard's work and play until 1922. It is a pleasure to remember her kindness, humour, and hospitality. It was she who arranged useful interviews with other contemporaries of Howard and Ethel in Adelaide, in particular Dr. Alan Lamphee, Dr. K. S. Hetzel, and with Mrs. J. R. Thomson, a friend of Howard and Ethel from their student days, and in England after their marriage. To my friends Professor and Mrs. G. J. Fraenkel I owe the most generous hospitality and much practical help in Adelaide, and information on the penicillin work in Oxford.

Professor R. D. Wright, of the Howard Florey Institute of Research in Melbourne, gave me a long and interesting account of his association with Florey in Oxford and Australia, followed later by detailed letters in answer to my further questions. My last interview in Australia was with Sir Keith Hancock in Canberra, who told me of his memories of Florey from his time as a Rhodes Scholar in Oxford and of his part in the complicated birth of the Australian National University.

In Oxford itself, my sincere thanks are due to Professor Henry Harris of the Sir William Dunn School of Pathology, and to all the members of his staff who recalled for me their memories of Florey and their work with him. Professor Harris, himself an Australian, one of Florey's pupils, and the successor to his chair of pathology, gave me the most valuable information, opinions, and advice, and access to his department's records and photographs. Dr. Norman Heatley, a key figure in the epic struggle to produce penicillin, lent me his notes and diaries covering this exciting period, helped me to convey a reasonably accurate picture of the team work it involved, and corrected my mistakes and misconceptions. Professor E. P. Abraham, whose biochemical work with Chain was of crucial importance, also gave me constant help and, most generously, the material he had collected for his own admirable Royal Society Biographical Memoir on Florey. I had several interviews with the late Professor R. A. Webb, a friend of Florey's from his Cambridge days, and his colleague in Oxford for many years; also with Professor J. L. Gowans (now Secretary of the Medical Research Council), the late Professor A. D. Gardner, and with my friend Dr. Gordon Sanders, who had done much personal work with Florey, besides his participation in the penicillin team. He, too, lent me his diaries and copies of his notes. No convincing picture of Florey could have been produced without the help of Mr. J. H. D. Kent, Florey's devoted and indispensable assistant for forty-two years—a service recognized by the award of an Honorary Degree by Oxford University in 1978. No one knew Florey's character as a scientist better than Jim Kent, and no one could have been more helpful to an author trying to capture it on paper. And Mrs. Kent, who had been the Florey children's nannie, has also been of great assistance. I wish to thank Mrs. P. J. Turner, for many years Florey's departmental administrator, for her information and the use of her notes and records, and also Miss W. M. Poynton, who had been his personal secretary. Mr. S. A. Buckingham, departmental photographer, has given me much willing and skilful collaboration in the finding, identification, and reproduction of photographs.

I owe rather special thanks to Sir Ernst Chain. After beginning the penicillin work in Oxford and sharing with Florey the credit for its final success, disagreements led to his departure for Italy in 1948. Despite the memory of these strained relations, he gave me much useful information in letters and an interview, and I trust that I have been able to strike a factual balance between his view of the past and those prevailing in Oxford. Others among Florey's Oxford colleagues have also helped me. Sir Rudolph Peters, Professor J. H. Burn, and Professor E. G. T. Liddell, contemporary heads of science departments, and Sir George Pickering, Professor L. J. Witts, Mr. R. G. Macbeth, Dr. A. E. Cooke, Dr. Alice Carleton, and Dame Janet Vaughan, have all provided parts of the picture. Dr. Charles Fletcher has written to me of his part in the famous first clinical trial of penicillin, and Sir Peter Medawar of his early work with Florey in Oxford. I am most grateful to Dr. T. I. Williams, who had also worked with Florey, and who read my manuscript and gave me valuable advice and help. I have a double debt to my friends Dr. A. H. T. Robb-Smith and Dr. H. M. Sinclair. They told me much about Florey from their own experience and, being learned in the history of medicine and of Oxford, helped to repair the lamentable gaps in my own knowledge in these fields, though they must not be blamed for any that are still apparent. In the matter of history, too, I wish to thank Brigadier and Mrs. Goadby of Standlake, Oxfordshire, for the information they kindly provided from their genealogical studies of local families, which included the Floreys, and to Mr. James Florey of Standlake, a distant cousin of Howard Florey, for reminiscences of this Oxfordshire family.

I particularly wish to thank Dr. Beatrice Pullinger, now in South Africa, for several long and detailed letters giving a vivid picture of the first few years of Florey's professorship at Oxford. Both she and Dr. H. E. Harding have described his time in Sheffield. I am grateful to Sir Alan Drury and Sir Ashley Miles for telling me of their memories of Florey in Cambridge, and of their work with him, and to the late Sir David Martin, while he was Executive Secretary of the Royal Society, for his views on Florey's presidency. Mr. N. H. Robinson, Librarian at the Royal Society, and his assistants have, during three or four years, made my task of working through the Florey archives as easy as possible by their willing co-operation and practical help. I am grateful to Dr. John Morris and Mr. D. Cawthron of the Medical Research Council for copies of official records and allowing me to use their summaries of voluminous material. I wish to thank the Registrars of Adelaide, Sheffield, Cambridge, and Oxford Universities for providing information and copies of documents; the Warden of Rhodes House for allowing me to consult records; the librarians of Gonville and Caius College, Cambridge, Magdalen College and Merton College, Oxford, for providing information, and to the Librarian of Lloyd's for research on various ships mentioned in this book and their movements. The Deans of the London Hospital and Guy's Hospital, London, have also been helpful.

I am personally grateful to the Bursar of All Souls College, Oxford, for receiving and administering the grants awarded for my work, and to Mrs. B. Clapton who has done most of my typing. I wish to thank the staff of the Oxford University Press for their unfailing courtesy, patience, and efficiency since the results of my labours were delivered into their hands. Last in the list, but by no means in importance, I record my gratitude to my wife, Dr. Hilary Macfarlane, for her constant help and encouragement, and her cheerful forbearance during the several years of a major preoccupation.

R. G. M.

Wester Ross, Scotland
December 1978

ACKNOWLEDGEMENTS

The author wishes to thank the following for permission to quote from their letters: Professor J. H. Burn, Dr. Alice Carleton, Sir Ernst Chain, Dr. C. S. Elton, Dr. Charles Fletcher, Dr. H. E. Harding, Sir Peter Medawar, Sir Rudolph Peters, Dr. Beatrice Pullinger, and Professor R. D. Wright.

Quotations from unpublished letters and documents by writers now deceased have been made with the kind permission of the appropriate literary executors, as follows: Lady (Margaret) Florey, widow of the late Lord Florey and also literary executor of the late Lady (Ethel) Florey; Dr. Joan Gardner, daughter of the late Dr. Hilda Gardner and niece of the late Miss Anne Florey; Mr. E. C. Poulton, son of the late Dr. E. P. Poulton; Mrs. E. M. Flew, daughter of the late Sir Arthur Hall; Sir Patrick Dean, son of the late Professor H. R. Dean; Mr. H. J. Barnes, brother of the late Dr. A. E. Barnes; Dr. Charles Fletcher, son of the late Sir Walter Fletcher; Mrs. Violet Gardner, widow of the late Professor A. D. Gardner; Mrs. J. B. Priestley, daughter of the late Sir Frederick Gowland Hopkins; Mr. G. M. Little, nephew of the late Sir Edward Mellanby; and Miss Unity Sherrington, granddaughter of the late Sir Charles Sherrington.

The author regrets that he has been unable to trace any surviving literary executors of the late Sir Phillip Panton and the late Professor E. H. Kettle, and he trusts that his quotations from their letters will be condoned by any legal owners of copyright, should they exist unknown to him.

The Medical Research Council has authorized the quotations from official papers and letters in their records which appear in various parts of this book, and Professor R. H. Ebert has permitted a quotation from the unpublished Dunham Lecture given by Lord Florey at Harvard University.

The author acknowledges permission given by authors, their agents or literary executors, publishers or their successors, and other holders of copyright to quote from published material as follows (the quotations being identified by the relevant chapter and reference numbers as listed in the bibliography): the Royal Society, London, 1 (1), 1 (14), 7 (11), 8 (6), and 16 (6); Oxford University Press, 3 (4), 4 (1), and 9 (6); Penguin Books, Ltd., London 1 (15); Cambridge University and Mr. L. P. Wilkinson, 7 (15); the *Journal of Physiology*, Cambridge, 7 (19); *The Listener*, London, 7 (26); Messrs. Curtis Brown Ltd., London, and Dr. Ronald Hare, 8 (5); the *British Journal of Experimental Pathology*, London, 8 (2) and 8 (8); the Royal Institute of Public Health and Hygiene, London, 8 (10); the Royal Society of Medicine, London 8 (9); the *Journal of Pathology*, London, 10 (7); the *British Medical Journal*, London, 10 (22); the Royal College of Physicians, London, and Sir Ernst Chain, 11 (5) and 12 (11); the Manchester Literary and Philosophical Society, 12 (12); Messrs. Angus and Robertson (U.K.), Ltd., Brighton, 12 (25), 13 (19), and 14 (2); David Higham Associates, Ltd., London, 12 (26); the American Medical Association, Chicago, and Dr. L. A. Falk, 12 (23); *The Lancet*, London, 13 (23); *Nature*, London 15 (15); Wayland Press, Ltd., Hove, and Dr. W. H. Hughes, 15 (6); Messrs. A. D. Peters and Co. Ltd., London, 15 (7); *The Times*, London, and Mr. G. J. Romanes (grandson of the late Sir Almroth Wright), the letter quoted in Chapter 15; the National Library of Australia (Hazel de Berg Collection) has permitted quotations from a recorded interview with Lord Florey (tapes 220 and 221), 2 (7), 9 (9), 10 (1), and 12 (15); and I.P.C. Consumer Industries Press, Ltd., London, 15 (4).

The author wishes to thank all those who have provided illustrations and have given permission for their reproduction. The frontispiece and Plates 2 (b), 2 (c), 4 (a), 4 (b), and 5 are from photographs in the possession of Lady Florey; Plate 1 (a) is reproduced with the permission of Messrs. Rigby, Ltd., Adelaide. Plates 1 (b) and 2 (a) are from photographs in the possession of Dr. Joan Gardner, Melbourne. Plate 3 is reproduced by permission of the executors of the late Mrs. Emmeline Brebner, Adelaide, and Messrs. Angus and Robertson (U.K.), Ltd.; Plates 6, 7 (a), 7 (b), 8(a), and 8 (b) are reproduced with the permission of Professor Henry Harris, the Sir William Dunn School of Pathology, Oxford. Figure 1 is reproduced with the permission of Dr. N. G. Heatley; Figure 2 by permission of Mrs. Ruth Duthie, widow of the late Dr. E. S. Duthie.

CONTENTS

LIST OF PLATES

Howard Florey   *frontispiece*

I

HISTORICAL INTRODUCTION

After Lord Florey died in 1968 there was an appeal for funds to create visiting fellowships between Britain and his native Australia as a memorial to him. In launching this appeal, Lord Blackett, then President of the Royal Society, and Dr. H. C. Coombs, Chancellor of the Australian National University, wrote: 'Millions of human beings have since owed their lives or their health to treatment with penicillin and related antibiotics whose production became possible as a result of his pioneering work. The consequences for the good of mankind even today have yet to be fully realised, and Florey is rightly honoured throughout the world as Jenner, Pasteur and Lister were honoured before him.'[1]

Many people may be surprised by this assessment because, though they will know a good deal about the three pioneers with whom Florey is compared, they will be unfamiliar with Florey's work or perhaps even with his name. During his lifetime he did receive the highest honours that society can give: a life peerage; the Order of Merit; and medals, prizes, awards, and honorary degrees from universities and scientific bodies all over the world. And he occupied during the latter part of his life important posts, becoming President of the Royal Society, Chancellor of the Australian National University (which he helped to found), and Provost of The Queen's College, Oxford. Florey brought to these offices the same qualities of vitality, inspiration, and sense of purpose that had made him a great and highly effective scientist. As an experimental pathologist and as the founder of one of the best schools of experimental pathology in the world, his work would probably have merited these honours, quite apart from his contribution to the use of penicillin. But penicillin therapy was, in terms of effort and organized research, his supreme achievement and certainly one of the greatest medical advances of all time. Why, then, has Florey not become a popular hero in the history of the long war against bacterial disease, when it was his work that ensured its most decisive victory? The answer is, of course, that the world has already chosen its hero for this role—Sir Alexander Fleming—and its deep gratitude for the gift of the antibiotics has found a single focal point. Those who know the facts of Florey's work have been puzzled and distressed by what seems to them to be a misdirection of credit. The facts themselves have been ignored or distorted in the popular version of the 'penicillin story', and it is one of the purposes of this book to set the record straight and to trace the elements of personality and circumstance that have led to the creation of a myth.

But its main purpose is to tell the story of Florey's early career and to follow the lines of thought and experiment that led him, not only to penicillin, but to other researches that are still yielding results of almost comparable importance in the hands of his pupils. In such a story the man himself must come to life as an actual person, whose feelings and character can be understood and whose efforts can be appreciated. Here there are two difficulties. The first lies in Florey's own character; the second is the problem of making the work of any scientist intelligible and therefore interesting to the general reader.

Florey was a man of deep reserve. His superficial manner was friendly, direct, humorous in what has been called a mordant way, and he had the ability to impress his seniors and to inspire colleagues and pupils by his imaginative drive and his immense capacity for hard work. But he had few, if any, close friends, and to none of them did he reveal personal feelings of any depth. Most, therefore, were quite unable to fathom the sources of his energy and dedication, and a general feeling was expressed by one of his close colleagues when he remarked, 'I never knew what made Florey tick.'

Luckily, Florey did reveal his deeper feelings—his interests and ambitions, his emotional isolation, and his attitude to the world in general—in the letters he wrote from England and America to Ethel Reed in Adelaide between 1922 and 1926. There are over 150 of these letters, which were carefully preserved by Ethel, and they contain factual details of his work and travels during this most formative time in his scientific career. Florey can therefore tell much of his own story in four of the chapters that follow, and set the stage for later activities that would, without this clarification, have seemed obscure.

The second difficulty is less readily resolved. Interest in the lives of great men is naturally related to the ease with which their achievements can be understood. Explorers, generals, artists, or politicians, for example, can be judged by familiar standards. Yet, though most eminent scientists are, as people, as interesting as anyone else who makes history, and their achievements can be more significant for humanity than those of any statesman, they seldom stir the public imagination. Partly this is due to the fact that most working scientists not only shun the limelight but avoid appearing at all on the stage of public affairs. Publicity is not their business and they regard it as distracting, or even as destructive of their work. Partly, too, it is the very nature of the work itself that discourages the layman's interest, however important its results may be. Progress in nuclear physics, for example, may well determine the ultimate survival or extinction of our species, but few people would strive to understand its technicalities on that account alone. Biology and medicine are less forbidding because they are, so to speak, nearer home. The working of living things and the struggle to defeat the diseases that afflict man and animals have a more immediate interest.

Scientific research has points in common with moun-taineering. There is the urge to explore, to surmount diffi-culties, to discover new peaks and to be the first to conquer them. There is the dependence on teachers and guides, on equipment, on teamwork and planning, and on the climate— for the scientist, the climate of thought and knowledge. Unlike mountaineers, scientists seldom begin at the bottom of their problem; they start from the point achieved by their prede-cessors. Florey's work can be properly appreciated only if it is seen in its historical context, against the background of con-temporary thought and knowledge in his field that had already been established.

The comparison of Florey with Jenner, Pasteur, and Lister is, in this context, an apt one because the work of these three men forms a logical sequence with his own that spans, in the course of about 150 years, the gulf between almost total therapeutic helplessness and the virtual defeat of most of the important bacterial diseases. Though the stories of these three earlier pioneers will probably be familiar to many people, a brief account of them will form a useful introduction to Florey's own career. And since Florey worked for most of his life in Oxford, and was greatly influenced by his teachers and colleagues there, and by the general attitude of the University to science, this chapter will include a short description of the stormy events that led to the eventual creation of some of the world's best science departments in the midst of an almost medieval scholasticism.

When Edward Jenner became a house-pupil of John Hunter in London, in 1770, the 'new philosophy' of the scientific experimental method had been gaining ground for nearly 150 years. William Harvey's *De Motu Cordis*, published in 1628, had been the virtual beginning of experimental physiology. The foundation of the Royal Society in 1660 was probably the most important event for science in Britain, since it became the focal point for the 'new philosophers' who had previously met informally at Gresham College and, during the Civil War and the Commonwealth, at Oxford. Its full title 'The Royal Society of London for Improving Natural Knowledge' originally included the words 'by the Authority of Experiments' and its motto *Nullius in verba* reflected a challenge to the dogmatic authority of religious and classical teachings that had halted scientific progress for two millennia.[2]

Clinical medicine, however, had not profited to any practical extent from this emancipated outlook. The classical texts remained deeply revered. The physicians were learned men and good observers of signs and symptoms who gave Latin names to recognized categories of disease, a fair indication of the probable prognosis, and comfort and confidence to their patients. But apart from alleviating some of the more distressing symptoms by the drugs then available, there was almost nothing they could do to cure the patient of any serious complaint. The prevalent practice of blood-letting, which persisted into the nineteenth century, is an example of the misplaced faith in the power of human reason held by most educated people, and the conviction that because an idea is reasonable it is also true. The practice of bleeding was based on a Greek theory of disease that was quite reasonable and also, as it happens, quite wrong.

John Hunter was not only a great anatomist and surgeon, he was also an acute observer of nature and a fervent believer in the necessity to experiment. 'But why think?' he once wrote in reply to a speculation by Jenner, 'Why not try the experiment?'[3] Jenner, already interested in natural history, responded to this sort of teaching, and he was employed by Joseph Banks to arrange and describe the specimens he had collected during Captain Cook's voyage to the South Pacific. Banks, a wealthy man and an eminent botanist, had financed the fitting-out of *Endeavour*, and had been the chief naturalist on board. He was pleased with Jenner's work, and suggested that he should be the naturalist during Cook's projected second voyage, but Jenner declined, preferring to take up medical practice in his native Berkeley, in Gloucestershire, where his father was a clergyman. There he remained for the next fifteen years; a good doctor, a fair musician and poet, and a popular figure in the sort of provincial society portrayed by Jane Austen. But he retained an active interest in natural history, and it was his careful and original study of the breeding habits of the cuckoo that brought him a measure of scientific fame. He presented a paper on the subject to the Royal Society in 1787, where it was well received.[4] Sir Joseph Banks was then President (he held the office for forty-two years) and Jenner was elected a Fellow in 1788.

Meanwhile the plagues, poxes, and fevers that had afflicted the general population in epidemic or endemic forms for centuries were as prevalent as ever, and the average expectation of life was about thirty-six years. Smallpox alone killed one child in ten under the age of four, and nearly half the adult population bore its facial scars. There was, however, a practical attempt to reduce its dangers, based on the observation that one attack of smallpox seemed to provide complete protection from a second one in the same person. Thus there was some justification for the hazardous practice of deliberately infecting people who had escaped the natural disease by inoculating them with fluid from smallpox blisters. The patient could at least choose to have his smallpox at some convenient

time and in congenial company; his induced attack was usually milder than a natural one; and having survived it he was
then free of an otherwise constant threat. Jenner, like most
doctors, was often called upon to perform such inoculations,
and it was in his rural practice that he made a most important
discovery, which can best be described in his own words.

Among those whom in the country I was frequently called upon to inoculate many resisted every effort to give them the smallpox. These patients
I found had undergone a disease they called the Cow Pox contracted by
milking cows affected by a peculiar eruption on their teats. On enquiry,
it appeared that it had been known among the dairies from time immemorial
and that a vague opinion prevailed that it was preventive of the smallpox
... I was struck with the idea that it might be practicable to propagate
disease by inoculation, after the manner of the small-pox, first from the
cow and finally from one human being to another ... The first experiment
[14 May 1796] was made upon a lad of the name of Phipps in whose arm
a little vaccine virus was inserted taken from the hand of a young woman
who had been accidentally infected by a cow. Notwithstanding the resemblance which the pustule thus excited on the boy's arm bore to variolous
infection, yet as the indisposition attending it was barely perceptible, I
could scarce persuade myself the patient was secure from the smallpox.
However, on being inoculated some months afterwards, it proved that he
was secure.[5]

Jenner repeated this admirable experiment on other
patients. He was careful to exclude a previous attack of smallpox as a cause of their immunity, and to show that the material
he used to test this immunity did produce smallpox in un-
vaccinated patients. In fact, he properly controlled his experiments. In his account he used two words that have since
passed into common medical language but with a changed
meaning. 'Vaccine' was, by his own definition, the infectious
material from cowpox (which he called *variola vaccinea*—
smallpox of the cow), but the meaning was widened by Pasteur. The word 'virus' was used by Jenner in the general sense
of a noxious agent. He was not, of course, aware of the existence of the submicroscopic organisms now called viruses.

Naturally excited by his discovery, Jenner wrote a report
which he gave to Sir Joseph Banks for presentation to the
Royal Society in 1797. But he was soon to learn that it is sometimes easier to make a great discovery than to persuade one's
colleagues to accept it. Banks read his paper and consulted
various supposed experts, but it seems that 'the perusal of his

cases and experiments produced no conviction whatever' and Jenner was advised not to injure the reputation he had gained from his former papers by presenting this one to the Royal Society.[6] He was told, in effect, to return to his studies of the cuckoo.

Jenner, without official scientific support but certain of the importance of his work, set out to fight his own battle. He published his paper privately, continued to vaccinate patients himself, and began to convince a highly sceptical medical profession. Public interest was aroused and the inevitable sides were taken. Jenner was hailed as a saviour and denounced as a charlatan. Violent opposition flared up, religious feeling somehow managed to be outraged, and the transmission of an animal disease to man was felt to be degrading to human dignity. There was much ridicule, too, expressed in the cartoons showing Jenner's patients growing horns or cow's heads. But, from 1800 onwards, the solid fact of the value of vaccination became so obvious that all except a few fanatics were gradually silenced.

Jenner lived to see vaccination made compulsory in several European countries and the virtual elimination of smallpox there. But he was acclaimed more abroad than at home, and he did not become a public figure. He died in 1823. His work illustrates the classic path of the scientific pioneer. He began with an empirical observation, recognized its possible implications, confirmed its importance by experiment, and then battled against every sort of difficulty to gain its general acceptance. But vaccination remained the only example of an efficient prophylactic against an infectious disease until Pasteur opened up the whole field of practical immunology.

Louis Pasteur was born in 1822 at Dôle, in the Jura, the son of a tanner. His school career was disappointing, and though he obtained a diploma in science at Besançon in 1842 his chemistry was graded as mediocre. Despite this, he was so inspired by the lectures of J. B. Dumas in Paris that he decided to make chemistry his career, and worked as a laboratory assistant while studying for his doctorate. Soon after he obtained his degree he accomplished a remarkable piece of chemical research on tartaric acid, a by-product of the wine industry. Pasteur showed that the commercial acid was a mixture in which one component rotated polarized light to the left while another rotated it to the right, and he conceived the novel idea that there were two mirror-image molecules, right-handed and left-handed respectively. He thus discovered the field of stereochemistry, one to which he might well have devoted the rest of his life. But his work on tartaric acid had led him to an even wider problem, the nature of fermentation.

In 1854 he was appointed Professor of Chemistry at Lille, and his interest in fermentation found a practical application in the brewing industry. On a commercial scale the process often went wrong and the local brewers appealed to him to solve their problems. This he proceeded to do and, as a result, discovered a whole new world: 'the world', as he described it, 'of the infinitely small'. To the orthodox chemists of that time any suggestion that fermentation could be due to living organisms would have seemed a return to the medieval superstition of 'vitalism'. Yet microscopists from Leeuwenhoek onwards had seen that yeast consists of microscopic cells capable of budding and multiplying, and Pasteur was prepared to use the microscope to study the troubles of the brewing industry. He found that these were associated with visible abnormalities of the yeast cells or the presence of other organisms. By simple experiments he proved that yeast is, indeed, alive and that during its growth it breaks down sugar into carbon dioxide and alcohol. He then moved from the particular to the general, showing that alcoholic fermentation was only one example of chemical decomposition produced during the growth of many different moulds and microscopic organisms, some of which produced the acids causing the souring of beer, wine, and milk.

Pasteur found that fermentation could be prevented by heating the liquid to a temperature just high enough to kill organisms and without changing its chemical nature. This process, now familiar as 'pasteurization', allowed a study of the origin of these micro-organisms. Sterilized milk became sour when exposed to air, but not if the air had been heated or filtered. Pasteur concluded that the lower atmosphere contains a host of organisms, but in a famous experiment he showed that the upper atmosphere is free from them, since fermentable liquids remained unchanged when exposed to the air on an Alpine peak. Such experiments captured the public imagination. Pasteur's success owes a little to his showmanship; he had something of the Gallic flourish that marked the engineering triumphs of his contemporary, Isambard Brunel.

Much of this work on fermentation had been done in Paris at the École Normale where Pasteur had been appointed Director of Scientific Studies in 1857. Having shown that the diseases of beer and wine were due to unwanted micro-organisms he began to speculate on a possibly similar origin for the diseases of animals and man. Such an idea was not new, but it had never before been given serious thought. Thus, when Pasteur was urged by Dumas, his old teacher, to investigate pébrine, a disease of silkworms, he was prepared to search for a living cause. And this line of approach yielded success where established experts in the field had failed to cure a disease that was threatening a vast industry in the south of France with ruin. Pasteur was able to discover that pébrine was due to a blood parasite in the mature moths. The infected stock could therefore be recognized and eliminated, and the disease stamped out.

In 1868, Pasteur's work was interrupted by a stroke. Then came the war of 1870 and the horrors of the siege of Paris, which left an indelible impression on his mind. In 1877, he began a study of chicken cholera, the disease that was causing serious loss to poultry farmers. He isolated the organism now known as *Bacillus avisepticus*, and showed by inoculating healthy birds with living cultures that it caused the disease. Then there occurred one of those 'fortunate accidents' that so often determine some great scientific advance. On this particular occasion Pasteur happened to use cultures that had been left, by chance, in the laboratory during the summer vacation. These cultures failed to produce the disease in the birds inoculated, and the natural conclusion was that they had become useless. Fresh cultures were made, but these, though they infected untreated chickens, failed to infect the birds already inoculated with the 'useless' cultures.[7]

In one of his most famous aphorisms Pasteur said that 'in the field of investigation chance favours only the prepared

mind'. On this occasion chance had favoured a mind familiar
with the work of Jenner and prepared to recognize the true
significance of an unexpected result. The old cultures had lost
their virulence, but not their power to immunize. Pasteur was
exultant. The birds, he said, had been 'vaccinated'—by accident. Jenner's vaccine was successful because the organism
of cowpox is naturally of low virulence but able to protect
against the far more deadly smallpox—a rare piece of good
fortune. But Pasteur saw that virulence might be reduced arti-
ficially without losing the protection, a principle that might
have wide application. Having produced attenuated chicken
cholera cultures for prophylactic inoculation, which he called
a 'vaccine' in honour of Jenner, he set out to make a similar
vaccine against anthrax. The bacillus causing this disease had
been isolated by the German bacteriologist Robert Koch, and
Pasteur, by culturing it under a variety of adverse conditions,
reduced its virulence. He then staged a demonstration at
Pouilly-le-Fort in which sheep, cattle, and goats vaccinated
by his method all survived an injection of virulent anthrax
bacilli that killed all of a similar group of untreated animals.

In August 1881 Pasteur attended the Seventh International
Medical Congress at King's College, London. It was an historic meeting between three men who were changing medical
science: Pasteur himself; Lister, who had enthusiastically
adopted his 'germ theory' of disease; and Robert Koch, whose
new method for growing bacteria on a solid medium allowed
the recognition and separation of individual colonies and their
subsequent propagation in pure culture.[8] Pasteur had as an
interpreter in these conversations a young Australian named
Watson, who was working with him in Paris. Watson had
studied medicine in Germany because there had been a price
on his head in Britain for suspected piracy.[9] Since he later
became Professor of Anatomy at Adelaide University and one
of Howard Florey's influential teachers, he will reappear later
in this narrative.

Pasteur's final research proved to be his most exacting trial
and his greatest triumph because, for the first time, he was
dealing with human patients. The subject was rabies, then
(as now) endemic throughout Europe. Pasteur was unable to
discover the causative organism (it is, of course, a virus), but

the infectious nature of rabies was obvious and he felt that the vaccination principle would apply to it. But the experimental approach was difficult. First, the time between inoculation and the appearance of symptoms might be weeks or even months, so that experiments became impossibly protracted. Secondly, the work was highly dangerous. The third difficulty is an emotional one. During the latter half of the last century a public outcry arose against animal experiments, and Pasteur's use of dogs aroused a storm of protest. Indeed, his work posed an ethical dilemma that is still unsolved, but he himself had no doubts and he was not deflected from his experiments.

The first practical necessity was to shorten the incubation time of induced rabies. Pasteur inferred that the infection did not produce the disease until it had invaded the nervous system. He therefore gave intracerebral inoculations in rabbits, and reduced the incubation period to six or seven days. Next, he must attempt to produce an attenuated vaccine. But the organism could not be grown in culture; it would propagate (like most viruses) in living tissues only. So rabbits became the living culture medium for the rabies organism, and a vaccine prepared from infected rabbit spinal cord was attenuated by drying for different periods. By 1884 Pasteur was able to demonstrate that a course of vaccine injections could protect dogs from rabies. He also suggested that, because of the long incubation time, prompt vaccination of a human being after infection might prevent the disease.

The opportunity to test this idea came on 6 July 1885. The patient was a nine-year-old boy, Joseph Meister, from Alsace, who had been bitten by a rabid dog three days before. There were no ill effects from the daily injections and the boy returned to Alsace. The next few months were an anxious period, but they passed with no sign of symptoms. Confirmation came from a second case in October. A fourteen-year-old shepherd boy named Jupille had grappled with a mad dog to prevent it from attacking five young children and had been badly bitten on both hands. Six days had passed, but again the treatment was completely effective. There was a limit to this success, however. In the cases that followed it became clear that the longer the time between infection and vaccination the less was the chance of preventing the disease.

During the next three years hundreds of patients from Europe and America came to Pasteur for anti-rabies treatment. Provided the time-limits had not been exceeded, it was almost always effective. In 1888 the Pasteur Institute was founded in Paris, and the first two patients treated by his method had their place in it. In the forecourt there is a statue that commemorates the bravery of the shepherd boy Jupille. Joseph Meister, when he grew up, became the gatekeeper at the Institute, and remained there until the Nazis occupied Paris in 1940. Then, rather than admit them to the crypt where Pasteur lies buried, he took his own life.[10]

Pasteur, who died in 1895, had lived to see almost all his work come to triumphant fruition. Almost every line of research had begun with a practical problem in commerce, agriculture, or medicine, but his supreme genius lay in his breadth of vision and imagination. Behind each practical achievement he could discern a new general principle. He was the 'applied scientist' *par excellence* and yet he may be said to have founded the 'pure sciences' of stereochemistry, enzymology, medical bacteriology, and immunology.

Joseph Lister was five years younger than Pasteur. His father was Joseph Jackson Lister, a Fellow of the Royal Society, who had developed the achromatic lens for microscopes. The Listers were a Quaker family, and when Joseph decided to study medicine he did so at University College London. There he was taught by William Sharpey, one of the world's great physiologists, and by Wharton Jones, a pioneer in the microscopic study of the capillary blood vessels and their reactions in inflammation. (Since the study of these capillaries was to be a major interest not only for Lister but for Howard Florey, it might be explained that they are minute vessels that permeate almost every tissue in the body. They are literally of vital importance since their thin walls allow essential chemical exchanges to occur between the blood and every living tissue cell.) Lister qualified in 1852 and became house surgeon to Professor Syme at the Royal Infirmary in Edinburgh. He too had done original work on inflammation and he encouraged Lister to continue it. This pleasant association was permanently cemented when Lister married his professor's daughter, a union that brought him a lifetime of happiness and certainly did nothing to hamper his career.

After his house appointment, Lister became Assistant Surgeon at the Royal Infirmary, where he studied inflammation as an approach to the problem of suppuration and gangrene that so often followed surgical operations. He observed the changes in the blood flow in the microscopic vessels of the bat's wing and frog's footweb, and saw that local injury caused a dilatation of these vessels, a slowing of blood flow, and a curious adhesion of the white blood cells to the vessel wall. The importance and technical elegance of this work were at once recognized, and he was elected a Fellow of the Royal Society in 1859 at the age of thirty-two. A year later he became Professor of Surgery at Glasgow University.

Physiological studies, which included an admirable study of blood coagulation, were only a part-time occupation for Lister, the surgeon. Anaesthetics by then allowed long and complicated operations, but the high post-operative mortality outweighed any technical advantage. However neatly major surgery was carried out by these excellent anatomists, the end-result in over 40 per cent of cases was death from 'hospital gangrene' or 'mortification' of the wound. Abdominal surgery was never willingly undertaken because fatal peritonitis was the usual consequence. It seems strange now that the idea of contagion as a cause of wound sepsis did not, apparently, even occur to surgeons who could see that it was a far greater danger in crowded hospital wards than in the patient's own home. The general opinion was that wound sepsis was local tissue death caused by the original injury. It might be reduced by improving operative technique, but never eliminated. Lister refused to accept this dismal conclusion. His observations on inflammation did not support the view that simple injury causes suppuration. He was inclined to believe that it was due to a decomposition of clotted blood brought about by exposure to air.[11]

In 1865, Dr. Thomas Anderson told Lister about Pasteur's experiments which showed the importance of micro-organisms in the decomposition of organic matter.[12] Lister at once saw the possibility that it was not the air but airborne micro-organisms that caused decomposition in wounds, and he began the search for some chemical that would be poisonous to bacteria but harmless to the patient. He tested a wide variety of substances, becoming an adept practical bacteriologist in the process. He was looking, in fact, for the ideal 'antiseptic'—a word that he had coined—but this ideal eluded him. It also eluded the countless searchers who followed him until a change of direction revealed the sulphonamides more than half a century later.

But Lister did not wait for perfection. Carbolic acid proved to be lethal for microbes at concentrations just tolerable by living tissues, and he chose it as a beginning. At first he applied it locally, then he introduced his celebrated carbolic spray for use in the operating theatre, which must have been almost as unpleasant for the human occupants of the room as for the microbes. But it became widely used and a greatly reduced incidence of post-operative sepsis followed. This apparent vindication of the ridiculed 'germ theory' created a new attitude of mind among enlightened surgeons that was probably more important than the carbolic acid, since an awareness of dangerous, invisible organisms focused attention on general cleanliness. During the next twenty years the emphasis shifted from the idea of antisepsis to that of asepsis: the exclusion from the operation site of organisms by heat sterilization of all the instruments, dressings, and clothing that might carry them there.

In 1869 Lister had stepped into the shoes of his father-in-law, and had become Professor of Clinical Surgery in Edinburgh. There, in his efforts to improve antiseptics, the field of his research remained the study of bacteriology. But he also became interested in the natural defences of the body against infection. He observed that the normal skin can inhibit the growth of bacteria, and that the white cells of the blood can ingest them. Thus he obtained a glimpse of the mechanism of immunity that can be stimulated by previous bacterial infection or by artificial vaccination. The idea of natural bacterial inhibition was already in his mind, therefore, when he observed that some organisms failed to grow in culture media contaminated by the mould *Penicillium glaucum*. His notebooks show that he began experiments on the antibacterial effect of this mould on 28 November 1871, so that he was probably the first investigator in a field that, some sixty years later, was to lead others to one of the most important advances in medical history. Lister's experiments were not conclusive, and he did not publish them, or follow this particular research much further. But he was sufficiently impressed by the antibacterial power of the mould to suggest to his brother that it might be used as an antiseptic. He used it himself in 1884 as a local application for a gluteal abscess that had resisted all other treatment. The result seems to have been so good that the patient demanded to know what the remedy was and Lister's registrar wrote the word 'penicillium' in her scrapbook.[13]

Lister had come to London in 1877 as Professor of Surgery at King's College, and in 1883 he was created a baronet. He had probably done more than anyone to promote the acceptance of Pasteur's 'germ theory' of disease. Robert Koch laid the foundations of scientific bacteriology, and immunology was also becoming a science. These advances inspired a move in 1885 to set up in Britain a laboratory modelled on the Pasteur Institute, and after six years spent in overcoming bureaucratic and financial difficulties, and the violent opposition of antivivisectionists, the 'British Institute of Preventive Medicine' was founded in 1891. It was a humble establishment by comparison with its Parisian model, being a converted private house in Great Russell Street, in which the bedrooms had become laboratories and the kitchen the animal room. Armand Ruffer, a protégé of Pasteur's, was the first Director.[8]

Almost from the beginning, developments in immunology made the animal quarters at Great Russell Street inadequate. From showing that germs cause disease bacteriologists at the Pasteur Institute and in Berlin had gone on to discover how they do so. In some diseases the damage is done by soluble poisons—toxins—produced by the invading bacteria. The often fatal symptoms of diphtheria and tetanus, for example, were found to be due to the action of such toxins. These toxins, which could be filtered from the bacterial cultures, could also be used as immunizing agents. Small, increasing doses given to animals led to the production in their blood of specific antibodies (antitoxins), proteins that would combine with and neutralize that particular toxin, so that the animal became immune to the disease. Moreover, blood serum taken from these immunized animals and injected into untreated animals conferred on them a 'passive' immunity, so that for a week or two they were also protected. The application to human disease was obvious, and the first such use of antitoxin was on Christmas Day 1891, when Dr. Geissler in Berlin used serum from an immunized horse in a case of diphtheria.

Thus the new British Institute almost at once became involved in the production of antitoxins as well as vaccines. But a converted kitchen does not provide good stabling for horses. The difficulty was solved by boarding them out at the Brown Institute in Wandsworth Road. This establishment is now almost forgotten but deserves a place in medical history. It was set up by London University in 1871 with a £20,000 bequest for 'investigating and curing maladies of quadrupeds and birds useful to man'. It was the first pathology research laboratory in Britain, and it had a series of distinguished men as its superintendents, including Burdon-Sanderson and Charles Sherrington, both of whom later became professors of physiology at Oxford. It was to Sherrington that Ruffer turned when he decided to produce diphtheria antitoxin, and he agreed to house a piebald pony named Tom at the Brown Institute for the purpose. Sherrington has a double claim on our interest. In later years he was the man with the greatest influence on the career of Howard Florey, and in the present context he was the first to use diphtheria antitoxin in Britain—the patient being his own nephew. The story is told in his own words, because it gives a vivid picture of the conditions of research in those days:

Ruffer and I had been injecting the horse—our *first* horse—only a short time ... We had from it a serum partly effective in guinea-pigs. Then, on a Saturday evening, about seven o'clock, came a bolt from the blue. A wire from my brother-in-law in Sussex. 'George has diphtheria. Can you come?' George, a boy of seven, was the only child. The house, an old Georgian house, three miles out of Lewes, set back in a combe under a chalk down. There was no train that night. I did not at first give thought to the horse and, when I did, regretfully supposed it could not yet be ripe for use. However, I took a cab to find Ruffer. No telephone or taxi in those days—'93 or '94. Ruffer was dining out. I pursued him and got a word with him. He said 'By all means you can use the horse, but it is not yet ripe for trial.' Then, by lantern light at 'The Brown' I bled the horse into a two-litre flask, duly sterilized and plugged with wool. I left the blood on ice for it to settle. After sterilizing smaller flasks, and pipettes and some needle-syringes I drove home to return at midnight to decant the serum, etc.

By the Sunday morning train I reached Lewes. Dr. Fawsett of Lewes— he had a brother on the staff at Guy's—was waiting in a dog-cart at the station. I joined him carrying my awkward package of flasks, etc. He said nothing, as I packed them in, but, when I had climbed up beside him he looked down and said 'You can do what you like with the boy. He will not be alive at tea-time.' We drove out to the old house; a bright frosty morning. Tragedy was over the place. The servants scared and silent. The boy was very weak, breathing with difficulty; he did not seem to know me. Fawsett and I injected the serum. The syringes were small, and we emptied them time and again. The Doctor left. I sat with the boy. Early in the afternoon the boy seemed to me clearly better. At three o'clock I sent a messenger to the Doctor to say so. Thenceforward progress was uninterrupted. On Tuesday I returned to London, and sought out Ruffer. His reaction was that we must tell Lister about it. The great surgeon (he was not Lord Lister then) had visitors, some Continental surgeons, to dinner—'you must tell my guests about it' he said and insisted—so we told them in the drawing room at Park Crescent. The boy had severe paralysis for a time. He grew to be six feet tall and had a commission in the First World War.[14]

Other successes followed, and the news of a cure for diphtheria was soon in all the newspapers. It was clear that the British Institute must expand. The Duke of Westminster provided a site on the Chelsea Embankment for a new building, the foundations of which were laid in 1894. But there was powerful and emotional opposition, largely organized by antivivisectionists, which involved petitions to the Home Secretary, mass parades, banners, and brass bands. Despite these protests a part of the new building was ready for occupation by 1898. Lister had proposed that it should be named the Jenner Institute, but when it was discovered that a small commercial laboratory already used this name, he was persuaded to allow his own to be applied to it.

Lister died in 1912. Like Jenner and Pasteur he had lived to see the success of his work. Though the ideal antiseptic still remained elusive, he had transformed surgery by the recognition and exclusion of bacteria. He was President of the Royal Society from 1895 to 1900, President of the British Association, and Chairman of the Lister Institute until 1904, when he became its President. He was raised to the peerage in 1896. In the Coronation Honours of 1902 he was the first recipient of the new Order of Merit. It could not have been a more appropriate award. If it had not been for Lister's work there would probably have been no coronation of Edward VII. In the pre-Listerian era, no surgeon would have dared to operate in a case of acute appendicitis.

As already mentioned, Lister made his researches in the Universities of London, Edinburgh, and Glasgow. These were, in fact, the only academic institutions in Britain where provision for scientific medicine existed during the first twenty years of his career. In the sweeping advances that were taking place in science, Oxford and Cambridge took almost no part until the last few years of the nineteenth century. Then, though late starters in the field, they came into the lead in the course of two or three decades and now their science departments rank with the best in the world.

The Faculty of Medicine in Oxford dates back to about 1330. From 1546 there was a Regius Professor of Medicine and several academic medical posts, but there were few candidates (sometimes one or two a year) who took the degree, and even these had usually studied the practice of medicine in London, Scotland, or on the Continent. During the Civil War and the period of the Commonwealth Oxford became for a time a centre for scientific thought. William Harvey's brief Wardenship of Merton College arose from his Royalist sympathies. The victory of Parliament resulted in the appointment of John Wilkins as Warden of Wadham and the arrival of Wallis, Goddard, and Petty.[2] These men had previously held meetings at Gresham College, London, to discuss and practice science, and they continued these activities at Wadham, forming the Oxford Philosophical Society. Boyle and Hooke were attracted to Oxford to join this group, and it no doubt stimulated Christopher Wren, a Fellow of All Souls, in his experiments in physics and in his pioneering work on blood transfusion with Dr. Lower, an Oxford physician.

With the Restoration, this brief encounter with science came to an end for Oxford. Most of the members of Dr. Wilkins's group returned to London, where they were soon to form the Royal Society. During the next two centuries Oxford declined as a great centre of learning. The colleges, with an autonomy unique to Oxford and Cambridge, grew richer from the Industrial Revolution but poorer in academic talent. In the world outside, science was challenging dogma and providing the basis for undreamed-of advances in technology. Oxford maintained a medieval isolation until about the middle of the nineteenth century, when it was forced to face the realities of change.

Until 1836 Oxford and Cambridge were the only English universities, and their insistence on 'religious tests' excluded nonconformists. From the point of view of science this proved to be no disadvantage. Indeed, Hill wrote: 'It was a great piece of good fortune for England that after 1660 the non-conformist middle class was excluded from Oxford and Cambridge where they would have learned to despise science.'[15] Many able young men had no choice but to go to Scottish or Continental universities, and the increasing demand for an English alternative prompted the founding of University College London in 1826 and of the University of London ten years later. From the first it offered practical instruction in basic medical science. The growing-point was physiology, the study of the working of the body as opposed to the mere description of its structure, and its most active exponent was William Sharpey, Professor of Anatomy and Physiology at University College. Of Sharpey's pupils, John Burdon-Sanderson (a Plymouth Brother) and Michael Foster (a Baptist) became the first professors of physiology at Oxford and Cambridge respectively.

The first decisive moves to improve science in Oxford were made by Henry Acland.[16] He had taken his Oxford B.A. from Christ Church and been elected a Fellow of All Souls when, in 1840, he decided to study medicine and went to St. George's Hospital, London. He was a young man of strange contradictions. He suffered—like some other eminent Victorians—from lifelong nervous troubles. But this apparent frailty concealed reserves of energy that allowed him to drive himself, and others, beyond the normal limits of endurance. He was profoundly religious and he developed an evangelical zeal for science as a revelation of the divine creation. In 1845, while he was completing his medical studies in Edinburgh, he was offered the post of Dr. Lee's Reader in Anatomy at Christ Church. After many doubts, he accepted. London and Edinburgh had revealed to him the scientific vacuum in Oxford, and he came to see this appointment as a call to spread the gospel.

Science in Oxford was indeed still embryonic, though various benefactors had tried to develop it. Dr. John Radcliffe had left his considerable fortune to advance science and medicine in Oxford, which thus gained the Radcliffe Camera and Science Library, the Radcliffe Observatory, and the Radcliffe Infirmary:[17] fine institutions housed in elegant buildings that did little to change the academic indifference to science. Elias Ashmole's bequest had produced a small museum and a natural history collection. Dr. Lee left estates to Christ Church to pay a Reader in Anatomy, and Dr. Aldrich gave about £130 a year each for chairs in chemistry, anatomy, geometry, botany, and natural philosophy. But the holders of these posts had few—if any—students, since undergraduates were not encouraged to divert their attention from their classical studies. As a result, science appointments were mainly nominal and several were often held at the same time by the same man. Dr. Kidd, for example, when he gave up his readership in anatomy to Acland and his chair of chemistry to Dr. Daubeny (who already held two chairs and a lectureship), remained Regius Professor of Medicine, Professor of Anatomy, and Professor of Clinical Instruction. What little had been done for the practice of science in Oxford had been done by individual colleges. Christ Church had provided two rooms for their Reader in Anatomy, and Magdalen had a laboratory for the chemistry and botany taught by Daubeny.

On taking up his new post Acland found himself lecturing to an audience of five or six at most. But if his lecture room was uncrowded, the room allotted as a museum was not. He had brought by sea from Edinburgh fourteen large crates of specimens collected by himself and, after difficulties with the Port of London Excise because many of them were preserved in whisky, he found the museum totally inadequate even for storage. He therefore appealed to Dr. Pusey, a canon of Christ Church and an old friend, who gave him space in his stable.

This displeased the Christ Church coachmen, who complained to Canon Faussett. Faussett ordered Acland to remove his bones and skeletons and, when he refused, told the stablemen to throw them into the street. Here the dogs of Oxford completed the havoc by running off with the choicer pieces. The tail of a giraffe was totally lost and the animal now in the Oxford Museum still has a false tail of plaster.

Despite such petty obstructionism Acland persisted. He had a microscope, a rarity in those days, and his small class came to include some of his senior colleagues. One of these—Dr. Kidd, no less—on seeing a morphological preparation through the microscope exclaimed, 'I don't believe it!' adding, on reflection, 'If it were true, then I do not believe that God means us to know such things'—a reaction that neatly sums up the contemporary attitude to science.

During the next two years Acland started a campaign. He had conceived a great idea, the creation in Oxford of a museum that would house the natural history collections and also laboratories. It was an ambitious dream for a young man with, as yet, no professional distinction and the certain opposition of seasoned university politicians. But he had advantages: an almost irresistible enthusiasm; a popularity which he could exploit quite shamelessly; a genuine piety that disarmed the clergymen who formed the most powerful University faction; and, finally, he was well-born, with friends in the Government. His first move was to prepare a memorandum and to persuade his colleagues to sign it. But the Reverend Dr. Buckland, Professor of Geology, refused to sign because he said that it was utterly hopeless to try to instil an interest in natural history in Oxford. Buckland, a fellow inmate of Christ Church, was the most influential of the Oxford scientists, and his defection was disastrous. He was, in any case, soon to quit the scene. 'Buckland is gone to Italy,' said Dr. Gaisford, Dean of Christ Church. 'Thank God we shall have no more of this geology.'

Acland then changed his ground to back a project by Daubeny for the creation of an Examination School in Natural Science, which would force the University to pay serious attention to science teaching. It was Acland's powerful friends in London who helped to win this attack through the

University Commission of 1850, which drove a reluctant Convocation at Oxford to establish an Honour School of Natural Science. Acland now had a more solid base for his projected museum and he began to enlist a wider support. He wrote hundreds of letters, nagged his friends, and succeeded in forming a committee of twenty influential members of Convocation, which included seventeen clergymen. They obtained an estimate of £30,000 as the cost of a suitable building, and a formal appeal for funds was made to the University, which at once pleaded poverty. Then it was discovered that £60,000, the profits made by the Clarendon Press from printing the Bible, was lying unallocated in the University Chest. But Acland's opponents were determined that, whatever this windfall should be spent on, it should not be wasted on a museum. Just before Convocation was due to vote on the museum motion, a pamphlet soundly denouncing the proposal, and probably written by Dr. Pusey, was circulated to the members. It was decisive and the motion was defeated by a large majority.

Acland did not accept this defeat. He set out to win back his lost supporters, in particular Dr. Pusey. His arguments, persuasions, and appeals to friendship finally succeeded, and in 1853 Convocation voted that a report should be made on the buildings, land, and money needed for a museum. The report was favourable. The University purchased from Merton College four acres of the meadows known as 'The Parks' and competitive designs were invited. From the thirty-two submitted the committee chose two: one was a Palladian-style building, the other was described by Woodward, its architect, as 'Rhenish Gothic'. Those who know Oxford will not need to be told which of these was finally selected. The Gothic revival was then in full swing and Acland's close friend John Ruskin had embraced the museum project with an almost embarrassing fervour.

The foundation stone was laid by the Chancellor of the University (Lord Derby) in June 1855. The building that arose is an Oxford landmark, a 'divine exhalation' to Ruskin and his followers, a monstrosity to later generations, and now an object of nostalgic affection. Certainly those who know the story of the Oxford Museum,[18] the labours and battles involved in its creation, and the hopes that it embodied have a good reason for appreciation. Ruskin combined instruction with aesthetics in its decoration. The glass-roofed hall has galleries supported on 125 columns each of which has a polished shaft made from a different British rock; their capitals are carved to represent various plants and animals, and the iron vaulting is fashioned in botanical forms. Ruskin hoped that Rossetti and Millais would design 'flower and beast borders' for the windows (a vain hope, apparently) and he himself insisted on doing some of the mason's work with his own hands, which, it is said, had to be professionally redone in secret. His enthusiasm was matched by that of the foreman mason, a wayward Irish genius named O'Shea, who could translate the most intricate design into carved stone with amazing skill. Unfortunately the monkeys and cats with which he decorated the external stonework offended the University Delegates, who told Acland to order their removal. O'Shea's response was to carve owls and parrots around the main doorway, each with the recognizable head of a University dignitary. His consequent dismissal brought his unfinished work to a sad end, and the literal defacement of the owls and parrots removed features that might have been among the historic sights of Oxford.

The first official function at the Museum was the meeting of the British Association in July 1860. This provided a confrontation with a new era that brought the mild, parsonical naturalists of Oxford to realize that the pursuit of scientific truth involves a ruthless disregard of the most cherished beliefs. Charles Darwin had published his *Origin of Species* a few months before, creating a storm that was to rage for many years. It was inevitable that the British Association should discuss this subject at its Oxford meeting. On the appointed day there was news that Samuel Wilberforce, Bishop of Oxford (affectionately known as 'Soapy Sam'), would speak and 'smash Darwin'. The excitement was intense, and the audience increased to over a thousand people. In his speech the bishop ridiculed Darwin and Huxley so savagely and with such eloquence that Acland was terrified that his beloved science would be permanently discredited. But Wilberforce made a fatal mistake when he resorted to a personal attack, turning to Huxley to ask him (in effect) to say whether it was his grandmother or his grandfather that was an ape. Huxley's devastating retort rolled round the academic world, though the reported wording varies from one witness to another.[19] There is no doubt, however, about its substance. Huxley replied that he would rather be descended from an ape than from a man who prostituted his gifts to pervert the truth. Thereafter, many in Oxford who had ignored science would be hostile to it. There was a closing of the ranks on either side of what was now a clear division, the beginning of the 'two cultures' that Acland had failed to foresee and would so much deplore.

Despite the rift, the opening of the Museum was the beginning of the growth of science in Oxford. Acland had become Regius Professor of Medicine and he set out to found a pre-clinical school where the student, after taking his B.A., could receive a scientific grounding before going to a London clinical School. In 1877, the second University Commission recommended that a new chair, the Waynflete Professorship of Physiology, should be created. Hitherto, physiology had been regarded as a branch of anatomy; now it was to be a subject in its own right. Acland had no doubts that Burdon-Sanderson, the foremost physiologist in Britain, would be the man for the post, but it was a choice that led to another and even more bitter conflict within the University.

John Scott Burdon-Sanderson[20] had studied medicine in Edinburgh, and also in Paris with Claude Bernard, one of the greatest of experimental physiologists. He had research ambitions, but he earned a living as a clinician in London, first at St. Mary's Hospital and as Medical Officer of Health for Paddington, and then as Assistant Physician at the Brompton Hospital where he was able to do some original work on respiration. When his father-in-law died and left him a fortune he gave up his clinical posts to study physiology with Sharpey at University College, and set up a private laboratory in Howland Street. He tried to encourage a scientific approach to medical problems and founded the Clinical Society in 1867, which used to meet in his Queen Anne Street house. Later, it was to merge with other groups to form the Royal Society of Medicine. In 1867, too, he was elected a Fellow of the Royal

Society and gave the Croonian Lecture in the same year on his respiration work. Four years later the Brown Institute was founded, and Burdon-Sanderson was appointed its first 'Professor-Superintendent'.

Sharpey retired in 1874 and Burdon-Sanderson succeeded him as Professor of Physiology at University College. He was interested in practical class work, and edited a new manual, the *Handbook for the Physiological Laboratory*, which became the standard work. It also became an object of horror for the antivivisectionists, since it described class experiments and was thus tangible evidence, apparently, of wanton cruelty. The strength of public feeling led to a Royal Commission, to which Burdon-Sanderson gave evidence. The result was an amendment of the Cruelty to Animals Act forbidding animal experiments except under a special Home Office licence. The antivivisection campaign made the physiologists close their ranks. In 1876 a meeting was held at Burdon-Sanderson's house, and the Physiological Society was formed, largely to present a united front.

When Acland asked him to come to Oxford in 1882 Burdon-Sanderson accepted on the condition that a new building and proper equipment would be provided. The estimate for equipment was £2000, and this Convocation cut down to £1500. But when it was learned that the new professor wanted £10,000 for a building there was the storm of protest that had become an almost reflex response to any request for funds for science. Apart from the unjustified expense there was now the highly emotional issue of the animal experiments that the professor would use in his classes. The antivivisection campaigners played on an understandable sympathy and painted a horrifying picture of utter callousness or deliberate cruelty. They were ignorant of, or chose to ignore, the fact that laboratory animals are treated as kindly and painlessly as any human surgical patient, and a large number of members of Convocation were swayed by appeals to prevent the creation of a 'Chamber of Horrors' in their University. Ruskin became one of the bitterest opponents of the science that he had once supported so strongly, and he was joined by Francis Dodgson (Lewis Carroll) and many other Oxford notables, who were shocked by the methods of experimental physiology. There

were letters to the newspapers, pamphlets, and petitions and
Burdon-Sanderson was deluged with abusive letters, some addressed openly to 'The Damned and Detestable Vivisector'.

The debate in Convocation on 5 February 1884, on a decree
to provide funds for the physiology building, is now a part
of Oxford history. Professor Freeman launched the opposition, maintaining with more eloquence than logic that a historian was as much a scientist as a man who operated on live
animals and asking if the historian would be justified in illustrating the siege of Jerusalem by reproducing the slaughter.
The debate proceeded along such extravagant lines, but the
restraint and good sense of Acland and his supporters prevailed and there was a majority of forty-one in favour of the
decree. But bitterness remained on both sides. Ruskin
resigned his Slade Professorship in protest, and many shared
his feelings. The rift between the two cultures had widened.

The new building, in suitably Gothic style, was completed
on a site just behind the Museum in 1886. Burdon-Sanderson
moved in, with two men from his London department,
Francis Gotch and F. A. Dixey, and he was later joined
by his nephew, J. S. Haldane. Since physiology had now
separated from anatomy, a full-time anatomist was appointed,
Arthur Thomson from Edinburgh, who occupied a tin shed
in the yard of the new department until he himself acquired
a new building and the rank of professor in 1893.[21] The shape
of scientific development in Oxford was becoming clear, crystallizing around pre-clinical teaching for medical students,
and with a strong emphasis on physiology. There were now
men of energy and international distinction in the Museum
departments, and Acland became disturbed by their very
strength and 'contentious specialization'. Behind this he saw,
in the words of his biographer, J. B. Atlay, 'the spectre of
that ... divorce of Science from Arts which he had always
dreaded'. In 1894 he resigned as Regius Professor and he was
succeeded by Burdon-Sanderson.

Francis Gotch became the next Professor of Physiology.
J. S. Haldane had hoped to be appointed, and Sherrington had
also applied.[22] Gotch seems not to have been a strong character, and Haldane (who certainly was) was able to develop a
virtually separate department in the field of respiration with

Douglas and Priestley. Two new branches of pre-clinical study appeared during Burdon-Sanderson's term of office as Regius Professor. One was pharmacology, which was taught by J. A. Gunn in a long attic room in the main Museum. The other was pathology. Pathology emerged in the form of morbid anatomy and histology—the description of structural changes in disease—and it was to develop along this line for many years to come. And with the recognition of the importance of micro-organisms in disease, bacteriology had become more aligned with pathology. Burdon-Sanderson was competent to teach both subjects, and until 1895 had done so. Then John Ritchie was appointed to teach bacteriology, and it soon became clear that the pace of discovery justified a new department where pathology and bacteriology could be housed under one roof. Another building therefore appeared behind the Museum in 1899 which, with new departments of anatomy, chemistry, and physics ('Experimental Philosophy'), formed the beginnings of the 'Science Area' that was to expand so widely into the Parks during the next seventy years.

In 1903, Burdon-Sanderson, then aged seventy-six, resigned as Regius Professor. As there was little endowment for pathology, he suggested that the Regius chair should be merged with a professorship of pathology, and that Ritchie should hold both offices. This suggestion was so strongly opposed by the Oxford medical graduates that he withdrew it and accepted the counter proposal that a former pupil of his own, William Osler, Professor of Medicine at Baltimore, should be invited to be Regius. Osler, a Canadian, was regarded as the most brilliant physician of his day. After some considerable doubts (which were ended by his wife's famous telegram: 'Do not procrastinate. Accept at once.') Osler came to Oxford to do for medicine what Acland had done there for science.

Though Ritchie received the title of Professor of Pathology when Osler arrived, he was a disappointed man and left for Edinburgh in 1907. His successor was Georges Dreyer, a Danish bacteriologist who had also been one of Burdon-Sanderson's pupils. With Ainley-Walker and A. G. Gibson, he built up his department into an active centre for research and teaching. The first few years of the new century saw the passing of the men who had begun the scientific renaissance in Oxford. Sir Henry Acland died in 1900, and Sir John Burdon-Sanderson in 1905. On the foundations they had laid, the two great men who followed them were able to build a school of medical science that could rank with the best in Europe. One of them was Osler, a teacher, writer, scholar, and clinician of such legendary eminence and personality that his mere presence in Oxford made it a place of medical pilgrimage. At the Radcliffe Infirmary his teaching created an unofficial postgraduate school, and his classic textbook *Principles and Practice of Medicine* remained the standard work throughout the world for over thirty years and went through sixteen editions. The other great man was Charles Sherrington, who became Professor of Physiology in 1913. Of his contributions to the development of scientific medicine and of his personal influence on his pupils, and in particular on Howard Florey, more will be heard in later chapters.

2

ADELAIDE

On the high ground of North Adelaide, where the terraced houses have an elegance suggesting Regency England, there is a statue of Colonel Light, the Surveyor-General who laid out the capital of the new Colony of South Australia in 1836. The bronze figure stands with an arm outstretched, commanding attention to a wide prospect of the city he created. On the plinth there is a quotation: 'The reasons that led me to fix Adelaide where it is, I do not expect to be generally understood or calmly judged of at present. My enemies, however, by disputing their validity in every particular have done me the good service of fixing the whole of the responsibility upon me. I am perfectly willing to bear it, and I leave it to posterity, and not to them, to decide whether I am entitled to praise or to blame.'

Light's confidence was entirely justified. To the south and east the suburbs of the city climb into the wooded foothills of the Mount Lofty range. A few miles to the west there is the seaport area on the Gulf of St. Vincent, and to the north some of the best fruit- and wine-producing country in Australia. The city centre is a mile-square gridiron of handsome streets, with gardens at the main intersections and tall, modern buildings that are not so high or crowded as to be oppressive. Around it is a complete ring of parkland with shady walks and avenues; the race course, the cricket ground, the botanical gardens, golf courses, and the river Torrens, which flows quietly from lake to lake. Beyond the park the streets of the city are wide and mostly tree-lined. Many of its houses date from the last century, with covered balconies and railings of delicate iron 'lacework'. There is a feeling of space and—a rare virtue in modern cities—a sense of time to spare. Adelaide may not be the most exciting of Australian state capitals, nor even the most beautiful, but it is certainly the most charming.

When Joseph Florey arrived there in 1885, Adelaide pos-
sessed many of the features that it has today. It was a pros-
perous place of about 100,000 inhabitants. Agriculture was
booming; copper had been found at Burra and Moonta. Adelaide was the commercial as well as the administrative capital
of South Australia, and by 1874 it had founded its university
and built its cathedral, hospitals, schools, and most of the
public buildings that exist at present. Joseph's wife had developed tuberculosis in London, and his decision to emigrate
may have been taken on her account. Adelaide has a sunny
climate without extreme heat, and it was also a good place
in which to start a business. Though the pace of life there
was less hectic than in London, Adelaide was very far indeed
from being the uncivilized 'outback' town of contemporary
English imagination.

Joseph was born at Bampton, an Oxfordshire village near
Witney, in January 1857. His father, Walter Florey, was a
bootmaker in the near-by village of Standlake,[1] where Floreys
had been farmers and craftsmen for generations and, after
Joseph's birth, he moved to London and bootmaking in
Islington. Joseph learned his father's trade, and obtained
employment at a local factory. In 1878 at the age of twenty-
one, he married Charlotte, the daughter of William Ames, one
of his workmates. Two years later, their first child, also chris-
tened Charlotte, was born, and in 1882 another daughter,
Anne. It must have been during the next two years that their
mother became consumptive. Pulmonary tuberculosis at that
time was a common and virtually untreatable disease. Sunshine and fresh air were generally supposed to be beneficial.
The well-to-do consumptive went to sanatoria in Switzerland.
Joseph Florey chose a new life for himself, and a better chance
of life for his wife, by leaving London for Australia.

Once in Adelaide, Joseph began to earn his living as a bootmaker. Footwear was then mostly handmade and, to judge
from the number of bootmakers listed in the Adelaide trade
directory of that time, was much in demand. His first
workshop was in Young Street, Parkside, a suburb just
beyond the southern parkland of the city, and his home seems
to have been near by. Soon, any hopes that he may have had
for Charlotte's recovery faded. She sank steadily into the typi-

cal consumptive decline, becoming weaker and more helpless. Joseph, with a home and two small children to maintain, engaged a nurse-housekeeper, Mrs. Wadham, an Australian-born widow with a twenty-three-year-old daughter, Bertha. Mrs. Wadham nursed Charlotte until she died in April 1886, and then seems to have stayed on as housekeeper for the next three years. In contrast to his personal loss, Joseph's business prospered. Adelaide was in optimistic mood. There were bumper crops for farmers, growing profits from the mines, and the celebrations to mark the jubilee of the city's foundation were gay and lavish. The small workshop became 'Joseph Florey and Co., Boot and Shoe Manufactory' and expanded in 1888 into the Enterprise Boot Factory in Pultney Street. Joseph was clearly a man of progressive ideas ready to exploit the commercial advantages of mass production, and to raise the money needed to do so. In November 1889, he re-married, and his new wife was Bertha Wadham, the daughter of his housekeeper.

Bertha proved to be a woman of strong character. She was an efficient housewife and her excellent business sense was probably a great asset to her husband. By 1890 the Florey family were living in a larger house in Fisher Street, Malvern, to the south of the city. The house, which remained their home for the next sixteen years, was a bungalow with the deep verandas and corrugated iron roof typical of Australian suburban houses to this day. Bertha's first child, Hilda, was born in 1891, and her second, Valetta, a year later when her stepdaughters, Charlotte and Anne, were twelve and ten years old respectively.

Joseph, meanwhile, was engaged in business manœuvres which were clearly to his advantage. In 1891 he sold his Enterprise Boot Factory to a large Adelaide retailer, but contrived that he should remain its manager. Within two years, he was Managing Director, having gained control of what was now a much larger concern with another factory in Halifax Street. In 1897, he was the proprietor of 'The Standard Shoe and Leather Company' of Grenfell Street, producing several stock lines of boots and shoes that were selling, not just in Adelaide, but throughout Australia.

On 24 September 1898 Bertha had her third, and last, child.

This time it was a boy, who was named Howard Walter. From the first, Howard, a late arrival and the only boy, was adored by his parents and by the four girls of the family. Charlotte was then a pretty girl of eighteen, tall and slender. She had studied cookery at school and seems to have had no other ambition than to be the household drudge. In this capacity she was shamefully exploited by her stepmother, who was careful to frighten off any masculine admirers who might have removed her. Anne was a little less malleable. She too was treated as a menial, but escaped to make her own career when nearly thirty years of age.

The first eight years of Howard's life passed relatively un-eventfully in the Fisher Street house, or at 'Nunkerri', the country cottage which Joseph had bought at Belair, high up in the Mount Lofty Hills, about 10 miles from the city. Charlotte and Anne did most of the housework. Hilda, obviously intelligent, was doing well at school. Valetta was taking a serious interest in the piano, for which she had an unusual talent. Howard seems to have been a quiet and well-behaved little boy who submitted to the alternate fussing and teasing of his feminine elders. They liked to dress him in the style of Little Lord Fauntleroy, and his hair, arranged in long, fair ringlets, hung to well below shoulder-level until he was over eight years old. Though physically well-grown, he was rather delicate, with a tendency to respiratory infections which later became serious.

Joseph Florey was deeply occupied with his business, which entailed a good deal of travel, and he could spare little time for his family. He was popular in the Adelaide commercial world, where he was becoming quite an important figure. There were articles about him in the local press, illustrated by sketches showing a stout and jolly-looking man with the walrus moustache in favour at that time. He was indeed prospering. A new factory had appeared in Freeman Street, with a warehouse and the company offices in Gawler Place. He was also branching out in other state capitals, with a factory in Richmond, near Melbourne, and warehouses in Sydney, Perth, and Brisbane. His products, which ranged from the heaviest working boots to the daintiest dancing slippers, included several brands of shoes that became famous throughout the Antipodes. Ironically it is the tiny replicas of these shoes made in porcelain for advertisement that still survive as collector's treasures long after their originals have been forgotten.[2]

By 1906 Joseph was a rich man. He had some hundreds of employees and a personal income of several thousand pounds a year. He was self-made and proud of it, but increasing wealth brought social aspirations, particularly to Bertha. The Fisher Street house and Belair were no longer good enough, and when the chance came in 1906 to acquire 'Coreega', a small mansion in Mitcham, he bought it for £3000. He also bought a Darracq motor car (Registration No. 16), then a possession of a very rare sort. It might be supposed that in the relatively young and democratic society of Adelaide, class differences would be unimportant. Nothing could be further from the truth. The big landowners of South Australia formed a natural aristocracy which, in many cases, had simply been transplanted from the British upper classes. Their often considerable wealth came from their land, or from the minerals found there, but mere wealth did not buy admission to their company. Below these landowners ranked the administrators and the professional and academic circles. People engaged in trade belonged to a still lower order, particularly if they were recent immigrants, and the ordinary workers had the sort of standing allotted to them in Victorian England.

These facts of social life became apparent to the Florey family when they moved to Coreega. Upper Mitcham was an old village around which were grouped the houses owned by a few of the richest and most influential families in the State. The private parks of these veritable mansions extended over hundreds of acres of the beautiful wooded foothills below Mount Lofty. The oldest of them is Rust Hall, built in 1846 by Sir Arthur Blyth, a Prime Minister of South Australia. Torrens Park, with scores of rooms, a private theatre, and 240 acres of land, had been the home of Sir Robert Torrens, also a Prime Minister. (Torrens Park later belonged to the Barr-Smith family and is now Scotch College, one of Adelaide's leading private schools.) Springfield House belonged to the Auditor-General, Sheriff Newenham; Strathspey (now

Mercedes College) to Sir John Duncan; and Carrick Hill to Sir Edward Hayward. There were several other mansions of similar size and distinction.[3]

Coreega should, perhaps, be classed as a large house rather than a small mansion among these palatial domains. It has only sixteen rooms, though these are, by modern standards, very large. (For example, the bedroom which had been Howard's is roughly 30 feet square.) There were 10 acres of land, with two tennis courts and a paddock, and a pleasant shady garden. It is a two-storey stone-built house, with a deep veranda on two sides, supporting a covered balcony on to which the main bedrooms open. There are three big reception rooms communicating to form a large area for entertainment. There are large and rambling kitchen quarters and the back stairs divide discreetly half-way up so that male and female servants go their separate ways to bed at opposite ends of the house.

The Floreys made few changes. They replaced the original main staircase in the vaulted hall with a more ornate and massive one, and installed a large and powerful mechanical organ in the drawing-room. Cows were kept in the paddock, and, as Howard grew older, he learned to milk them. But the reception rooms remained empty of guests. The Floreys did not entertain. The reason, probably, was their non-acceptance by Mitcham society. It was a village of the very rich, long established on their estates, and the very poor—people employed on these estates or in the neighbouring quarries. The Floreys fitted nowhere. Joseph Florey was not unpopular, but he was a busy man with little time or inclination for country life or social occasions. Bertha in her new environment became embarrassingly ladylike, overdressed, and ostentatious, and she was positively disliked.

One family, however, made friends with the Floreys, and in particular with Howard. The rectory garden adjoined the grounds of Coreega, and Archdeacon Clampett, an extrovert Irishman, who was a Canon of Adelaide Cathedral and Rector of Mitcham, had an English wife and three children. Howard Florey used to escape from the rather forbidding atmosphere of Coreega to play in their garden with the rectory children, and Mollie Clampett, who was eighteen months older than he was, became a firm friend from that time and for the rest of his life. Mollie was a day-girl at a local private school, Unley Park, run by a Miss Thornborough, and Howard, too, became a pupil there. The journey entailed a two-mile ride in the horse tram that plied between Mitcham and Adelaide. Mollie was charged with Howard's safe collection and delivery. She used to hold his hand on the way to the tram and called him 'Dowie' because his long hair reminded her of the flowing locks affected by a well-known Mormon preacher of that name.[4] But at school he was called 'Floss'—a nickname that he retained for the rest of his life. Its origin has been variously explained. Some of the Floreys still living near Witney were also called 'Floss' at school because their surname was pronounced 'Florrie' which, like 'Floss', is the common diminutive of 'Florence'.[5]

It seems that Howard's doting family did their best to prolong his infancy, and that at school he found himself the butt of small boys, who seldom refrain from personal remarks. Perhaps this experience established the reaction to criticism or personal difficulties that remained with him for the rest of his life—a determination to do better than his critics, to excel in everything that he undertook. At the beginning of 1908 Howard, shorn of his long hair but not of his nickname, transferred from Miss Thornborough's school to Kyre College, also in Unley Park. Kyre College was a private school for boys, which later merged with Scotch College. When Howard entered it there were over a hundred day-boys and boarders, and the school was gaining a good academic and athletic record, having won the Intercollegiate Cricket League. He very quickly made his mark in this larger world. Within a term he was top of his form, and had won the under-14, under-12, and under-10 flat races in the school sports. His school reports, carefully preserved by his parents, were uniformly excellent. He worked hard, played hard, behaved well, and was obviously both athletic and clever. But he was said by those who remembered him there to have been a very quiet boy, shy and reserved.

Meanwhile Hilda Florey, an intelligent and hard-working girl, had entered Adelaide University to study medicine in 1906. The University had none of the sex discrimination then to be found in Europe, and, almost from its inception, had admitted women on equal terms with men. The science faculty was well served, with Horace Lamb, the mathematician, William Henry Bragg, the physicist, E. C. Stirling, the physiologist, and E. M. Rennie, the chemist, among the teachers. But, although the Clinical School had begun in 1887, only about a dozen women (including Helen Mayo) had graduated from it when Hilda became a medical student.

It is not recorded why Hilda made this choice of a career, though she is more likely to have been influenced at school than by her family. It was a decision that required strength of character to carry through. Though the University had liberal ideas, these were not always shared by the medical teaching staff, who found a solitary female among a class of twenty or so Australian youths somewhat inhibiting. An extreme example was that most remarkable man, Archibald Watson, Professor of Anatomy, who had once acted as Pasteur's interpreter. His expositions were considered too lurid for feminine ears, but rather than modify them he refused to admit the one girl in a particular year to his classes. The girl, very properly, complained to the authorities. At the next class, Watson lined up all the young men in the dissecting room, turned to the girl and said, 'Choose any one you fancy, marry him, and I'll see that he passes all his exams. Then you needn't bother me any more with trying to do medicine.' The girl burst into tears and locked herself in the lavatory until the Professor had apologized, which, in the end, he did.[6] Hilda Florey was the only girl in her year, but there is no evidence that she was the heroine of this episode.

After three years of success at Kyre College, Howard was entered as a day-boy at St. Peter's Collegiate School in 1911. This had been founded by Dr. William Wyatt and Dr. George Mayo, both eminent not only in medicine but in the early affairs of the State of South Australia. St. Peter's was modelled on the lines of an English public school and soon became recognized as one of the best schools in Australia. Its Gothic buildings, which now accommodate about 700 boarders and day-boys, are set in about 75 acres of playing-fields and park within two miles of the centre of Adelaide. The Headmaster at that time was Canon Girdlestone, who had graduated from Magdalen College, Oxford. In this even larger world, Howard, then aged thirteen, continued to succeed through the total application of his considerable mental and physical abilities. During his first three terms his reports were glowing, stressing quite unusual progress 'in so young a boy', and the necessity not to force him unduly. He acquired an enthusiasm for chemistry from an inspiring teacher, J. S. Thompson, whose habit of padding about in rubber-soled shoes had earned him the name of 'Sneaker'. It was probably Thompson's ability to make his subject come alive and to give an inkling of the excitements of research that determined Howard's choice of a career. He decided that he was not interested in entering his father's business, and one day he told Hilda that, when he grew up, he was 'going to do research'. Towards the end of his life he recorded his memory of her reply to this remark and his reaction to it: 'She said "You mean you'd like to be another Pasteur?" At the time I did not know what that meant.'[7]

The year 1912 was an eventful one for the Florey family. Hilda, having been first in her class in her fourth year, won the Davis Thomas Scholarship at her medical school and graduated M.B., B.S. in November. She then embarked on her house appointment and finally specialized as a pathologist. In the same year, Howard, who was liable to colds and bronchitis, had his first attack of pneumonia. He was seriously ill for several weeks, and his parents engaged a trained nurse to look after him. Anne and the nurse were soon friends and Anne, although she was now thirty years old, became determined to take up nursing. She had always had more spirit than Charlotte, who remained completely under her stepmother's domination, and decided that she must leave Adelaide. She enrolled as a student nurse in Launceston, Tasmania, and three years later she was with the Queen Alexandra Nursing Service in the Middle East war zone, where she was awarded the Royal Red Cross Medal. Meanwhile, Valetta, Howard's younger sister, was showing real gifts as a pianist. She was studying at the Elder Conservatorium in Adelaide, and was commended in the press for her concert performances. But she was handicapped by deeply webbed fingers which restricted her span.

It was in 1912, too, that Joseph Florey's business affairs took an important, and in the end disastrous, turn. In this year the Standard Shoe Company became a public company. Previously Florey had been the sole proprietor, but the manager in each state capital had a partnership agreement, and had to be bought out by the new company. One of these managers, O. M. Harry, contrived to become co-director with Florey. Florey, as Managing Director and with more than 50 per cent of the ordinary shares, apparently retained control, but a few years later he realized that he should have been much more careful about Harry. Harry, when a manager in Perth, had risked (and lost) the firm's money in starting a chain of retail shops in which he was later found to have installed his own relatives. But in 1912 Florey seemed quite content with the arrangement and for all manufacture to be transferred from Adelaide to a new factory in Melbourne under Harry's sole management. Meanwhile the prospectus painted a glowing picture of the commercial future, and Joseph had the satisfaction of now drawing a large salary from the new company as well as the income from his 29,000 ordinary shares, which he expected to earn about 15 per cent per annum.[8]

Looking back on his schooldays, Howard Florey often remarked that his greatest pleasure came from roaming through the country near his home. The bush (or the desert) is never far away in Australia, and Mitcham is on the fringe of the native forest that clothes the deep valleys and lower hills of the Mount Lofty range. Howard had only to climb the paddock fence to be free to wander for as many miles as he wished, and often with Mollie. Wildlife was not then the popular object of concern and interest that it is today, but the possums that still install themselves in convenient attics, the flights of brilliant parakeets, the maniacal dawn screeching of the kookaburras, and the musical song of the Australian magpies must have formed a part of these memories. The Floreys' house at Belair was an ideal base for picnics and explorations. The road there was steep and rough and the Darracq car, and even the new Coventry Lotis that Joseph bought to replace it had to be pushed by its passengers up the worst inclines.

Back at school after his illness, Howard soon resumed his

steady, dedicated progress. He was almost always top of each form within a term of entering it. He won prize after prize: five in chemistry, and seven in physics, mathematics, and history—a subject in which he retained a lifelong interest. He also won four scholarships, which more than paid the cost of his education. His parents were naturally delighted, though money at that time was not important. One day he said to his mother, 'What will you give me if I win the Farrar Scholarship?' Bertha in an unguarded moment replied, 'Fifty pounds.' 'Hand it over,' said Howard, 'I've won it!'

On the games field he was equally successful. Though not robust, he was a natural athlete and made every effort to excel. He played cricket, football, and tennis for his school, and won event after event as a sprinter and long-jumper. But, though he was liked and respected, he was not a very popular boy. He was reserved and withdrawn and seems to have formed no close or lasting friendships at school. He was responsible and well behaved, qualities more appreciated by his masters than by his less responsible or well-behaved schoolfellows. His dedication to success was a little too intense; the young tend to admire excellence when it is achieved with an appearance of careless ease. But there is an episode that reveals a human streak of mischief. At an annual school concert, with the hall packed with boys and parents, Howard was stage manager. One of the items was a comic ballet. In the absence of adequate dressing rooms, the prima ballerina had to doff his trousers and don his *tutu* on the stage behind the curtain during the interval. Howard chose the moment of maximum exposure and then raised the curtain. The effect, one must imagine, was dramatic.

St. Peter's, modelled on the typical English public school with an English headmaster and a fair proportion of Englishmen among his staff, probably dispensed the precepts of Dr. Arnold which had made the British Empire what it was. But St. Peter's did not produce the typical English public schoolboy—it produced Australians. The difference, which is considerable, is important because Howard Florey was, and remained, essentially an Australian and as such his reactions were often misunderstood and sometimes resented during his

life in England. Like most people, the Australian is conditioned by his early environment, and this must, for a
moment, be considered.

The prime factor in the Australian environment is the bush.
The vast areas of wilderness, the hundreds of miles that may
separate the scattered homesteads and townships have, as well
as their obvious physical significance, a profound psychological impact. South Australia, though one of the smaller states,
is nearly twice the size of France. In 1912, it had a population
of less than 400,000 people, of whom more than half lived
in Adelaide. The day-boys at St. Peter's came, of course, from
these city homes, where the life was superficially not unlike
that of a large English country town. But the boarders, who
tend to make the character of a school, had a very different
background. Many were the sons of the fruit- or stock-farmers
or the owners of outback stations the size, perhaps, of an
English county and where the nearest neighbour might be
fifty or a hundred miles away. These people were often the
pioneers who had carved their holdings from the bush, and
to whom survival meant a constant battle against the facts of
nature. They had to be tough, determined, resourceful, and
independent. Their welfare came, not from the State, but
their own efforts. They learned to be self-reliant, to prefer
action to discussion, and to meet and overcome their difficulties head-on. They had little patience for the pretences of
polite society or its reverence for tradition, and none at all
for official pomposity and procrastination. And if transplanted
to such an environment they were apt to feel insecure and
thus aggressive.

Something of this refreshing 'outback' vitality infected the
Australian city-dweller and even the recent immigrant. When
Howard Florey was at school, rural travel was difficult and
sometimes dangerous. Adelaide lost its horse-trams in 1909
in favour of electric ones, and was linked by rail with the larger
towns of the State. But in the country, roads were little more
than tracks, and the only practical alternative to walking was
the horse. Today, almost every Australian has a car and he
will cheerfully drive hundreds of miles in a day over roads
that may be as bad as they ever were. His idea of an outing
is a day on the beach or to take to the bush, setting his car.

at natural obstacles as if it were, indeed, a horse, in order to grill his steak and boil his billy over an open fire. The old pioneer spirit lingers; and this attitude of mind, obvious to the visitor as quite un-European, does not strike the Australian as peculiar—until he himself visits Europe. Howard Florey absorbed it naturally and the more readily because it harmonized with his pleasure in overcoming difficulties.

In spite of a second attack of pneumonia in 1914, Howard continued to excel at school. He had made up his mind on a research career in science, and Joseph Florey, though disappointed that his son would not eventually take over his business, discussed the matter with Canon Girdlestone. Howard's best science subjects were chemistry and physics, but he himself recognized that he had not the mathematical ability needed for physics research, and Girdlestone advised that there was no future in Australia for a research career in chemistry. An obvious alternative was medicine, which can provide a living while offering research opportunities. So it was decided that Howard should enter the Medical School in Adelaide, as Hilda had done ten years before.

It was a troubled period. War had been declared, the war that everyone had expected to be over by 1915. But it was not over. The battles were raging with greater fury in Europe, and the campaigns in Mesopotamia and the Dardanelles brought increasing numbers of Australian troops into action. Young men from the cities and the outback enlisted in their thousands, trained for a few weeks, and vanished overseas. And, by the time the war was over, 60,000 of them would have been buried there. From St. Peter's, most of the leavers were going straight into the armed forces.

Howard would be eighteen in September 1916, and with no sign of an end to the war, he naturally considered his own position. Enlistment was voluntary, but he would have preferred conscription—he argued in favour of it at a school debate—since it obviated the need for painful personal decisions. His parents were bitterly opposed to his joining the army straight from school. His father's health was failing, he was their only son, and he himself had suffered from two attacks of pneumonia and repeated bouts of bronchitis. These arguments were used whenever the subject arose, and they

were probably repeated in a letter (from (Melbourne) written
by his father in July 1915, to which Howard replied as follows:

Dear Pa:

I received your letter today at school. It came as a complete surprise....
However, I reckon I ought to do something at the end of this year or beginning of next, not because a lot of balderdash that's talked about 'King and
Country' and other patriotic rubbish, but because its the right thing to
do. If a chap stayed home and pleasantly followed his ordinary routine,
what a fool he would feel when all the other chaps came back and, as I
told Anne, he'd want a prolonged holiday in some very secluded nook ...
It may appear that it reduces the business to one of public opinion, and
in ninety-nine cases out of a hundred that's what it is, for I don't think
anybody *wants* to go, but they reckon they ought to go because mainly they
think that other people point them out as having no 'guts'. Doubtless it
is very high falutin to talk about no notice being taken of public opinion,
but it is a pretty potent factor nevertheless. Whenever I hear a chap saying
he'd like to go to the war, I feel like calling him a liar or a d. fool, for
no-one in his right mind would want to go, but still he ought to go. I don't
want to go, but I ought to, and that's the whole position ...

                                        Your affec. son,
                                        Howard

P.S. That was pretty good about Annie. She's pretty flash. [Probably a
reference to Anne's decoration on active service.][8]

Howard's clearly stated attitude to what he considered his
duty was not accepted by his parents, who refused to consent
to his enlistment. Also, the fact that Howard was entered as
a medical student, and that doctors were needed as urgently
as soldiers, did something to salve his conscience. But his self-
respect suffered, and on later occasions he would have to
explain the reasons why he spent the war in Adelaide.

Joseph's failing health was a real issue, and one accentuated
by ominous signs of failure in his business. Orders were dropping; the Melbourne factory, under Mr. Harry's management, was producing less than half its planned output, and
this of poor quality. In 1915, the company that was to have
paid a dividend of 15 per cent made a loss of £41,000 and
paid nothing. Joseph Florey became convinced, first that it
was Oscar Harry's mismanagement that was to blame, and
later that Harry was deliberately wrecking the company in
order to oust its founder and take it over at a rock-bottom
price. There were angry meetings of the shareholders, and the
troubles of the company attracted the attention of the Inter-
State Commission from whom the directors got, in Joseph

Florey's words, 'a severe raking'. The Commission found that 'the method of book-keeping was arranged more for concealment than exposition'. Florey wrote a letter of dismissal to Harry, who, far from accepting it, brought an action against Florey. Counter-actions followed and the company proceeded to tear itself to pieces.[8]

Joseph Florey as its principal shareholder was well on the way to ruin. In 1916, he put Coreega up for sale at £4200, but there were no buyers. He then raised mortgages on both his houses and on 950 acres of land at Myponga, bought as an investment some years before, and managed to stave off the final collapse. Though Bertha was fully in his confidence, he seems never to have discussed these troubles with his children. But the endless meetings and recriminations, the frequent visits to Melbourne and worry about the future wore him out, and it was clear to everyone that he was becoming a very sick man.

Howard, in his last year at St. Peter's, was top of the upper sixth form and head boy of the school. He was preparing himself for his medical studies. One of his schoolfellows remembers how, when the form-master was unavoidably absent, Howard gave an extempore lecture on human anatomy with such knowledge and assurance that the class was spellbound. In his last term he was awarded the Young Exhibition to the University and a State Bursary that would mean that his training there would cost his hard-pressed father nothing.

Howard became one of the twenty-two students to enrol at the Medical School in March 1916, the beginning of the antipodean academic year. He was now to study in earnest the subjects over which he had already skimmed: organic chemistry and pharmacology, zoology and botany, anatomy and physiology. The teaching staff was depleted; many of the younger men were away at the war. But Howard worked hard, made the best of his opportunities, and was top of his year when it came to examination time.

As a scientific introduction to medicine the pre-clinical course at Adelaide was below the contemporary standard of European or American universities. There was little reference to research, nor had any significant research been done there. Among the teachers, Archibald Watson stood out by virtue of a rebellious independence and an inquiring mind that at least inspired his pupils to ask their own questions. Watson was an amazing man, one of those legendary figures that add colour to the most prosaic background. He was a native of New South Wales and, as a boy, had won prizes for divinity. To escape becoming a minister he had run away to sea and became 'cook's helper' on board Captain Armstrong's ship, *Carl*. Only then did he discover that Armstrong's trade was 'black-birding'—kidnapping natives and selling them in South America as slaves. 'It was all a mistake my being a pirate,' he maintained. When the Royal Navy caught up with the *Carl*, Watson escaped, but with a price on his head. He went to Bonn, 'the only German town I could spell', and studied medicine, obtaining one M.D. in Göttingen and a second in Paris, where he worked with Pasteur. After studying cholera in Egypt in 1883 he was appointed to the staff of Adelaide hospital as pathologist and Lecturer in Anatomy. He attempted to rid Australia of rabbits by importing animals from Germany infected with *Sarcoptes cuniculi*, but they died on the way.

A photograph of Watson shows him in riding breeches, a bush shirt, and a wide-brimmed hat and looking very like Colonel Cody. But his mount was a fearsome motor cycle on which he would roar through the streets or over the open country. One day he pulled up by one of his students outside the Anatomy Department. 'Jump on!' he commanded. The youth, unwilling to offend, mounted the pillion, and Watson rode at terrifying speed out of the city and up a track into the Mount Lofty hills. There, in a field, was a sick horse. 'Take off your trousers,' said Watson. 'Why?' asked the student, now seriously alarmed. 'Because I'm going to operate on that horse and a pair of trousers makes the best anaesthetic mask,' said Watson. When the student indignantly refused, the professor removed his own trousers, soaked them in chloroform, and performed the operation.[6]

Watson survived rows and scandals that would have ruined a lesser man; he survived by sheer effrontery and because it was recognized that he was a superb comparative anatomist. He was accused, quite justly, of body-snatching, but his only regret was that he had not weighted a coffin with enough coke to avoid detection, and he was unabashed when a special law had to be passed through the State Parliament to make such things illegal. In the hospital his habit of taking his anatomy class into the gallery of the operating theatre in order to expound loudly on the anatomical ignorance being displayed by some unfortunate surgeon made him powerful enemies. He was dismissed at least once, but as Joseph Florey had found with Oscar Harry, there is a difference between dismissal and departure. Watson was so much of a nuisance outside the hospital that he was soon reinstated.

Both Watson and his motor cycle had their effects on Howard Florey. Despite his flamboyant eccentricities, Watson was an able teacher, not only of his official subject but of physiology and pathology as well. He gave the study of anatomy, therefore, a breadth of application and a scientific interest not to be found in a more orthodox approach, and it was this refusal to be bound by the artificial confines of a 'subject' that Howard found so effective in his later career.

The motor cycle was, perhaps, less important, though it provided lessons in practical mechanics. When it failed him, Watson would wheel it into the dissecting room and announce that anyone who could put it right would be guaranteed a pass at the next examination. It was not surprising that several members of the class bought motor cycles of their own. Howard already had one, having acquired a second-hand Triumph while still at school in order to avoid the long, slow tram-ride to and from Coreega. He made expeditions to Victor Harbour, or nearer bathing beaches, often with Alan Lamphee, a fellow-student from St. Peter's, who had a Douglas. The roads were rough, punctures were frequent, and any country journey was an adventure. Mollie was sometimes a passenger on the Triumph's pillion, even to dances, though as a form of transport it must have been the ultimate in discomfort. Howard crowned his career as a motor-cyclist in an unusual way. It happened that a famous anatomist was visiting Adelaide and the Dean of the Medical School asked Professor Watson to arrange an interesting demonstration. Watson expressed his opinion of such visitations with his usual pungency, but finally agreed. The visitor and his entourage arrived in the dissecting-room. Watson blew a whistle, and through the double doors roared six motor-cyclists, including Howard on his Triumph, to perform intricate manœuvres between the tables.

This was almost the last of Professor Watson's official misdemeanours at the Medical School, because he retired a year or two later. But his extravagant adventures and amorous intrigues continued until he died in 1940 while making an expedition to Thursday Island at the age of 91.

The year 1917 closed, in the Australian high summer, with no end to the war in sight. Joseph's business troubles were deepening, but life at Coreega went on. Charlotte was still the household drudge. Hilda had met and married Dr. John Gardner in Ballerat and then gone with him to live and work in Melbourne, where she was to become one of the best clinical pathologists in Australia. Valetta had given up her ambitions to be a concert pianist, because of her restricted hand-span, and seemed content to be idle at home. Howard disapproved of this capitulation. For him, difficulties were to be surmounted and talents were to be used to the full.

Though the Floreys seldom entertained formally, there were parties in the garden. To one of these parties Howard brought Tom Bowen, who had been a schoolfellow at St. Peter's, and introduced him to Mollie Clampett. For them it was a momentous meeting. Tom Bowen had enlisted and was just about to be sent overseas, but Mollie immediately captured his heart and when he returned safely after the war the first thing he did was to marry her. Howard, too, was beginning to find girls interesting. He was particularly attracted to Nance Lamphee, Alan's sister, but he was anxious that his family should not discover this. When Alan and Nance came to Coreega, Howard would take them to the drawing-room, ostensibly to play the mechanical organ. Alan was then deputed to keep this monster at full blast while Howard and Nance slipped out into the garden. Alan was surprised and sorry when this affair came to an end, and he remained convinced that they would have been well suited.

In 1918 Howard passed his pre-clinical examinations with honours, and started on his clinical work in the hospital. He was an adept examination candidate, setting himself to read the questions carefully, to make notes on each, and then to write clear and concise answers. This simple, but rarely applied, formula ensured that he made the best of his excellent memory and unusual critical judgement and also pleased the examiners. He usually finished his papers long before the other candidates, who were apt to view his almost unvarying success without much enthusiasm. As at St. Peter's, he entered into every sort of student activity. He played tennis and football for Adelaide University, becoming the equivalent of a 'double blue', and he kept up his athletics. Tennis was, perhaps, his best game and he was able to give Davis Cup Players a run for their money.

After the relatively pure science of his pre-clinical years, the teaching of medicine was something of a shock. The formal teaching was straight from standard textbooks (such as Osler's on medicine and Rose and Carless's on surgery) which the students were expected to absorb—even memorize—but not digest. On the other hand, it became obvious in the wards that patients persist in being individuals, and that their ailments can rarely be fitted exactly into neat categories. There was a tendency therefore for teachers to gloss over or ignore misfit facts, while the best clinicians who relied more on personal judgement and experience than on the textbooks were not considered the best teachers. It was clear that the practice of medicine was still an art rather than a science. Howard would have accepted this situation, though he was more of a scientist than an artist, if it had been honestly presented. What he could not accept was an unquestioning attitude to authority and the concealment of ignorance by long words and pompous phrases.

In the clinical laboratories he was happier. Blood cells could be counted, haemoglobin measured, the chemistry of certain body functions determined with fair accuracy. The recognition of the different blood groups was making blood transfusion a practical proposition. Bacteriology had reached the stage at which almost all the organisms (except the viruses) responsible for disease could be cultured and identified. Histological techniques were allowing the cellular patterns of disease to be recognized. Radiology was opening up entirely new methods of examination. Such information, properly used, was of immense value in diagnosis.

But, apart from blood transfusion, treatment had scarcely changed since Lister had transformed surgical practice thirty years before. The incidence of infection had been reduced by better hygiene but, once established, it was as lethal as before. The mortality from tuberculosis, pneumonia, peritonitis, meningitis, osteomyelitis, and septicaemia was as high as ever. Lister's 'ideal antiseptic' had eluded discovery. Only for one organism had a specific and effective antibacterial chemical been found. Paul Ehrlich's 'magic bullet', Salvarsan, discovered by him in 1909, could kill the spirochaetes of syphilis within the body with not too much harm to the patient, but it remained an isolated example that could not be widened into a general principle. Great advances were soon to be made by the application of physiological knowledge to the treatment of such fatal diseases as pernicious anaemia and diabetes, but in 1917 these were still in the future.

In September 1918, Joseph Florey travelled to Melbourne in his last effort to shore up his collapsing company. He stayed, as he usually did, at the Tasma guest house in Parliament Place, and he was taken ill with a severe heart attack soon after he arrived. His son-in-law, Dr. Gardner, was called and a telegram was sent to Bertha, who immediately went to his bedside, but he had died on 15 September, before she could get there. He was sixty-one years of age. Bertha wrote a grief-stricken letter to Howard, and arranged for the funeral at Mitcham. There is no record of the reactions of the children to their father's death. As a family they seem to have been undemonstrative, but it is a strange fact that nowhere in all the many personal letters from Howard that have survived does he ever mention his father.

On the financial side, the shock was a severe one. As the tangled affairs of the Standard Shoe Company began to unravel, the business was found to be insolvent, and went into liquidation. The ordinary shareholders got nothing, but Bertha's acumen had prompted her to convert 10,000 ordinary shares standing in her own name into preference shares, and on these the liquidator paid nine shillings and sixpence per share. When Joseph's debts were settled, probate stood at £7405, the sum of Bertha's inheritance. She was not penniless, but it was clear that the family would have to give up

Coreega. The dispersal of Joseph's proudly acquired possessions took time. Coreega was sold in 1920; the house at Belair and the Myponga land had already gone. The Lotis car, which had cost £800, was sold in 1921 for £100, and in that year the family moved into No. 12 Glenunga Avenue, Glen Osmond, a suburban bungalow very like the Fisher Street house in which Howard had been born twenty-two years before.

His father's death did not interrupt Howard's studies. The scholarships he had already won made the completion of his medical training secure, and he continued his clinical work, but in a changing environment. With the signing of the armistice in November 1918, those who had survived their war service began to return to their civilian occupations. Young war veterans appeared among the first-year students at the University, teachers who had been on active service came back to their posts. At the hospital, the old men who had carried on beyond retiring age made way for those returning. Professor Watson was succeeded by Wood Jones, less of a character, but a distinguished and cultured man. H. S. (later Sir Henry) Newland, returning with a D.S.O. and a C.B.E., resumed his position as Honorary Surgeon. J. B. Cleland became Professor of Pathology, and Dr. C. T. C. (later Sir Trent) de Crespigny came back, also with a D.S.O. and the acting rank of Colonel. De Crespigny was a striking figure. Though basically a physician, he had been appointed pathologist to the hospital in succession to Watson in 1911. He was also Medical Superintendent, and Director of the Australian Government Laboratory of Bacteriology and Pathology. He returned to become an Honorary Physician. His wide experience made him a good teacher, and he was much admired, particularly by his female students, who found his aloof manner and immaculate appearance impressive. A fair number, it is said, found him less than aloof if they took his fancy.

In 1920, in his fourth year, Howard began to consider his future. His experience of clinical medicine had convinced him that his place was in the laboratory and his heart was still set on research. He could probably make his living in Adelaide as a hospital pathologist, or as a teacher of physiology, but research was then hardly possible there, with little encouragement and almost no facilities for original work. A Rhodes Scholarship would, however, give him all the opportunities he wanted, and he would be a strong candidate. His school and university record was excellent. He had all the qualities required by the Rhodes Trustees: industry, determination, leadership, and outstanding academic and athletic success. The Scholarship would maintain him for three years in Oxford, where he would be able to study in what had recently become, under Sherrington, the finest physiological department in Europe. Against this course of action was the financial plight of his family. During the time in Oxford he would have money only for his bare subsistence and could do nothing to help them. If he stayed in Adelaide he would soon be earning a reasonable salary. He had doubts then, and regrets later, but he decided to apply.

His application was made in August 1920. It contains an impressive list of achievements, and a personal statement in which Howard set out his interests and his hopes for the future. He also explained, at some length, why he had not enlisted in 1916. There were excellent testimonials from a number of schoolmasters and medical school teachers and a certificate from the Vice-Chancellor of the University.[9] The president of the Adelaide Medical Students Society commended Florey's work in many student activities, including his editorial work for the Society's *Review* and the *Adelaide University Magazine*. But it was not until December that Howard was told the result.

It was Howard's editorship of the *Review* that had introduced him to Ethel Hayter Reed. She was a medical student, three years his junior, and, as she was one of the very few girls then studying medicine, he felt that she might care to contribute an article on 'Women in medicine'. He wrote from Coreega on 21 January 1920, concluding his letter with an invitation to tea, so that they could discuss his suggestion. It was the first in a series of 153 letters that he was to write to her during the next six years and which she carefully preserved.[10]

Ethel was the daughter of the manager of the Bank of Australasia in Adelaide. Her mother was a member of the du

Ve family and prided herself on her French ancestry. The Reeds lived in a large and handsome residence, The Red House, overlooking the park in North Adelaide, the most fashionable part of the city. But money was not plentiful in the Reed household. Mr Reed religiously gave a tenth of his income to charity, but was less bountiful to his family. Though all his four children had expensive educations, he made them conscious of his generosity. Ethel, in fact, had achieved her ambition to study medicine against parental opposition.

At the medical school Ethel was popular. She was pretty and vivacious, and being the only girl in her year she was the centre of attention. She was fond of dancing, parties, and playing tennis. She enjoyed light music and the lighter sort of literature. She was intelligent and ambitious to do well in medicine. She admired her teachers, in particular Dr. Trent de Crespigny. She insisted on being regarded as 'one of the boys' and was generally thought to be a 'good sport' and 'great fun'.

Alan Lamphee, who was very fond of her, said that he detected a steely determination beneath her charm and that he would not have cared to marry her. It is clear that Howard soon became her most serious admirer. They met frequently at the hospital and parties, but never at Coreega. Howard, years later, wrote to her recalling a special occasion in the Lamphees' garden, when he decided that he must be in love. 'Do you remember how we ate peaches from the wall? I couldn't believe that so small a girl could eat so many!' But despite his warmer feelings he continued to address her as 'Dear Ethel' in his letters.

In December of 1920, Ethel developed pleurisy. Howard wrote to her at her home. He prescribed rest, 'as I do for all my patients, it's the only treatment I'm sure about'. He sent her books—*The Poet* and *The Professor at the Breakfast Table*. He himself, he told her, was reading Osler and Mackenzie, thirty pages of each per day, and was 'fed to the back teeth with the lot of them'. But he had been allowed, in a case of fluid in the chest, 'to wield the knife. After about 12 shots I struck oil—I was thrilled!' The patient was less so—she accused him of experimenting.

On 8 December 1920, Howard was named by the Rhodes Committee as the successful candidate for South Australia. The award had great prestige as well as practical value. Cecil Rhodes had told his trustees to select 'young colonists' and students from America who were not mere bookworms but showed 'a fondness of and success in manly outdoor sports and qualities of manhood, truth, courage, devotion to duty, sympathy for and protection of the weak, kindliness, unselfishness and fellowship, moral force of character and of instincts to lead'.[11]

Howard had therefore acquired great distinction simply by being elected. But he was by no means overawed. He insisted on making his own conditions, as he was so often to do in the future. In his application to the Rhodes Committee he had stipulated that he wished to take his final examinations in December 1921, in Adelaide. But the Scholarship required that he should start in Oxford in October, the beginning of its academic year. There were two ways in which this condition could be met. He could defer his final, qualifying examinations until he had completed his time in Oxford, or he could defer his scholarship for another year. Previous Adelaide medical scholars had done one or the other. Howard, however, insisted that he should start in Oxford a term late, in January 1922. This insistence led to considerable difficulty and letters and an exchange of cables between the Dean of the Medical School and even the State Governor on the one hand and the Rhodes authorities in Oxford, but Howard got his way.[9]

Later in December, Howard went to Mallee Cliffs Station, Mildura, in New South Wales, for a working summer holiday. He had travelled by steamer for several days up the Murray River, the great waterway that is navigable for 600 miles above its sand-blocked mouth near Victor Harbour, and he wrote to Ethel about the voyage.

'I saw the most wonderful sunset in the world one night. The river was in flood, and it seemed to stretch for miles beneath the trees. At the time of the sunset it became silvered like a mirror—but I won't describe it, I'm going to write a sonnet about it in the Italian style.' It was an experience that he remembered all his life. He was to see many fine sunsets during his world-wide travels, but none, he used to say, came up to 'that marvellous sunset on the Murray River'.

From Mildura, Howard went to Broken Hill Hospital as an unqualified clinical assistant, a post reserved for the most promising of the fifth-year medical students. He wrote of his change of status, 'I'm now "Doctor" to the patients and I have to cover my ignorance by waving my arms about and looking grave.' He put Ethel right on several technical points in her letters, advised her on her choice of textbooks, and asked her to write more often. It is sad that almost all Ethel's letters in the long correspondence that ensued should have been destroyed by Howard. Something of what she wrote can be gathered from his replies, and it is clear that she had an amusing turn of phrase and definite—often critical—views on life in general and Howard in particular.

On returning from Broken Hill, Howard settled down to work for his finals. He complied with the custom of his medical school and endeavoured to memorize large tracts of textbook medicine. In May 1921 Ethel was again ill, and this time a more serious view was taken of her condition. Repeated attacks of pleurisy in a young person suggest tuberculosis and her doctors advised a year of recuperation in the mountains. Howard was deeply concerned. He rang up The Red House, only to be intimidated by Mrs. Reed. In a letter to Ethel he sympathized with her on the loss of a year of her studies, but reminded her that she was 'approximately young' (she was nineteen). He, himself, he said, had been suffering from enteritis and 'with my hypochondriacal nature was certain it was typhoid'. He wrote to her every two or three weeks and gauged the passage of time by the number of pages of Osler he had absorbed. He sent her books, but did not call at The Red House. 'Your parents appal me', he wrote, 'I don't think I shall ever have the courage to visit your sick bed.' He advised her to read, not textbooks of medicine but of physiological principles.

Howard's social timidity where the Reeds were concerned is in contrast to the professional toughness he later displayed. The Reeds probably felt themselves in a higher stratum than the Floreys, and even Ethel seems not to have met Howard's family until two years after he left Adelaide. Perhaps Howard, at this stage, was anxious not to commit himself as an official suitor. Or perhaps he simply disliked the rigid gentility of the Red House ménage. But, by the time he had done another 200 pages of Osler, he had worked up courage to call to see her.

They clearly had the sort of exploratory and semi-philosophical discussion that young people need in getting to know each other. Howard wrote to her later to say that he was very touched to find that she appreciated the beauties of nature. 'Anyone who could do this would never lack for interest and enjoyment.' In the same letter he describes a melancholy Sunday spent with a friend [Charles Muirhead], looking at old photographs. 'We were weeping in each others arms before long, to think of all the people we knew in our carefree days. He and I seem to be islands left; most of the others are married and we'll never see them again ...' He tells her that he will be 'pushing off, D.V.' in December, and that time grows short. 'Can I come round to hoe in your garden? And don't play cards—d'you know I much prefer to talk.'

Despite his impending trials, Howard found time to read, and discuss with Ethel some non-medical literature. *The Mirrors of Downing Street* and *The Glass of Fashion* by 'A Gentleman with a Duster' led him to become temporarily 'a thundering moralist'. He goes on to say, 'I have just read a dreadful book, 'The Harvester' [by Stratton Porter], and had severe abdominal pain in consequence. I prefer the solid ideas of Rose and Carless to these wretched novel writers.' By the end of September he is 'looking out for a job as carver on a ship'—in other words, for a free passage to England as a ship's surgeon. He succeeded and signed on with the S.S. *Otira*, a 6000-ton Shaw-Savill freight liner due to leave Port Adelaide about the middle of December 1921.

Howard sat for his finals during the last two weeks of November, which involved a long sequence of written papers and oral and practical examinations. He passed, but not with the expected honours. In consequence he did not gain the Everard Scholarship which had confidently been predicted for him. The reasons for this lapse from the highest achievement are not clear. Perhaps he did not find the subject of clinical medicine as taught in Adelaide congenial and his letters give no evidence of an interest in his patients as sick people. Also, his departure for England within a few days was unsettling; Adelaide and its affairs were already losing importance as his ambitions shifted into an exciting future.

At the beginning of December he wrote a long letter to Ethel. He began: 'I'm not going to write sentimental slush. I don't think either of us are built that way.' But it was, for him, an emotional letter—the nearest approach to a love-letter he was ever to write to her. They had said good-bye on the previous day. Ethel was going away to the mountains for her convalescence. Howard would, soon after, be leaving for England. He was in grave doubt about their futures. 'If either of us fail it is going to be myself. I shan't reach your height, I may not be able to throw off my selfishness and domineering manner.' He said that his feeling for her went beyond physical attraction. 'I see more in you—you may not be brilliant at your work, but what is that? You ... have a deeper understanding of men and things than falls to the lot of the average girl ... a level of mental purity and serenity that I could not hope to reach.' Ethel, however touched she may have been by this tribute to her character, was not likely to have appreciated Howard's assessment of her academic abilities. She was, in fact, intensely ambitious, and rated her medical career as more important than affairs of the heart.

On 7 December Howard wrote his last letter to Ethel before sailing. He thanked her for her photograph, which he had collected from her father's office.

It's been rather amusing the last few days to hear the different things people say to me. 'Lots of luck' is the usual—I must have said 'Thanks very much' a million times. 'Make lots of money'—I feel like pushing this type somewhere near the umbilicus and they usually have a fair expanse in that neighbourhood. 'Marry a millionairess!' 'How many broken hearts are you leaving behind?' None—is it kid?

My steam ship, S.S. Otira, arrived in this port, so I met the Captain at Dalgety's and went down to see the Chief Officer on the boat. Scotch; I catch about 4 words out of 5. Had a look at the stores of medicine. The tools for carving consists of some scalpels which might cut tobacco, some needles and a pair of artery forceps. I ordered a few extra but feel inadequate for anything big, e.g. disarticulation of head from neck. I was tickled to death by the quantity of purge carried—enough to purge Port Adelaide. The boat's made of tin throughout, decks and all, so we won't burn.

I'm sad at heart, and I'm dreading the family goodbye tomorrow. I shan't see my mother again probably, and I've been a beast—that's the rub. If a chap could dispense with a conscience, or be like Elija and push off in a chariot.

I've collected lots of good letters. Wood [Wood-Jones] gave one to Sherrington, Thomson and Keith; and Cresp [de Crespigny] one to C. J. Martin, of the Lister. The Head of Saints [St. Peter's] one to the Countess of Harrowby.

S.S. *Otira*, with Howard on board, sailed from Port Adelaide on 11 December bound for Cape Town and Hull.

3

OXFORD

By Christmas Day, *Otira* was half-way across the Southern Indian Ocean, a vast and lonely sea perpetually ridged by the huge rollers that run up from the 'roaring forties' to the south. After a dinner of turkey and plum pudding, Howard wrote to Ethel. His worries about possible seasickness had proved baseless; he seemed to be immune. He had had few patients to deal with—one or two stomach upsets he had treated with bicarbonate and purgatives, and a possible case of cholecystitis had responded to 'mustard plasters and nature'.

He was not impressed by the ship's company. 'The Captain is an absolute dud. His only conversation is ships and women, and he has soon said all he has to say on both subjects. An Irishman who dislikes Ireland; small-minded—everyone dislikes *him*. The First and Second Officers are Scots—good chaps. The chief Engineer is a Cockney who swears very heavily. There are 3 cadets, and I play deck-tennis every afternoon.' The only other entertainment was cards and anecdotes. 'Most people have been everywhere—any seaport you like to mention ... I dislike their attitude to women, which is very loose.'

What had impressed him most was the albatross that had followed them for several days. The ease with which these great birds glide with motionless wings on the air currents just astern of a moving ship is truly astonishing, and Howard spent much time in trying to photograph it. His efforts led to an invitation to photograph the ship's jazz band in the crew's quarters, and thence to his first experience of naval rum. 'It brought tears to my eyes, but I kept up a forced laugh and between gasps uttered some inanity. No more rum *whatever* happens!'

His next letter described his arrival at Cape Town, with the sun rising above mountains that looked black against the sky. He spent the day exploring the city and the slopes of

Table Mountain by tram, appreciating the Cape Dutch architecture and the centuries-old furniture and pictures he found in the public buildings. In the evening he had a night out with his fellow officers, but he did not, apparently, emulate the traditional excesses of the sailor ashore in a foreign port.

The ship left for England next day, and soon there was a serious row with the Captain. Howard had taken one of the crew off duty for medical reasons. The Captain was furious, accusing him of incompetence and the sailor of malingering. Howard stood up for himself. 'I didn't like to receive *all* the abuse, so I hoed in with a few remarks of my own. He hasn't spoken to me since—I'm quite glad, because he is an awful bore.'

Otira steamed on. Off Portugal engine trouble developed. 'Some piece of tin attached to the engine broke, and if the engineer hadn't been leaning on the throttle at that moment the whole thing would have collapsed. He shut if off just in time.' It seemed that the ship might have to limp into Lisbon but, to the disappointment of Howard, who would have liked to see the city, the engineers repaired the damage.

Off Finisterre they met the British Home Fleet, '24 destroyers and the battleships *Hood, Queen Elizabeth, Warspite, Valiant, Royal Oak, Resolute* and several others—a magnificent sight.' Then across the Bay of Biscay, 'rolling like a cow', and into the channel. 'It was a gorgeous night, clear and cloudless with ships passing every quarter of an hour. Next day it was foggy and cold as the deuce. The first of England was Dungeness Light, where we picked up our pilot. He told us that 'flu was very bad in England.'

When *Otira* docked at Hull on 24 January 1922, the Captain tried to take his revenge. He waited until Howard had packed his gear into a taxi, and then ordered him back on board for the day. Howard wasted no time on arguments about the legality of deserting ship. In his own words, he simply 'pushed off'.

He went to London by train on the coldest day for five years. The first snow he had ever seen was falling and his railway carriage was an ice-box with the condensation freezing on the inside of the windows. His sister Anne (who was working in England) met him at King's Cross, as in a letter to the family in Australia she described:

Our dear laddie has arrived in the land of snows. It was sleeting—horribly, and he felt very cold. I've been looking forward to his arrival for so long that I can hardly yet realise that he is here. He has altered a great deal—he knew me first. Well, Mother, my dear, apart from the brains which I know are a considerable quantity, you have given to the world the making of a fine man. I like his open, honest countenance, his natural reserve and quiet manner. He'll be liked here. Australians who brag are detested, and rightly so. He was looking very well. That sort of voyage was best for him. There's lots of temptation when one is on a big liner for many weeks.[1]

Next day they went to Stoke Newington to see James Florey, one of Joseph's four brothers and the only one who had not emigrated. Howard did not find him very interesting, but, to do his uncle justice, he was in poor health. Anne had bought the various domestic things Howard would need in Oxford, and next day he went to Paddington Station and took a ticket to Oxford. 'Needless to say the significance of the occasion did not fairly sink into me then, or indeed yet,' he wrote to Ethel. 'Well, it was still most damnably chilly. We passed at rapid rate through delightful misty country—it must be glorious in summer. In due time we arrived in Oxford.'

In 1922, Oxford was still a city of buildings and institutions that seemed to have remained unchanged for centuries. But deep changes both in the University and the town were already well under way. The town, which had for long resented the opposition of the University to any commercial development (it had, for example, prevented Brunel's main railway line from coming nearer than Didcot), was beginning to assert itself. William Morris, a strange contrast to his Pre-Raphaelite namesake, was building one of the largest motor works in Britain within two miles of Magdalen Bridge. The colleges, though aloof from commerce, profited from the commercial development of their large holdings of land, and within a few years the University itself would profit from the munificence of an ennobled William Morris.

In the year or two before Howard's arrival the University had reluctantly made three concessions: the admission of women to most of its degrees and departments; the abolition of compulsory Greek; and the surrender of seven centuries of independence by the acceptance of a Government grant. All were considered by conservative academics to be disastrous capitulations to modern barbarism, the only point of

debate being on which one was the worst. But a more fundamental undermining of University traditions had been taking
place within its walls during the previous half-century. This
was the growth of science. Acland's museum, with its various
extensions, housed fifteen active professorial departments by
1922.

Having arrived in Oxford, Howard's first duty was to report
to Francis Wylie, the Secretary of the Rhodes Trust. When
Cecil Rhodes died in 1902, he left a complete plan for scholarships at Oxford for twenty 'young colonists' and 32 young
students from the United States in each year. The tenure was
three years, the stipend £300 per annum, and foreign travel
was encouraged. This scheme was a surprise to people who
had known Rhodes as a man of action with no great interest
in academic pursuits. Of the seven trustees he appointed, only
two had any first-hand experience of Oxford; Lord Milner,
who had been a Fellow of New College; and Lord Rosebery,
who had been sent down from Christ Church because he
refused to give up his racing stud. The scheme caused misgiv-
ings. Oxford dons were dismayed at the prospect of invasion
by young barbarians. And how could they pass the compulsory Greek in Responsions? In the 'Colonies' and United
States heads were shaken over the effect an ancient and effete
university might have on their virile youth.[2] The worst fears
were not realized. Oxford gained vitality from its young barbarians without noticeably emasculating them; it was stimulated by the demand for postgraduate study and if, as it did,
compulsory Greek became the casualty its loss proved not to
be the ultimate disaster.

Rhodes Scholars had the choice of the college to which they
would be attached. Howard had chosen Magdalen, probably
because his former headmaster had been a Magdalen man
and because its President took a personal interest in Rhodes
Scholars. But, in some respects, Magdalen was not then the
most welcoming of institutions for an outsider. Oxford in 1922
was in a strange mood. The war years had seen the virtual
cessation of University life. But in 1919 there were over 4600
undergraduates—more than ever before. A number of these
had come back from the war bringing with them a mixture
of relief and sadness. Many were maimed for life, and so many

more of their friends had *not* come back. The undergraduate society determined to forget war. The languid aestheticism of Oscar Wilde's day returned, being something as far removed from the crude horrors of the trenches as could be imagined. By 1922, when Howard arrived, almost all the young war veterans would have gone down, their places taken by the boys who had been only fourteen when the war had ended. But the atmosphere remained, and perhaps as strongly at Magdalen as any other college. The caste system there was rigid; to be accepted by the élite, one had to be a member of the English upper class, and to have been educated at one of the four top English public schools. To be Australian and of humble birth, though tolerated in a Rhodes Scholar, meant exclusion from that particular herd.

After reporting to the Rhodes Trust Office (Rhodes House was not built until 1928) Howard went to Magdalen.

Needless to say, the buildings are wonderful and I could scarcely realise things as I walked past all these ancient places. Magdalen is undoubtedly the most beautiful college here, that is, taking it as a whole—the view from Magdalen Bridge is magnificent, and the High! The most beautiful street in Europe!

I interviewed the crowned heads of the place—the President who asked sundry questions and said 'Oh, really!' when I answered. Saw Professor Dreyer the pathology man, he seems very decent . . . Saw Arthur Thomson, the anatomist. Delightful old chap—gave me much advice. He was a friend of Wattie's [Archibald Watson]. Wood evidently told him I was thinking of taking up a scientific life. He advised me to do physiology, and then read in pharmacology—in which so little is known. He also advised doing an M.R.C.P., as it gives so much standing. The most comical thing of all was that if I decided on a scientific life it would be a big struggle financially, and if I were married I could only just carry on. However, if I did what he recommended, and if I were a dud or didn't wish for the hardship of scientific life, then, with an M.R.C.P. I'd have a good standing as a physician. So I'm to carry on for 3 years and then decide.

Howard had enrolled in the honour school of physiology. It was the second term of the final honour course, which made difficulties. As he would have to wait two terms for the next course, he found himself doing elementary work which he had already done at Adelaide. He wrote that, in college, he had a

very decent pair of rooms. There are three other South Australians, Girdlestone, Wells and Coombs. Hopkins, who went to Saints, is quite a lad and has introduced me to numerous fellows. It is a fact that one tends to meet only colonials. The Englishmen are a queer lot. The chap on the same

landing as myself went to Winchester ... and has spoken, but the majority
preserve a frigid silence. I am assured that it is their manner and when
you thaw it they are very decent. I'm also told it takes patience. They all
seem to take themselves so seriously that I want to laugh ... It snowed
fairly well last night and this morning there is a frost. All very beautiful.
Took some photos. Damnably cold. I have to sprint roughly 200 yards
across the quad and through cloisters to have a bath. Breakfast in Hall,
lunch can be had there but I have a snack in my rooms. Dinner in Hall
at 7.15. Afternoon tea is the great meeting time. I've started to play
hockey—just smacked the pill when it came my way, not knowing any
better.

About the middle of February, Howard received letters
from Ethel which had arrived together by the same mail. One
of the difficulties of their correspondence was that any question raised by either could not receive an answer for nearly
three months. Howard was not the most patient of men; his
changes of mood were frequent, and Ethel found herself argu-
ing or consoling, to his intense frustration, on situations that
had ceased to exist when her letters arrived. But she had sent a
photograph of herself, which vastly pleased him. She was, he
wrote, looking at him with a half-amused, half-critical expression. The picture had the place of honour in his room and
he often referred to it later, saying that he would try to read
from it her approval or disapproval of the decisions he was
to make.

His letters during these first months in Oxford, and the reflections in them of Ethel's own, reveal something of their
relationship. It is clear that they did not know each other well.
They questioned each other about their likes and dislikes,
ambitions and ideals. Howard was greatly attracted by Ethel;
he did not ask her to marry him because, having chosen the
'scientific life', he was in no position to marry for several years.
But he wished to maintain their correspondence and to build
up a picture of her as his ideal woman—something that can
be done at a distance of 10,000 miles with fatal ease.

Ethel, for her part, found Howard attractive (though not
to an extent that precluded other friendships), and she realized that he was likely to do well professionally. She had no
immediate intention of marrying anyone, because she herself
wanted a successful career in medicine. But she had told
Howard, before he left Adelaide, that she did not think that

'there would ever be anyone else'. He reminded her of this in a letter, adding 'and if our plans—of which we have none!—go wrong, I won't forget that.' Howard, lonely and homesick, repeatedly begged her to write longer letters and more often. Without this constant dependence on her, Ethel's attachment to him might well have lapsed. He told her that he wanted a relationship based on more than sexual attraction.

Sex, as can be demonstrated by recording drums and electricity, is our greatest stimulating force. The struggle of youth in our civilization is to suppress it and make other interests loom large ... Some minds—and I hope ours are among them—can perceive beauty in biology or chemistry—but anything romantically or 'humanly' beautiful will probably stir us more.

The other issue that Howard discussed was religious faith and morality. Ethel told him that she was an atheist and a materialist. She rejected beliefs that could not be scientifically proved, including the existence of God and of mind except as a 'secretion of the cortical cells of the brain'. She rejected, too, any abstract ideals of morality. Behaviour, she maintained, should be determined by its practical effects on other people and society. Howard replied: 'I am in a transition stage. I don't believe in a personal God but—this is undemonstrable, all such things are necessarily so—I do think there is something ... immeasurably superior to our best thoughts.' On the question of morality, he had this to say:

If there is no inner incentive then one is only moral because of mutual benefit. It is simplicity itself to lead a life free from restraint doing what pleases one, but it is exceedingly damaging to our Ego. If one thinks that this Ego stops with the last ventricular systole, why worry? But I prefer to look on the brain as an instrument of mind, and not on mind as a mere secretion of the cortical cells.

A moral life is an ideal because we have to strive—and damnably hard sometimes—to attain it. You agree on being moral, but deny the ideal. For me, the ideal would not be a great driving force if I thought it was all going phut. Possibly my love for my fellow men is not of a high enough order.

These exchanges establish two things: Ethel believed that she was quite emancipated from what she regarded as the old superstitions, and Howard was an idealist.

The greater part of Howard's letters described his everyday

doings. He made some social progress. There was a 'Wine and After' party at Magdalen, in which he joined.

Prolonged dessert and port in the Junior Common Room until the higher cerebral activities are dulled and everyone is happy and talkative. Too much is drunk by many. Afterwards we adjourned to run round the cloisters uttering discordant cries. Altogether a very splendid show and an excellent method of getting to know people.

He also received an invitation from

some old duck from North Oxford to a tea party. As I'm equal to all experiences I said 'Right-o' expecting amusement. My suit with a crease in the trousers smelt horribly of moth balls as I arrived. I was the only young male of the party (the curate stunt). The man of the house was a pleasant old chap among all the old gossips and dowagers. I endeavoured to look like the hardened criminal of the bush everyone expected, and talked about the weather to a don's sister and a woman with deformed teeth. The old man was something learned at the Bodleian, and was writing a book on Oxford. He said he didn't know how to describe Magdalen without offending Christ Church. I was very bland and cheery and said 'Ah, a little thing like that doesn't matter' and the old chap took this to mean that it was his book that was the little thing, saying that of course he carried no weight. I collapsed under explanatory efforts and finally fled.

(Howard's 'old man' was probably Falconer Madan, formerly Bodley's Librarian, who published his *Oxford Outside the Guidebooks* in 1923. If so, he had solved his problem. He described Magdalen as 'the most beautiful of Oxford Colleges' and Christ Church as 'like nothing else in the world'.)

The teaching in physiology did not greatly impress Howard. 'The majority of Englishmen seem to have an almighty liking for themselves. In fact, they conceal a great amount of ignorance under a noble exterior. I was thoroughly over-awed at first. I said "yes, sir; no sir", with great vigour until I found them saying absurd things with a convincing air.' But, about the middle of February he wrote, 'the great Sherrington himself came and gave me a word the other day. It gives me quite a thrill to think that these are the men who are altering all physiology and they themselves don't seem to realise it. The people who really are top notchers couldn't be more humble.'

Sherrington, then aged sixty-five, was still at the height of his scientific powers. He was President of the Royal Society. His research was doing for the nervous system, it was said, what Harvey's had done for cardio vascular physiology.[3] But he was more than a great research worker and technician. His book *Man on His Nature* is a work of philosophical insight. He was deeply interested in the arts, particularly in painting. And he was a poet, the author of a published volume of verse.[4] Sherrington's career, with its wide variety, its almost leisurely pace, and its tremendous accomplishments forms a contrast to the hectic and often non-productive hustle of today's science.

He was born in 1857, the son of James Sherrington of Great Yarmouth. When his father died, his mother married Dr. Caleb Rose of Ipswich, a physician with interests in the classics, art, and archaeology. Charles grew up in a house frequented by well-known painters and crammed with their pictures. In this atmosphere of arts and science, so prized by Acland, his versatile mind could flourish. Like Howard Florey, he was good at games. He played rugby for St. Thomas's Hospital and for Caius College at Cambridge, and rowed in the Caius boat. As a pioneer winter-sports enthusiast he was one of the first few who provided Switzerland with a new industry.

Sherrington entered St. Thomas's Hospital Medical School in 1876, but interrupted his medical studies in 1880 to go to Cambridge as a demonstrator in Foster's department of physiology. The event that determined his career took place at that momentous International Medical Congress held in London in 1881. Following the demonstration by Goltz of Strasbourg of a dog from which parts of the cortex had been removed, a heated discussion arose and it was agreed that the brain should be examined by a panel of experts. One of these was Langley, Foster's lecturer, and Langley brought in the youthful Sherrington as a collaborator in work that revealed to him the potentials of the experimental approach to neurophysiology.

Sherrington's years at Cambridge were occupied by the mixture of unhurried research, teaching, and study that often marked the liberal scientific education of a century ago. He demonstrated at Cambridge and taught histology at St. Thomas's in London. He took his M.R.C.S. examination in 1884, and went to work with Goltz in Strasbourg during the winter. In 1886 he passed the L.R.C.P. examination, thus qualifying as a doctor ten years after enrolling as a medical student. But he had gained First Class Honours in the Cambridge Natural Sciences Tripos, published his first neurological papers, and his original work was already attracting attention. He was a member of a research team sent by the Royal Society to Spain to study the 1885 cholera outbreak, and there met Ramón y Cajal, a young neurohistologist, whose methods were to play such an important part in his own work. A year later Sherrington visited Venice and Puglia, again to study cholera, though it was Venetian art that claimed much of his attention. From Italy he went to work with Virchow and Robert Koch in Berlin, and in 1887 he returned to England as Lecturer in Physiology at St. Thomas's Hospital and Fellow of Gonville and Caius College, Cambridge.

In 1891, Sherrington was appointed Professor-Superintendent of the Brown Institution in London. The previous holders of this post were all men destined to achieve great distinction: Burdon-Sanderson; W. S. Greenfield, the Scottish pathologist; C. S. Roy, who became the first Professor of Pathology in Cambridge; W. H. Welch, the American pathologist, and Victor Horsley, the neurosurgeon. And it was from there that Sherrington carried through the historic first use in Britain of diphtheria antitoxin, described in Chapter 1.

This clinical diversion did not distract Sherrington from his commitment to neurophysiology. He was elected a Fellow of the Royal Society in 1893. Two years later, when Burdon-Sanderson vacated the chair of physiology in Oxford, Sherrington applied for it, as did J. S. Haldane (Burdon-Sanderson's nephew) and F. Gotch. On Gotch's appointment, Sherrington accepted the Holt chair of physiology in Liverpool, where he developed his original ideas on practical teaching and began a lifelong friendship with a young American visitor, Harvey Cushing. In the research field, his methodical experiments led him to a brilliantly conceived picture of the integrative action of the nervous system. An example was his work on the scratch-reflex in the dog. By using an 'electric flea' he showed how the reflex brought into action nineteen muscles beating rhythmically five times a second and seventeen more that kept the dog upright during the process.[3]

In 1913 Gotch died and the Waynflete chair at Oxford again became vacant. Once more Haldane and Sherrington were obvious candidates. Both were physiologists of great distinction, and it was Sherrington who was appointed. Haldane resigned from the Department of Physiology and retired to a private laboratory in the grounds of his home in North Oxford.[5]

Physiology in Oxford had suffered from a lack of firm direction for some years. Gotch had been overshadowed by Haldane—a man powerful in body, mind, and character—whose experiments on the respiratory effects of deep-sea diving, mining, and high altitudes had often involved personal danger and earned him great respect. He had built up a sub-department of his own, with his collaborators Douglas and Priestley and, later, his son John Burdon Sanderson Haldane, a man very like his father. Such a division of interests and authority within a department leads to uncertainty, and it was one of Sherrington's first tasks to create a sense of unity. Haldane's departure made his task possible, and without losing the work on respiratory physiology, since that was still continued by Douglas and Priestley. But Sherrington's reorganization of his new department was interrupted by the war. He himself was involved on Government committees, and in his study of industrial fatigue he worked for a time in a shell factory. It was not until the war was over that he could put his plans for teaching and research into effect.

As a lecturer, Sherrington was apt to be led away from his topic by his own enthusiasm, but he was an inspiring teacher able to convey difficult concepts with a sense of discovery exciting to young people. He enjoyed their company. His warmth of manner and a genuine interest in their opinions and problems won their affection. He did not overawe them. His humility was that of a great scientist, and he never allowed his immense prestige to be a barrier between himself and his students. By 1922 when Howard Florey joined it, Sherrington's laboratory was already providing the best neurophysiological teaching and research in the world.

Howard wrote to Ethel in March that he was coming to like Oxford.

If one is sympathetic, it grows on one. The beauty of it all—the old grey walls—the associations. It's a wonderful place and I'm very glad not to have missed it. It makes the past live—one is humbled by one's own insignificance. I got very little, intellectually, from the chaps at Adelaide Varsity. Quite the reverse here. I feel I have all to gain and nothing to give.

But he was still out of tune with the English undergraduates.

The great majority annoy me excessively. The Aesthetes, who wear gilded ties and other odd things are nauseating. The few I have liked–probably they don't like me—all play games. But even some of these are insufferable. On the other hand some of the older men are nice. I'm rather unfortunate in being much older than the general fresher. The attitude towards women here—and in England generally—is amusing. They are tolerated. They should be stud-ewes, warm their lord's slippers and cook the meals or be diverting in the butterly style. The women dress extraordinarily well here and look very natty—they are, on the whole, a much more good looking lot than in Australia.

He replied to a letter from Ethel in which she must have told him that she did not want to be committed to marriage. He felt the same, he said. He wanted a scientific career and a bachelor could achieve this more easily than a married man. 'But, dear girl, give me the chance to write to you and know you better—you are practically my only friend.' And he continued: 'The trees are just beginning to bud here—it is now much warmer. Magdalen will be simply a paradise in summer. If we were married, we ought to live here. Qui sait?'

Howard was falling under the spell of Oxford. How much has been written, and with the skill that Oxford both inspires and teaches, on the charm of that ancient city. The timeless sense of history there does indeed revive the past, as if the centuries are somehow arrested in its stones and customs. Even the casual visitor feels the presence of that ageless other-world of culture behind the physical beauty of some college building. Those who know Oxford better, but are not *of* it, may find the pedantry and studied verbal cleverness disenchanting. But to the young Oxford undergraduate of past generations the impact of such a university was overwhelming. For the first time, perhaps, they were free; free of the constraints of school or home, free to study, or to play or to idle, surrounded by companions from whom they could choose sets or cliques or lifelong friends. This was the springtime of their lives, and they would always identify Oxford with the magic of those years. Henry Tizard, the scientist who later became President of Magdalen, wrote: 'I remember the joy of coming over Magdalen Bridge when I was an undergraduate. One never quite gets over that.' For the rest of his life he could never judge Oxford quite by those same critical standards which he used for all else.[6]

Howard's response to this atmosphere was at first uncertain. When Ethel wrote about her stay in the sunny mountains of Tasmania, he suffered an attack of homesickness which was aggravated by a return of winter to Oxford at the end of March. He told her that he found the bare trees damp and depressing. He had changed his rooms and was then in the Grammar Hall,

put up somewhere around 1300, very comfortable and quaint ... I have spent 3 days in bed with asthma and bronchitis, the first decent attack in two years—and that's enough. This blasted climate is about the dead end. Couldn't read much because of the quaintness of the buildings and their poor lighting. The only conversation is ten minutes per day with my scout, and that dull ... Knowing myself what a pseudo-person I am, I wonder what you, a genuine person, can see in me. I'm always pretending, damn it, even to myself. Also, large numbers of people pretend too. If one does, oneself, one can see it in others.

When he recovered, he went to tea with the Sherringtons. Lady Sherrington was well known for her hospitality, particularly towards young overseas visitors. Howard described her as

rather an enigma. Her complexion fades from pink to brown *suddenly*. She obviously assists nature by art ... I have a prejudice against this. It seems to me, if you don't own your complexion you may be slightly false elsewhere. I was also surprised to find women as well as men smoking cigarettes. Didn't see much of Sherry, who was late.

His first term at Oxford ended, and he was free for the vacation. When he next wrote to Ethel it was from London, where he was staying with his uncle. He examined St. Paul's ('excessively grand in the Italian style'), Westminster Abbey ('crumblingly ancient'), and the Tower of London ('quite awe-inspiring'). He met Hugh Cairns. Cairns, a previous Rhodes Scholar from Adelaide, was then holding house-appointments at the London Hospital, before specializing in surgery. He and Howard seem not to have known each other in Adelaide, but Cairns welcomed his younger compatriot, showed him over the London Hospital, and was to play a considerable part in his later career.

Howard found London very wonderful and fascinating—crowds, traffic, historic places. He spent much time at the British Museum, and discovered the little-known John Soane Museum where he enjoyed the pictures by Hogarth, Turner, and Watteau. He also went to a Promenade Concert to hear 'Henry Wood tickle an orchestra'. But he saw two sides to London: 'there is a dreadful contrast between the East End and the West End. The West End is beautiful: after seeing the buildings one realizes that there aren't any in Australia. The slums are dreadful ... The people are pitiful, poorly dressed and in multitudes, but full of life. But colour is lacking. All seems to be grey.'

When he next wrote, it was from Salisbury. His sister Anne, who was then matron at a boys' school in Abingdon, near Oxford, was spending her holiday with a Dr. Livingstone in Salisbury, and Howard had been invited too. 'I like my sister,' he writes. 'We see eye to eye in numbers of things. I never could with the rest of the family.' Howard was being taken up by influential (and usually elderly) ladies. In London he had been entertained by Lady Francis Ryder, the daughter of the Countess of Harrowby, to whom he had a letter of introduction. He also spent some time with Miss Kerr, an interesting old lady, who had been a Lady-in-Waiting to Queen Victoria. And after his visit to Salisbury there was a plan for him to stay with another rich old lady in Sherborne, Dorset. 'Somebody I don't know, but little things like that are beginning not to affect me. I wear a smile, say inane things and hope for the best—they probably think I'm quaint, if nothing worse.'

In Salisbury Howard was staying in a house 600 years old, and was unable to put on his shirt without hitting his bedroom ceiling. He was greatly impressed by the Cathedral, with its tall, graceful spire, and he explored the surrounding country on a bicycle. In Sherborne, the rich old lady had a motor car, and drove him to see Stonehenge, the races at Wincanton,

and the meet of the hunt at Sherborne Castle. Her house was
only 500 years old, he wrote, 'but it is haunted—both the old
lady and her dog see the ghost'. He went to several dances,
but found the girls 'very stodgy, one has to draw words like
corks'. He was enchanted by the countryside, and it is interesting to see it through Australian eyes:

> Dorset is fascinating and extremely beautiful. We have nothing I've
> seen in Australia at all like it. The paddocks—fields, they call them—
> are small and each is surrounded by well-clipped hedges—just think of
> clipping those miles of hedge. The fields at this time of the year, just as
> the grass begins to grow are absolutely like lawns. Nothing is wild here. The
> roads are ... narrow and sunken ... the villages have just happened, and
> then a road put through. Nearly all the houses are centuries old, with
> thatched roofs.

He was astonished to find that the agricultural worker's wage
was only 35 shillings per week.

Howard returned to Oxford in April to start another term.
He was now in rooms in Long Wall. He wrote that he was
still uncertain about the future. He would rather do research
than general practice. 'I'm no good at top-dog stuff, depending on one's smile and method of holding the fair patient's
hand and her mother's. But, as for research, I believe I'm a
dud after all—but I'll do a year and see.' He was playing a
good deal of tennis. 'They are under the delusion that I'm
a good player ... so you can gather the rotten players that
exist here, and they're all so damned polite about it too!' He
was in the tennis team that played the Staff College at Sand-
hurst. 'We had a glorious motor drive. The King happened
to be there too.' He was also spending time on the golf course
at Cowley—'the place where they make the little cars. You
may have noticed them buzzing about Adelaide.'

Summer had come to Oxford and with it the delights of
the river. Howard took up punting and found it more difficult
than it looks. But he was enchanted. 'The punting on the
river—ah, the River! To sit in a gently rocking punt, under the
magnificent elms, the sun setting and the birds warbling, is as
pleasant a situation as I've ever been in—only you are needed
to complete my bliss!' And the river at night—gliding along
the moonlit waterways—made him still more sentimental.

Punting is, indeed, a most romantic occupation of which Sherrington had written:[4]

> And now below through shadows starred a boat
> Steals by me laden with singing and young laughter
> And, higher, a wide-flung casement casts afloat
> Pulses of waltz the which white robes sway after.

Howard told Ethel that she was always smiling in her photograph. 'Englishmen never smile except under provocation. I hope I don't get like that.'

Howard was perturbed by the next letter from Ethel, who had been ill again. She complained that she was lonely, and was finding correspondence with Howard a poor substitute for companionship. He replied that he, too, had no intimate friends.

Sometimes I think I am more or less liked here. At others I get most depressed about it, and revile the English. But I don't feel justified in chucking England. I'm sure I will do something as good here as elsewhere—making due allowance for being a dud. I told you how I stand financially. In three years I'll be in precisely the same position, and I couldn't ask you to marry me then, even.

The more I see of women here, the more I realise what you stand for. They seem so artificial—faces, manners. It is hard to believe that they could ever be sincere ... I'm glad you appreciate old things and buildings. Most Australians have an amused contempt for them I, myself, feel a sense of awe when I stand in an ancient place and realise that 500 years before people also stood there with much the same feelings. I'm thankful that you and I like the same things. I get most awfully lonely for someone I can really talk to—you're my only friend.

By this time he was taking more interest in physiology. After a lecture by Bayliss of University College London, Sherrington, Haldane, Priestley, and Douglas took part in the discussion. Howard wrote: 'It was the best collection of scientific heads I've seen. It's a wonderful thing to see these men in their native habitat—one can appreciate that they are men and also supermen at the same time.' He mentioned what was to prove an important event. 'I'm going to meet Harvey Cushing tomorrow. My American friend is doing the trick—ought to be rather interesting.' Cushing was then the world's leading neurosurgeon, and Howard's American friend was John Fulton, a fellow Rhodes Scholar at Magdalen.

The rest of his letter was unlikely to have pleased Ethel, if she took it seriously. Howard dilated on the inferiority of women in every field of art (including 'rubbishy women novelists'). In science he felt that they lacked the necessary logic though in medicine he allowed that they might make 'good average general practitioners'. There was an inevitable emotional clash between their domestic, maternal instincts and career ambitions. Why bother, he asked, when men were so much better at such careers? Howard's ideal partnership was 'a good and beautiful woman drawing out the best from a great man'. But, on the subject of her own future, he would not expect her to give up everything, should they marry. A vow of celibacy was not essential for her medical career.

The Long Vacation came. Howard had planned a continental tour, preceded by a visit to Wales. He stayed with 'an old duck and her elder brother' near Fishguard, and they motored to Dale, in Pembrokeshire, through beautiful, wooded, damp country. Milford Haven reminded him of Sydney Harbour, hence, he supposed, the name New South Wales. His next letter was from Brussels. He was travelling alone, third class, and worried by his lack of languages. Fortunately, he had 'met up with an old duck from Cambridge who knows French and German like a native'. He appreciated the Flemish architecture and, particularly, the pictures. The girls of Brussels he thought pretty, but suspected that many were immoral. However, he found *Le Manikin qui pis* amusing. 'I can't imagine an English city putting up such a thing—they won't even allow Sunday games! The continental attitude is superior to our mock modesty.'

He visited Cologne, the castles of the Rhine, Bonn, and Nuremberg, where he stayed for a few days at a hotel that charged him one shilling and sixpence a night. Then on to Linz and Vienna, where he planned to spend several weeks learning German and working in the medical clinics. 'Perhaps you don't think I do much work', he wrote, 'Well,—I don't!' But though not much impressed by Viennese medicine he was enthusiastic about the opera, to which he went every night at the cost of sixpence or a shilling. 'All the Austrians seem very nice, polite, and friendly. The only people who are rude are the English ... Like you, I get irritated by most of the people I meet. Does this mean that we think, unconsciously, that we are superior?'

Howard settled down in a *pension* to work at his German and enjoy Vienna. The hills and woods surrounding the city reminded him of Adelaide, and he found the buildings magnificient—particularly by moonlight. He visited palaces with marvellous art collections, but felt something of a trespasser, since they had only recently been taken by the State from their private owners. He went often to symphony concerts and it seems that his real appreciation of music dates from this time, though he was more aware of the quality of performers than of composers. He told Ethel that he disliked jazz. 'It fills one with sordid thoughts—but some people like the sordid.' Nevertheless, he has been having a gay time with a party of youngsters—including two Czech girls—dining, dancing, going to the theatre and opera, and 'becoming a thorough-going Viennese, lazy and café-loving'. His living expenses, including all such amusements, were about thirty shillings a week.

There was a grimmer side to Vienna which distressed him. Thousands of its inhabitants were on the verge of starvation. 'There was a bit of a stir with the out-of-works yesterday', he wrote, 'and about ten had the Politzei's swords stuck into them. Poor devils—they're hungry. A loaf of bread costs 4,000 K. They are getting 2,000 K. a day for wife and all—it's dreadful. I was having some food in a café today. A man with the saddest face I've seen came round selling matches … He was starving—the intelligentsia get about £10 a year to live on!' He had also been hearing of the malnutrition in Germany that resulted from the Allied blockade. 'Don't let's be so damned self-righteous about our part in the war and the German atrocities.'

Then in Vienna, Howard received a letter from Ethel in which she said she wanted their correspondence (and relationship) to be ended. He wrote, quite distractedly, to ask why. 'Is it that you prefer medicine to marriage? Or is it your mother? Or is it that you don't love me?' If it were the first, he would feel hopeless. If either of the others, he might change things. He was prepared to wait years for her. He begged her to carry on with the letter-writing in a quiet way. 'I can't let you fade out of my life without a bit of a struggle. Do you think you could write to me once a month without unduly straining things?'

But their correspondence did lapse. He sent her a Viennese picture for her twenty-second birthday, with a brief note saying 'if you don't want this, give it to your sister, she can decorate her kitchen with it'. He did not write again until the Michaelmas term had passed, and with it his first year in Oxford. Then he told her that he had been doing 'a small amount of work—cats and respiration stuff, and biochemistry. I play hockey for the college, ran in the relay races, and get the odd game of golf. The Prince of Wales came down and stayed here a night—quite a gay scene, several people even got drunk.'

Ethel, perforce, had to thank him for the picture, and her letter broke the three months' silence. He replied with an account of his doings during the Christmas Vacation. He had been back to Dorset and wrote 'I've never been bored so stiff in my life. Met the local Nabobs—Colonels and Majors whose cortex could have been removed without any perceptible effect. I was forced to go to a village dance where, while waltzing with a six-footer, I backed her into one of the card tables, upset it and then fled.'

After Dorset, he went on a continental tour with Hopkins and Ride (the Rhodes Scholar from Melbourne). They visited Brussels, Louvain, Cologne, Koblenz, Frankfurt, Heidelberg, and Freiburg, and Howard maintained that he was becoming a real connoisseur of ancient architecture. He had time for sport, however. He bought skis and boots for thirty shillings and had five days skiing in the Black Forest for ten shillings, all inclusive. Just before he left, the French occupied Essen. 'They may get their coal but they have stirred up a hatred and enmity which is dreadful. The Germans state openly that they're going to have another slap at France and won't wait a century to do it.' He went on to Berlin, where there was an intense hatred of the French and indeed all foreigners. He understood enough German to overhear the sneering and insulting remarks made about him as a supposed Englishman. He found Berlin 'rather an ugly place. The Cathedral is more like a grand opera house and the Unter den Linden is nothing striking. Also it is under snow, which probably makes it look better than it really is.'

Back in Oxford he found no letter from Ethel, and resigned himself to a solid term's work and the looming prospect of his examinations in June. He planned to go to the Black Forest in the Easter Vacation, but when the time came he was ill with influenza and had to stay in Oxford which, in vacation time, was 'like living in a vault'. Then he heard from Ethel, who wrote that she found his description of his travels unsettling. She, too, would like to see the world. Adelaide now seemed small to her. 'True,' replied Howard, 'yet one could be perfectly happy there—if ignorant. Happiness comes from within ... I don't think anyone can be happy for more than a short time in any one period. I'm talking about myself, as usual. I've got opportunities most chaps would give a lot for, and yet I'm not satisfied. It's a frightfully lonely business living among strangers.'

Ethel also mentioned a wish to come to England. 'I'd give quids if you could!' wrote Howard. He would love to act as her guide, demonstrate Oxford, and introduce her to Sherrington. She had taken him to task for being too enamoured of cleverness. Howard replied, 'I reserve the right to call people bores if they bore me. You always thought me selfish—which is true; and swollen-headed—which is true.' He tells her, in passing, that he has burnt all her previous letters. His examinations are 'only a month ahead, all physiology and I trust I've got a smattering of it now'.

At the beginning of the summer vacation Howard went to Bath to stay with his old headmaster, Canon Girdlestone. He had done the written part of the examination and thought it 'not too bright an effort—lack of knowledge, principally. There's still a week of practicals to wade through and I'm fed up with exams.' He had been to the Rhodes dinner. 'Baldwin—the Prime Minister—and several other hot-air merchants spoke, but it was quite a good evening.' He then went on to describe a discovery made by Dreyer and his co-workers at the School of Pathology in Oxford of a supposedly effective vaccine against tuberculosis. 'It holds forth, I believe, an extraordinary promise of success, though at present, owing to its limited trial, they are very cautious. If it should be so, it is the greatest discovery since Lister. One gets rather lost in a maze at the prospect of stopping the appalling thing of seeing young people maimed or wiped out while one can do nothing.' (Unfortunately Dreyer's 'diaplyte' vaccine was soon found not to work in human cases.)

Howard next writes on 11 July from a Tudor manor-house on the Welsh border. 'I'm sitting under a cherry tree writing letters like the heroine of a novel.' The house belonged to the family of a Magdalen friend. 'He's in love too, poor chap, and has told me about it. He gets extremely agitated about the thing. He has asked me away, I think, because I didn't get obviously bored (though I am) by his outpourings.' But Howard is more obviously bored by his other companions there. 'This English girlhood—their chatter is intolerable. Cries of "topping" and "priceless"—they're utterly unsophisticated, even girls of 23.' In Oxford he had been to a meeting of the Physiological Society.

All—or rather some—of the big guns were there. Sherrington in the chair. Haldane, Douglas, Priestley, Banting of insulin fame was there—a most poisonous looking fellow. Benedict (of the sugar test), Barcroft, Leonard Hill, also A. V. Hill. I sat next to Thomas Lewis ... I've arranged to work on some research in pharmacology with Professor Gunn next year. Should be O.K.—he's an awfully nice chap with a very pleasant wit.

I suppose really the main reason I'm writing is to tell you about the exam. I managed to raise a first in the Physiology, I don't quite know how it happened—all the others couldn't have known too much. As a point of fact I may as well say I'm rather pleased. There were 64 in for the show and only 5 firsts [another of them was Fulton] ... It may mean that I can raise a more or less decent job when I leave here. If you can excuse some more egotistical talk, what mainly pleased me was to be told by the Pres. of the R. S. [Sherrington] that I'd done very well and he was pleased. I hope you're pleased too.

Howard's achievement was, indeed, considerable. Sherrington would now pay particular attention to him. To obtain a first in the Honour School of Physiology in Oxford is not a guarantee of success in scientific medicine, but it is a guarantee of interest on the part of medical scientists.

After these results were known, Howard set off on his third continental tour. He had been invited by the two Czech girls he had met in Vienna to stay with them in the Carpathians, a prospect he thought likely to be interesting. He planned also to visit Hungary and Italy, and spend the last two to three weeks in Vienna. 'I think I'm wise to do this travelling rather than working in hospitals—the chance will probably only occur once, and I think it broadens one's outlook.' He travelled via Bruges, where he bought a lace handkerchief for Ethel made by an old woman on a doorstep surrounded by children, dogs, spools, and pins; Ghent, which was *en fête* with fireworks and bands; through Germany to Prague and then into Hungary. 'There was a young Czechoslovak lass sitting next to me in the train. Well, as the night proceeded, she thought she would like to sleep on me. I did my best, thought "Well, I'm in Bohemia anyway"—and she slept!'

He gave, unfortunately, no account either of the Carpathians or of the two Czech girls. But he described the colourful scene in a market town, with the people in national dress, and bands of gypsy musicians. He found the Hungarian music romantic. 'If you were here it might make you want to hold my hand or something equally absurd.' He returned by way of Italy. 'Venice', he wrote, 'is in some respects as good as the pictures of it ... St. Mark's and the Doge's Palace are incomparable. One was in Venice! Across the lagoon, gondolas black and the buildings standing out in relief—quite wonderful!' During the next two weeks he visited Ravenna, Florence, Siena, Pisa, and then Florence again for a stay of six days. 'I often wonder if the Italians who created these marvellous things are like the local inhabitants who spit on the church floors. They're a dirty crew.' On Lake Garda he took a boat journey to Riva. 'The mountains rise sheer from the lake which is an unbelievable blue—odd, colourful villages hang on the sides, boats with chocolate sails. Then by train to Innsbruck via the Brenner—the best journey I have done.'

But he was disturbed during his travels by the evidence of nationalistic feelings. 'Germans, Austrians, Italians, Czechs, Hungarians—all fussing about their national honour, damn them ... I've come to the conclusion that it's the menace of the future. I've been very distressed to see the beginnings of a national pride being stirred up in Australia.'

On his return to Oxford for the Michaelmas Term, Howard changed his plan to do research in pharmacology. He still had four more terms of his Rhodes Scholarship, and he was free

to choose how he would spend his time. The choice was made
for him—by Sherrington. Before the summer vacation he had
obviously encouraged Howard to apply for a post as demonstrator in his department. In a letter dated 12 July 1923 he
wrote:

Dear Florey,
    Let me congratulate you on the excellent work you have done. In reply
to your note, I shall be very pleased for you to come on in October, as
a Demonstrator for the mammalian class.[7]

Howard was delighted by this offer. He wrote to Ethel:

Sherry has seen fit to let me work under him—that is, I'm not doing phar-
macology research. This is God's own opportunity. He's the President of the
Royal Society and probably is recognised as the greatest living physiologist.
Also, I might get a D.Phil.—Oxford's best research degree—in the time.

Howard had applied for a Fellowship in physiology or zoology at Merton College. 'Have done a pretty stiff exam—languages, essays, all the trappings. I didn't get it, but I was considerably elated when my tutor told me that the examiners
sent their congratulations and would like me to put in for the
next vacancy in physiology.' The examiners were Sherrington, J. B. S. Haldane, and Professor Fordisch. They
recommended three candidates for the prize fellowship
(Howard being one of them) and the final choice rested with
the Fellows of the college.[8] They chose G. R. de Beer (later
Sir Gavin de Beer), a zoologist who became world-famous.
    Ethel was still discussing, in her letters, the possibility of
coming to England. Howard encouraged her in this idea,
though not her suggestion that she should bring her mother.
He thought Ethel should get away from Adelaide, if only
because of the narrowness of the teaching, which had the
'definiteness and dogmatism that saps the souls of medical
students and makes them mere senseless blocks of absorbent
wood'.
    During the following weeks Howard became a hard-work-
ing demonstrator, arriving in the department at 7 a.m. on
class-days to prepare the practical work for the students, and
dealing with their difficulties and questions. Sherrington had
thought that Howard and John Fulton, also a demonstrator,
might work together in a joint research on the nervous control

of muscle. He had proposed to the University that each young man should be eligible to take a D.Phil. on his contribution to the joint research.[9] This scheme did not materialize. Apart from any University objection, Howard did not take kindly to the subject. His attitude to neurophysiology was summed up in a typically laconic 'Electrodes—who cares?' Sherrington, with an equally characteristic tolerance, found him something else to do.

Howard tells Ethel what this is.

I'm now researching with Sherry, doing a bit at present on capillaries. I can recognise them now, and think I've seen a Rouget cell [Rouget cells were supposed to be contractile and to control the bore of the capillary.] Am going to try to observe the circulation in the cat's brain by means of a microscope. Will be rather difficult, but I think it can be done ... Also, incidentally, I raised a thing called the Gotch Memorial Prize the other day—5 quid and a medal—for physiology ... I've been to hear Kreisler fiddle—quite good, in fact he must have played before!

On Boxing Day Howard wrote to Ethel in answer to an obviously critical letter.

I'm glad you shriek at me. I want someone on the spot to keep on 'putting in the boot'. I can see myself developing into a rather nasty product ... When you hoped I wouldn't be lonely at Christmas you just about hit the nail. Kid, I'm that most damnably lonely I don't know what to do, sometimes. I'm not complaining against anyone or anything—people in Oxford have been exceedingly decent to me—but the fact is I really haven't got a friend as I conceive the term, and am quite incapable of having one ... It must be something wrong with me.

But he was being received into Sherrington's home. He stayed there 'to keep the burglars away while Sherry was messing about at the Royal Society' and he remained there for two days after Sherrington's return. 'I could sit and lap up his conversation for hours. It's quite thrilling at times to hear him talk of mythical figures.' With a letter of introduction from Sherrington, Howard had then 'pushed off to Copenhagen to see Krogh. Pretty rough from Hull, two days round the N. of Jutland. Krogh is an awfully nice old chap—gave me the run of his lab—*the* authority on capillaries. Spent a fortnight there, quite invaluable. Copenhagen quite a nice town, about the size of Melbourne. The Danes are the nicest Europeans I've met.' Then he returned to London, to take his sister to Paris for Christmas. This 'was a bit of a failure.

You see, she's much older than me and this place shrieks with joie de vivre. But I like my sister. She's also most frightfully on her own, but good tempered.' 'Drifted into Notre Dame on Christmas Day, with lights everywhere—just points in great dim aisles—most impressive.' He was also working on 'some nerve and muscle stuff' at the Sorbonne with Professor Lapigne.

During the following term at Oxford Howard went on with his capillary work, with the idea of a D.Phil. He used a binocular microscope to study the cerebral circulation in cats, which, he told Ethel, had never been done by this technique before. He is thinking of writing a paper about it. 'But don't tell anyone—I'm a very secretive person, you know.' He had been staying with J. B. S. Haldane (then Reader in Biochemistry) at Trinity College, Cambridge, and he had been introduced to Rutherford, J. J. Thomson, and Barcroft—'about the most brilliant scientific collection in one room' he had met. He had heard that a lectureship in physiology was to be established in Adelaide, and that he was being considered for it. 'Should I take it, if offered? But I might pick up something here, which means that one lives in the thick of things.'

In April, after another term, he wrote to Ethel: 'Time flies. So much to learn, so little done. The same thing is being borne in on you I suppose. By September you'll come to the conclusion you know very little, and by January, having safely negotiated your exams you'll be frank and realise you know nothing.' He had done a locum at the Radcliffe 'in the fog of ignorance and helplessness in which we live. Case after case of T.B., encephalitis and so on—one can diagnose them but one might as well fold one's hands for all one can *do*.'

He had been 'on the move' again, staying in Cambridge with Dean, the Professor of Pathology. 'He's a very decent old boy—showed me his lab and gave me dinner in Trinity Hall.' Then to London, where he went to the theatre to see Marie Tempest, and also George Robey who was 'positively vulgar'.

I like London immensely—it just pulsates with life. Strolling in Hyde Park I have been watching the élite riding in Rotten Row. I wonder if these people serve any useful purpose? They have infinite educational advantages, but the main result is that they don't eat soup with a knife, wear delightful clothes and can talk a fair issue of rubbish.

From London he went to Holland with a Canadian friend, largely to see the Vermeers and Rembrandts at Sherrington's suggestion. He was entranced by the pictures in the Rijksmuseum. 'The Frans Hals, 400 years old but astoundingly fresh and brilliant. They give one a slight idea of the immensity of the men who can create; artists, architects, musicians—they stand alone, they create beauty from nothing.' He was able to laugh at his former depression. 'Here we are, well fed, living a patrician life, seeing more of Europe than most English people do in a lifetime, doing congenial work—the work I've always wanted since I was about 15—and yet we become melancholy.'

Then, in May, he writes with momentous news. 'My luck seems to be hanging on pretty well ... in spite of my efforts to detach it. At Sherrie's suggestion I put in for a Studentship at Cambridge—Sherry wrote a letter and I went to see the man I mentioned. Well, they've been foolish enough to give it to me. I've got a frightfully swelled head. It's the John Lucas Walker Studentship. £300 p.a. + £200 to buy apparatus. The thing that tickles my fancy is that it is pathology and I don't know any. However, they think I know some physiology and rather like my idea—not mine really, of course!—of doing pathological physiology along Cohnheim lines. As the thing was open to anyone in England, I feel a bit bucked. The thing is really a great honour, and Sherrie told me I couldn't possibly get a better launching into experimental medicine. It means if I do any good at all that I'll get a decent lectureship, and possibly in a few years' time a professorship.'

In June Howard received a letter from Ethel in which apparently she, in her turn, complained of depression. Her health was poor and she was losing weight. She felt that they were wasting the best years of their lives writing letters. But, if they were to marry, they would be very poor. She thought that Howard cared more for his work than he did for her. He only wanted her to minister to his comfort and she would have to cook all day. He did not want children. She did not want to leave her mother, and Howard hated her parents. Howard did his best to deal with each of these points in turn, though not very convincingly. About his own doings, he wrote that he had been out motoring, through the Cotswold country, with

Lady Sherrington—'Some of the most beautiful country in England—Bibury said to be the most beautiful village there is'. He had also dined at High Table at Magdalen, where his health was drunk by the President and Sherrington. 'The appalling thing about going to Cambridge is that I shall have to start all over again to know people. I'm just beginning to know some in Oxford, and now I pass on.'

His next letter to Ethel, dated 14 June, was a bare announcement. He had been offered the post of medical officer to the Oxford University Arctic Expedition which was due to leave in a few days' time. 'They're going to collect birds and whatnots, tap the rocks and do a bit of sledging. We sail from Newcastle; Tromso, N. Norway and hence to Spitzbergen, coming back via the fjords of Norway about the beginning of September.' He had accepted.

So ended Howard's first years in Oxford. Ended too, when she got his letters, were Ethel's hopes that he would come back to Adelaide as a Lecturer in Physiology, or at least on a brief visit during the Long Vacation, when their personal and family problems could be discussed. Howard seems not to have even considered such a visit, but it transpired later that Ethel had somehow persuaded herself that he would come, and that she had cause for disappointment.

Despite Howard's constant protestations of his friendless condition at Oxford he realized, when the time had come to leave, that he would miss his companions. Though he did not, apparently, class them as friends as he 'conceived the term', John Fulton, Keith Hancock, Aldrich Blake, and Lindsay Ride among several others had all enjoyed his company. Apart from the elderly ladies who had obviously enjoyed befriending him, he seems to have had no feminine companions. He found the English girls boring and irritating; and he told Ethel that they gave him the feeling that he always said the wrong thing. Howard deeply respected the scientific leaders of his day— in fact, he was frankly dazzled by them. They, in their turn, responded with a personal interest in this ebullient young Australian. He had made a lifelong friend of Sherrington, and he had acquired the interest of Haldane and Dean at Cambridge, personal contacts that determined his future.

4

CAMBRIDGE VIA THE ARCTIC

The Oxford University Arctic Expedition which Howard joined in June 1924 was the third in a series. The object was to explore and to study the natural history, geology, and glaciology of Spitsbergen and North East Land. Oxford provided eight members (including Howard as medical officer); Cambridge three, with nine from the services and the National Physical Laboratory. There were also four Norwegians to take charge of the dog teams, including Captain Hanssen, who had been with Amundsen to the South Pole. The expedition leader was George Binney of Merton College, and the chief scientist C. S. Elton of New College. For the first time in arctic exploration it was proposed to use an aeroplane. The expedition ships were the *Polar Bjorn*, a 300-ton Norwegian whaler, and the 45-ton trawler *Oiland*.

Howard wrote to Ethel of their departure after many delays from Newcastle. The first thing to impress him was their wireless.

While in a Norwegian fjord we listened to a church service from Aberdeen—sermon very inadequate—and dance music from the Savoy Hotel. It's all well-nigh incredible! ... The voyage past the Lofoten Islands was very beautiful, with snow-covered mountains, blue seas and the sun reaching the horizon at night—a perpetual sunset ... There has been a considerable storm and our little ship tossed about like a cork. My innards are still a bit annoyed. But I've seen my first pack ice and icebergs ... Now we're anchored off the old whaling station at Green Harbour. Lots of snow and a glacier ... Very difficult to arrange sleep and working hours in the perpetual daylight. The weather is perfect—warm and sunny with marvellous colours, shades of blue reflected from snow, ice and sea. I'm getting to know the members of the team. Some very nice and intelligent. Some 'bone-heads'—typical Englishmen without a spark of imagination and a conviction of English superiority in everything. They have a habit of getting angry when one challenges their beliefs.

He describes the wild life. 'The birds up here are quite beautiful, but we have some egg clutchers who go about robbing nests—a most detestable thing to do. Then there are some people who shoot "specimens". I detest shooting for "sport". If I want something to eat, yes.' Hanssen had a boil on his face, which Howard feared might become a carbuncle (a staphylococcal abscess particularly dangerous on the face). He was using 'hot fomentations and hope'.

A week later the expedition set up camp at Reindeer Peninsula. Howard was based on board the ship and had little to do. Hanssen's boil was subsiding, but he was still unfit to take charge of the dog-sledge parties. The aeroplane had taken off and then it failed to return. Sledge parties had to be sent out to search for it and it was finally found by its radio signals two days later, having made a forced landing. The crew were, in Howard's words, 'as well as could be expected after having been wet through for 18 hours'. Then the radio operator, E. Law, developed pleurisy. After six days with no improvement, Howard felt that he must be taken to hospital. The nearest, a small miners' hospital, was at King's Bay, and *Polar Bjorn* was therefore sent back through pack ice that made the voyage a hazardous one. Law was landed at the hospital, and eventually sent home by collier. Hanssen's abscess had not quite resolved, and Howard had to open it. *Polar Bjorn* then began her return voyage northwards which, in the words of the published report,

was full of contretemps. A thick belt of ice had formed north of Wahlenberg Bay. By repeated charging and taking advantage of ice channels, open water was reached. The *Polar Bjorn* then grounded heavily on a reef. To refloat, it was necessary to jettison all the ballast, and from that day (July 23rd) till we returned to Tromsö, we were without ballast.[1]

The ship struggled back to North East Land. There sledge parties set out, leaving Howard on board. He described a period of boredom, fog, wind, and cold. When the ship sailed to Liefde Bay, he lived ashore for two weeks, and spent his time collecting specimens and 'growing a splendid beard'. Charles Elton was in charge of this base camp, and wrote that he and Florey were alone there.

It was this that founded a life-long friendship. He was a wonderfully reliable and, in his mordant way, an amusing companion. We did one 16 mile voyage, rowing a dinghy across Liefde Bay, but otherwise explored the flat tundra country. The base camp consisted of very large wooden crates that had housed seaplane parts, and our only source of heat was a camp fire kept going with chopped driftwood that had floated from Siberia. It was rather cold.[2]

Howard describes to Ethel how all the sledge parties had been picked up safely 'but have accomplished very little. The aeroplane has crashed again on landing. It has been a nuisance from beginning to end.' He is coming back on board the little *Oiland*, and expects to be 'damnably seasick'. However, he reached Tromsö safely, shaved, bathed, 'even had a haircut and became respectable'. He plans to return to England via Trondheim, Christiania, Stockholm, and Copenhagen.

Despite his laconic descriptions and disparaging comments, the expedition had a considerable impact on him. He was deeply impressed by the beauty of the Arctic wastes and their wildlife, to which he often referred later. And, perhaps for the first time, he experienced something of the companionship and the shared sense of purpose that unites any group of people faced with physical difficulties, hardships, and danger and utterly dependent on each other. For years he attended the reunion dinners of the Arctic Club to relive those few weeks of adventure.

Ethel, in her letters, probed Howard's taste in literature. It was, in fact, surprisingly varied. He read Sheridan, Marcus Aurelius, Hugh Walpole, Katherine Mansfield, Oliver Wendell Holmes, Wells, and Shaw. He particularly admired Shaw, whom he found amusing as well as instructive, while Wells was 'merely instructive'. What, she had asked, did he think of *Hassan*? Howard had seen it, with Fay Compton and Henry Ainley. 'It was the man who faltered, you may note,' he replied, adding, 'I know myself that if there's any good in me, you could draw it forth. . . . In a book I've just read, "The Opium Eater", he [De Quincey] states that the hardest burden man has to bear is the burden of the incommunicable. He's quite right. During this last trip through Norway, I've felt it more and more, this loneliness of spirit. I can't write about it. It's something intangible and beyond me.'

Ethel had read and enjoyed *Jurgen* and asked Howard's opinion of it. In due course he gave it.

It has made me very cynical again. I thought the book lacking in taste and sensibility. It's a different thing, it seems to me, to face 'facts-as-they-are' delicately and with due appreciation, to doing so in a blatant way. Most well brought-up young women should be shocked by it. As you weren't, you can't have been well brought-up.... I knew a Jurgen in Oxford—an American, totally unmoral, quite cynical and quite human. I often envied him, but never had the strength of mind to follow the thing out. I must be one of those people with a conscience and shall, of course, look forward to my specially constructed flames.

(*Jurgen*, James Branch Cabell's fantastic 'Comedy of Justice' had then just been published, with a preface by Hugh Walpole, after having been banned as obscene in the United States. To describe it as obscene or even blatant would now be thought laughable, but its lightly veiled symbolism is certainly erotic—more so, perhaps, than the crude and often inaccurate expositions of sexual anatomy, physiology, and morbid psychology which betray the failing powers of many modern writers and their readers.)

On his return in mid-September, Howard packed up in Oxford, with genuine regrets, and took up his new post in Cambridge. Dean had not only provided him with laboratory space and equipment, but found him lodgings at No. 12 Priory Road, and Sherrington had recommended him to his old college, Caius, as a member of Common Room. Caius had long associations with medicine. It had been founded by John Kees, a successful physician, who had been a pupil at Gonville Hall and, in 1557, refounded the college at his own expense, with himself as Master, incorporating a Latinized version of his name in its new title, Gonville and Caius College. At the time of Howard's arrival, the Master was Sir Hugh Anderson, F.R.S., the physiologist.

The development of science had taken place in Cambridge some years earlier than in Oxford, and with far less turmoil. The Natural Science Tripos was introduced in 1848, and in 1870 Michael Foster came from University College London to teach physiology in Cambridge, and build up a department. In 1883 the University established a chair of pathology, and C. S. Roy, then Superintendent of the Brown Institute, London, was appointed. In 1886 a University Demonstratorship in Pathology was created, and it was filled, temporarily, by Almroth Wright, who was to become an almost legendary figure and one indirectly concerned with later events in

Florey's life. Roy died in 1897 and A. A. Kanthack, Director of Pathology at St. Bartholomew's Hospital, London, was appointed. He brought with him T. S. P. Strangeways, and his laboratory assistant W. A. Mitchell, who later became Superintendent of the Cambridge Pathology Department. But Kanthack held his professorship for only a year. In 1898 he died of cancer at the age of thirty-six. He was succeeded by G. S. Woodhead, who drew up a scheme to provide accommodation for all departments of medical teaching, which resulted in a new building, opened in 1904. Much of the credit for these developments was also due to Clifford Allbut, Regius Professor of Physic. And it was largely Allbut's efforts that established the 'Cancer Research Hospital' for Strangeways, whose tissue-culture work was opening up a whole new field of research.

In 1921 Sir German Sims Woodhead died and Henry Roy Dean succeeded him as Professor of Pathology. Dean was an immunologist, who encouraged the experimental approach in his department. He had been educated at Sherborne School, New College, Oxford, and St. Thomas's Hospital, London. He had a remarkably extended academic career, holding the chairs of pathology at the Universities of Sheffield, Manchester, and Cambridge successively, the last for almost forty years. Though not himself a research worker of much originality, he had the great gift of choosing men who were, and the power to place them in high academic positions when they left his department.

One highly significant development in which Cambridge played a part should be recorded. This was the appearance of biochemistry, a subject, it has been said, 'begotten out of chemistry by an unknown father'. When Foster invited Frederick Gowland Hopkins to come to Cambridge in 1898, it was to develop 'chemical physiology'. Hopkins had started his career in 1877 as a chemist's apprentice. Then, at the age of twenty-eight, he became a medical student at Guy's Hospital, qualifying in 1894, five years later. He was recommended to Foster by Starling, the physiologist. It happened that in his practical class at Cambridge one of the students, J. Mellanby, failed to get a test to work. Hopkins investigated the reason for this fortunate mischance and discovered tryptophan in the process.[3] This led, by stages, to the discovery of the vitamins, for which he received a Nobel Prize in 1929. Meanwhile, he had been elected a Fellow of the Royal Society in 1905, and became the first Professor of Biochemistry in Cambridge in 1914. In this, however, Cambridge was behind Liverpool, since the first chair of biochemistry in Britain was established there in 1902. In 1924, the trustees of the Sir William Dunn Bequest provided a fine new building for biochemistry in Cambridge as part of their policy to further medical science. Meanwhile in the physiology department, Langley, the neurophysiologist with whom Sherrington had done his first research, had succeeded Foster in 1900 and he was still Professor of Physiology when Howard came to Cambridge as John Lucas Walker Student in Pathology.

On 28 September, Howard wrote to Ethel and suggested that they might become 'publicly engaged'. He considered that neither of them knew what to expect from marriage.

You think, so you say, that if you can cook me a decent meal and know when I want to be chatty that's all one can want. But I expect you to be my companion. What should an ideal husband do? Not nag—that's for certain. As a point of fact I've not had much of a chance this last few years. Anyone I chose to nag at would just tell me to push off, so perhaps I've lost the habit.

He also wrote of his research. 'I now have a cat with a glass window in its head running round the lab quite happily.' (This was something of a technical triumph and the beginning of a method for the observation of living tissues that he was to use on a wide variety of problems for the rest of his life.) He would try the technique in a study of the brain changes in epilepsy by using thujone (the active principle of absinthe), which produced fits. He intended to write up his experiments for an Oxford B.Sc., and to work for a Cambridge Ph.D., not an Oxford D.Phil.

In her next letter, Ethel told him that two of their Adelaide Medical School friends had developed tuberculosis. 'Absolute tragedies,' wrote Howard, 'and all on account of a little saphrophytic organism—it's appalling!' Moreover, Ethel was worried about her own tuberculous tendency and he continued: 'When you surround yourself with morbid forebodings it's unbearable. I've known all along precisely the situation, hence my entreaties to you not to work too hard. To be quite frank, I hate the idea of you doing medicine at all. We've enough to contend with without manufacturing an illness for you.'

In October he was less optimistic about his work.

My experiments have mostly been going wrong of late, and my usual re-action of gloom sets in. And this sensation of being alone. I know only a few people at Cambridge, and those I do don't excite me. It's an awful business getting to know congenial people again. Haldane I know. He's the chap I stayed with in Trinity. Quite a brilliant person who does and says extraordinary things. Much disliked here on account of his booming voice, but one can put up with him for other qualities.

He told Ethel that he spent much time imagining their meeting. 'I've already taken you round Cambridge and seen your eager face gazing at things. And, of course, I've met the boat several times—it's my favorite pastime. We've been to Paris, too—though of course you don't realise it.' But he mentioned an ominous reference in her letter to something wrong with her hearing and the possibility that he would have to shout. 'I refuse to marry you under those conditions. There's nothing, I think, could be worse than two people to live together with raucous voices.'

In November he wrote:

It's all over now, at any rate by the time you get this—the thing that's been making your life hideous—exams. I'm sure you've done well. You'll be a full-blown M.B. looking indulgently back at 'my student days' and thinking now how easy it really was. You are 24 and I am 26, and the best years of our life are passing away at an amazing pace. What a simplification of life it would be for me to be a general practitioner or something equally useful instead of a pretty mediocre pathologist. If youth has to be lost one may as well bargain it for experience. That's what is so depressing—age. I have a most unmitigated fear of death, not so much for myself, but in others. If you were to die, you'd be blotted out for the people you leave behind—appalling!

On the credit side he had been to a magnificent concert by Backhaus, Hislop, and Florence Austral. His golf was improving, too.

I've been round in 121! By the time we're married I'll be down to 100 and then I'll retire. I'm very pleased that the M.C.C. are being taught that they're not such good cricketers as they think. Of course the wail goes up here about unfairness—they're really very poor sports here in general.

Ethel wrote complaining that 'rain leaked out' of all Howard's letters, and that the gloom had made her weep before her examinations. The financial plight of his family worried her, and she felt that she could not contemplate marriage if he had to support them. She had met Mrs. Florey (apparently for the first time) in July, and she and her daughters had moved to a smaller house on the Fullerton Estate in South Adelaide. Howard replied that he now had facts and figures from them, and that they were living partly on capital. 'I feel that they ought to do something for themselves, but they lack driving force.' Howard's feelings for his family were complex: he was more concerned by his own lack of affection for them than by their actual situation. This attitude can be illustrated by quotations from some of his previous letters.

I'm disgusted with myself for my feelings for my family. I'm just not interested. Yet they've done a hell of a lot for me—possibly in the wrong way—but they did it. Is it because I'm most selfish? I don't know. I feel I could do a lot for other people. [25 June 1922.] It is a terrible thing to be the only son, and the youngest at that. Everyone puts you under the obligation to do well. [8 August 1922.] I'm glad you liked my mother. It's truly horrible that I don't give her anything. I've tried, and the failure will probably haunt me all my life. I really think it is because I've never been able to tell her anything that mattered. She has never really known or could take an interest in the things I did. Externally she has sacrificed everything for me, but in all the decisions that counted I've had to make up my own mind. She was strong-willed and I was obstinate. [1 September 1924.]

He had heard that the Adelaide lectureship had been advertised. 'Life seems to be a series of sudden decisions—decisions that may turn the whole current not only of my life but yours. I've been tossing up whether I should apply, but on the whole I think it's better not. I'd have to say farewell to any chance I've now got of doing some decent research.' A new project was opening up for him. He had been to Oxford, staying with Hancock at All Souls, and then with the Sherringtons. 'He has given me a line to go on—the permeability of membranes in inflammation—Sherry is a good egg, and I hope you'll like him.'

In January 1925 Howard wrote to congratulate Ethel on passing her examinations. She was going to do a house job at the Adelaide General Hospital, and he gave her a list of

medical books and papers to read 'to counteract the bad influence of the staff'. He had written his thesis for the Oxford
B.Sc. He had another eighteen months at Cambridge

and might get a Ph.D. By then I'll be sick of degrees. My ultimate aim
is to try for a Fellowship at Caius in October 1926, and get a grant from
the M.R.C. If that happens I'll have about £700 p.a. which perhaps you
might consider adequate. If I get the Fellowship, I might go to America
with Mr. Rockefeller's bounty, and as they pay more to a married man
I can see no adequate reason why you shouldn't come too.

In his next letter he tried to deal with Ethel's worries about
what she will find to do if they marry. She was not, she told
him, interested in laboratory work. He replied that she could
not know this, since she had never done any. He had been
suffering from periodic attacks of indigestion since his return
from the Arctic. 'I don't know if it's an appendix, or gastric
ulcer, or simply "dyspepsia". You know how it's possible to
imagine things ... I think I should be a bad patient. It's quite
terrifying to realise that if one's health goes there's nothing
left but a large spot of morphia.'

Ethel's friend and adviser de Crespigny warned her on
health grounds not to come to England. He also told her that
he thought Howard should be doing clinical work. Howard
explained (once more) that he did not want to be a clinician.

There's an immense field on the purely experimental side without dabbling
in clinical research. My erstwhile strictures against clinicians I still adhere
to, but I've a deal of respect for the clinical men of the type of T. Lewis.
But the muddle-headed clinician with his mal-diagnoses and ridiculous
theories and fish-mouth that swallows anything that's put there is a lamentable product—especially if it teaches!

Howard heard from his mother that Ethel had been to see
her. His mother 'thought Ethel very nice'. 'But she'll get a
shock', he wrote, 'when she gets my next letter which will
tell her of our engagement. They'll ask you out. My mother,
I'm sure, will kiss you and weep a little. I'm sorry for you.
I remember the tortures I went through with *your* mother,
when I was beastly rude and she ticked me off in the drawing
room. I was pale pink with fury, largely because I knew she
was right.'

Howard's official duties included teaching. 'I've been
demonstrating bacteriology. As I haven't done any since my

third year you can gather it's a considerable mental strain. It's very flattering but most depressing to be called "sir". It's indicative of age and decrepitude. Do people call you "doctor"? We'll soon be referred to as "The older generation".' He had been playing golf, and improving. He reminded her of their games together in Adelaide.

I used to be largely in a mental twitter in those days. You always seemed so calm and impersonal—forbidding. I've been looking at your photo. From where I sit you look wistful, as if you'd just missed the metaphysical bus all your life, but when I look closely, there's a firm set about your mouth. Then, sometimes, you smile. Altogether a remarkable photograph.

Ethel was wondering if she could do medical work in England. Howard wrote: 'I suggest that while you are waiting to get a job in London you should share my digs in Cambridge—if your views of the proprieties are as lax as mine. I have a couple of spare rooms and my landlady would be charmed. It would be a good idea to get to know each other and remove four years of separation.' Sherrington had read and approved of his thesis and suggested that it should be published in *Brain*. 'I've now started, at his suggestion, to find out why capillaries leak, e.g. in inflammation. It's quite fascinating and fundamental when you come to delve into it, but very difficult technically . . . They're rather amused in the lab by my demands for animals—cats, bats, dogs, rabbits and now a goat.' He had been combing the area for bats (the blood vessels of their wings are easily seen), but it had been denuded by some enthusiastic physiologist, and he would have to go to Ely Cathedral for them. He suggested that they should marry before he applied for a Rockefeller Fellowship. 'Then I'd get £415 instead of £300.' He asked her if she is an early riser. 'If so, you'll have trouble with me. I have the utmost difficulty in struggling down by 9.30. I'll never be a great man—I sleep too long.' He upbraided Ethel for overworking in her hospital job. 'I suppose they expect you to look after 60 people and do all the latest lab fandangos, only half understanding their significance.' She seemed to have lost her colour and become thinner. 'That's sad news, you were so very pleasantly coloured.'

'Had a moderately interesting experiment today,' he wrote in February. 'In inflammation it is usually described that the capillary walls become sticky, and hence the leucocytes stick. I think it is the leucos that are sticky, and not the capillary walls. I produced inflammation in a frog's web—leucos sticking well—then shoved into the heart a suspension of carmine particles. These were carried round alright, but did not stick to the capillary walls, but did to the leucocytes, which became studded with them. Will repeat with a normal frog. All accounts date from Cohnheim, and he said the caps were sticky, and everyone let it go at that.'

He was sorry that Cresp (de Crespigny) thought that he was bound to have got the Adelaide job. He felt that he had let her down. The salary, £750 p.a., was tempting.

But if you state that you wouldn't want me back if I didn't like the job, it eases me to say the coin was the only thing about it I liked. Clinical Physiology, in my opinion, is a bastard subject. Prof. Thomson of Oxford put me right off it at first, and I'm afraid his words sank in deep. As a counterblast which gave me encouragement, the Master of Caius dropped in the other day and told me I'd been elected to a rather swagger club here. It's composed of 12 members, J. J. Thomson, Rutherford and folk of that calibre, and there are 6 associates. I'm one of these. They give two dinners a year.

Ethel had been telling him about her likes and dislikes. For instance she did not like watching sporting events. 'You needn't be alarmed,' he wrote, 'I can't stand it myself. I go occasionally to a good football match such as an Intervarsity. I shan't drag a charming wife about in the old mediaeval spirit.' Ethel feared that marriage would make them bored or indifferent. 'It all depends on whether we let it. I do hope, oh, so much, that we won't be disappointed in one another when we meet. Sex is not much foundation for lasting happiness, but we have common interests and the same way of looking at the world's futilities.' He begged her to come to England as soon as her year's appointment at Adelaide was over. She had told him she would like to have children, and this pleased him. 'To be frank, married women messing about in labs always appear to be slightly pathological to me, but I was frightened you would get bored after living a pretty active life and not take kindly to domesticity. I want children ... they will be to you, I hope, what my work is to me.'

He had given a dissertation to his new Club (the Rayleigh

Club) on Spitsbergen—his maiden effort in Cambridge, and he was due to talk to the Physiological Society in London on the cerebral circulation. His abdominal troubles were still in evidence and he had been examined by a surgeon, who thought the condition was not organic and put him on a diet. He had also been having headaches, and had his eyes tested. He had developed astigmatism, and would have to wear glasses, which he would dislike intensely. He was very worried by the possibility of getting ill.

He next wrote that he had read his paper to the Physiological Society.[4]

I think it went down pretty well, as a matter of fact, as Dale congratulated me and said it was very interesting. I hope you don't mind me telling you such things, as it would buck me up if you were pleased. Had lunch with Adrian and his wife on Sunday. Dale is considered by Sherrington to be England's foremost physiologist. Adrian is also a physiologist, and an F.R.S. He hinted that he could do something for me with the M.R.C. as he is on the committee dealing with head injuries.

If he got his Fellowship at Caius and some tutoring they would have £650 p.a. to live on and he could give his mother £100 per year. Would Ethel consider marriage on this? His research was opening up.

I hope soon to knock out old Colnheim's story about the capillary walls becoming sticky in inflammation. It is, I am almost sure, a property of the leucocytes which are under the influence of the substances released by the inflamed tissues. They become more phagocytic, and more valuable as a defence. But perhaps you'll think this academic and of no practical value.

He reminded her that two months of the twelve-month appointment had gone and he hoped that they would be married early in the following year, and then go to America on a Rockefeller Fellowship.

Writing about their relationship he said:

I'm sure you would not like me to pour my sticky sentimentality over you. But aren't you a bit that way? Would it make you feel just a bit wild to sit out in the moonlight? In Venice would you feel the thrill of the thing that has shaken so many thousands in their time? You are a bit of a contradiction. I once thought you were so cold you were above all this world's emotions, that was when I didn't know you.

Ethel complained of her disappointments during the past year. Howard wrote, 'I have a horrible feeling I'm the cause

of many of them. You were disappointed that I went to Spitsbergen instead of coming home, and that I didn't take the
Adelaide job. I think I was a fool over that. I feel a perfect
beast in the way I am treating you.'

In his research, he had been 'looking at sections showing
that the capillaries leak through the cells and not through
spaces between them. I perfused the gut of a dog with a solution of pot. ferrocyanide and iron ammonium citrate, which
is non-toxic. Then Prussian blue stain for the iron, which
shows as a blue haze in the endothelial cells, indicating it was
going through them when they were fixed.'

A week later, he wrote: 'Another term has gone. Very little
to show for it. I'm working out several techniques at the same
time, none of which yield to the first breath. If they don't
work out pretty quickly, I give them the go-by, which is foolish, as I know.' He was trying to trace the path of colloids
through the capillary wall by injecting starch into the blood
stream and then staining the tissue with iodine. *Brain* had
accepted his article on the cerebral circulation.

Howard said he had been feeling rotten during the past
three to four months with indigestion.

Did a test meal and was shocked to find complete achlorhydria [lack of
the normal acid in the stomach] and buckets of mucus. Ran through the
possibilities—cancer, of course, or pernicious anaemia—and worked myself
into a sweat. Then went to Cairns in London, who passed me onto a radio-
grapher who did a barium meal and only found I emptied rapidly. This
morning I saw Robert Hutchison, who said I had undoubtedly a chronic
gastritis, and has given me a diet. Very relieved it's nothing worse, but
I've got achylia, which is annoying ... I don't know whether the flesh silk
stocking and high skirt craze has reached Australia, but it's in full bloom
here. I have a theory that, as women are in excess here, sex is an important
matter to them, and hence fashions stress to the uttermost limits their biological attractions, even a display of thigh on occasions.

In her next letter, Ethel complained of sinusitis and
anaemia. 'What a pair we shall be, with your nose and my
stomach,' Howard replied. He also dealt with her dislike of
his views on clinical medicine. 'If it interests you, they are
going to try out a suggestion of mine at the London Hospital,
for the treatment of tetanus. I'll tell you what it is, but please
don't tell anyone else. I'll tell you the reason for this last pro-
viso. I had one of my rare ideas the other day. I told a man

working in the Path Lab about it, and he got all the information out of me and is now calmly pinching the whole thing. To say that I'm peeved would be mild. Hence I don't want any of my ideas pinched again.' Howard's idea was to inject tetanus antitoxin into the spinal fluid and then to give hypertonic saline intravenously. The salt would cause shrinking of the nervous tissue and the antitoxin should be drawn into the spaces around it and neutralize the toxin *in situ*.

He had received a pathetic letter from his mother, and told Ethel that he was considering arranging for his family to come to England, where houses were cheaper and he could provide for them more easily. 'My conscience is a Banquo's ghost. I feel I'm making a beastly mess of the whole show, and my mistakes rebound on so many people.'

Only two letters out of all the correspondence that Howard had with his family in Adelaide seem to have survived. One is a letter written to his mother in April, on their finances:[5]

I received your reply to my rather ill-considered epistle of a month or two ago. I have forgotten exactly what I said, but it was apparently rather scathing. Well, perhaps I did not mean all I said, but you must realise that I am approaching a bog from which there is no apparent exit. That being the case you will have to excuse a little exasperation. The whole point is that my material prospects are of the slightest . . . and all I can do is research which is extremely ill-paid, so you will understand that I would be very relieved to see at least part of the family self-supporting, for when it comes to the point I will only be able to look after you and that somewhat inadequately. I don't know whether you have ever considered the possibility of coming over here to live, but that seems the solution of the difficulty. As I live at present it costs me for two small rooms as much as I could hire a complete house of seven rooms. You must think this over and put it to the rest of the family. You need not fear that you would be bored stiff and that you are leaving the friends you have as, if I recollect, there were very few people that had truck with us before I left. It has also become apparent to me that if I wish to do any decent work I must remain in this country. And there is the complicating factor that I wish to get married just as quick as I can and I don't feel that can be done till something in regard to your welfare is settled. Well, I hope that you are all well and that you at least approve of Ethel a bit. Your enthusiasm appeared of the doubtful type, that doesn't worry me in the least.

Howard also wrote to Ethel to tell her that he is thinking of taking a house in a near-by village. 'It has 7 rooms, not big, certainly, but still a house. And a garden with a tennis court. No furniture, but there are deferred payment plans. The rent is £50 p.a. My friends tell me it costs £1 to 30/- per person per week for food. £50 for your clothes, £25 for mine—it all adds up to about £300 p.a. I could scrape another £80–100 for odds and ends, holidays, etc. I wonder if you would mind doing without a maid for a bit, until I can get something more remunerative?' (It seems that he was suggesting that his bride should share the house with his mother and two sisters. This highly impractical scheme got no further, since none of the ladies concerned even considered it.)

Ethel said that she thought it would be a good thing to quarrel occasionally, for the pleasure of making it up. 'But we are both hot tempered,' Howard replied. 'I can say the most beastly things and I should think that you would not be far behind. The main problem is likely to be mutual boredom.' They must have a gramophone and a good collection of records—he had been listening for hours to 'wonderful records' and his current favourite was Tchaikovsky's Sixth Symphony, which he played once a day. 'What else can married people do in the evening? Just sit by the fire and talk or read?' But she might want to talk when he did not.

'My last fortnight's work has been quite useless,' he told Ethel. 'However, I'm getting used to that, as it seems a necessary accompaniment to any research till you are on the track and have solved your technical difficulties.' Nevertheless his work on capillary permeability was progressing. Starch could traverse the capillary wall, and then be stained with iodine. He now had to learn some colloid chemistry, and felt the lack of mathematical knowledge. He believed that increased permeability is due to a decrease in the viscosity of the endothelial cell interior. He had devised a method for measuring this viscosity by injecting iron particles into the circulation and then drawing them through the endothelial cells by a powerful magnet, and measuring the rate of movement.

As in almost every letter, he had something to say about his own failings. 'I wish we were all without any illusions about our personal attainments and the necessity to go through the painful process of losing them. It makes life disappointing. One of the chief errors in our make-up is that we *must* be thought well of by our fellow fools, and it is this absurd feeling that is the cause of so many of our futilities.' He also seemed to be incapable of enjoying things alone. He recalled the sunset on the Murray River and his continental travels, saying that half of his pleasure had been lost because he had no one to share it with. 'I could have brained the happy *pairs* I saw enjoying the beauties of Venice.'

In his letter to Ethel of 29 April, he said that he must write formally to her father for his permission to marry her. But he did not know his initials! He had been tracing her various moods in her letters, 'from gentle acquiescence to beating me with a no uncertain bludgeon. I suppose you'll be like that in the flesh. I expect that, with a moistened finger, I shall find out which way the wind is blowing.'

His paper had appeared in *Brain* and he was justly proud of 'my somewhat mis-shapen first offspring. It gives me a frightful thrill to think I'm probably about the second person to see the blood running about in the cerebral vessels.' He was going to Oxford on 2 May for his B.Sc. examinations and he wrote again a week later. 'I think I put it across 'em for the B.Sc., as they told me it was an excellent thesis. The exam was a farce, and I think it merely means passing over more coin to get it.' (Howard's examiners were J. S. Haldane and J. G. Priestley.)

Saw Sherrie in Oxford, and told him of results of expts. He thought them interesting and important. Next day, Sherrie asked about Dean and the other pathologists [in Cambridge]. I told him that no pathologist I'd spoken to knew what I was talking about. He said those he'd seen had said that the twentieth century had nothing to find out about inflammation. Well, I'd already got enough to show that's rubbish. Sherrie advised spending my middle year away from Cambridge, gaining technique in America. He'll do all he can to get me a Rockefeller. I just loathe the idea of being in America by myself. Please do say you're prepared to come.

He wrote that his mother thought that Ethel looked very nice and not the kind of girl who would let him slop around untidily. 'I hope that's not too true. I prefer old clothes, as long as they're clean, and I take my collar and tie off in the evening . . . I hope you won't be a bridge-mad or a Mah Jong wife—I'll divorce you. I'd sooner spend the evening with my feet on the mantelpiece.' And, on his foreign trips, he preferred to travel cheaply and simply. 'Do you think you'd like messing about with me in low Albergos in Italy and small Gasthauses in Austria? I find it intensely amusing.' 'I showed a specimen of a bat's wing with contractile veins at a Science Club meeting. In these days it seems incredible that some anaemic old men go into ecstasies if they find another kind of bug crawling about. It's miserable, to me, this fictitious scientific activity. However, I may be wrong and quite prejudiced against morphologists in general.'

He had seen Dean about his plans. There might be a job for him teaching the tripos class in the new building, when it was ready. But he advised Howard to go to Spain to learn neuro-histology rather than to America. In his indecision on this point, Howard went to see Sherrington again, who maintained his original view on an American visit, to which Dean reluctantly agreed. Howard described his stay in Oxford. 'Beautiful weather and Oxford was looking magnificent. The river was lovely; still nights, but undesirable females to share it with. You've probably never seen a punt—a most luxurious thing with cushions on which you recline. Lady Sherrington says her son is married on £400 a year, so it can be done, apparently. Lady S. says modern young couples want too much.' Ethel had told him that she 'went gay at the Palais', but didn't enjoy it much as the people irritated her. 'I entirely sympathise,' replied Howard. 'I feel too it's an awful nuisance getting interested in everyone. I hope we dislike the same people!'

In June, Howard wrote about experiments he was doing with Harry Carleton, the Oxford histologist, who had come to stay with him in Cambridge. The experiments were on the omentum, the apron-like fold of peritoneum that sticks to any point of injury or infection in the abdomen, like a living adhesive dressing. 'No-one knows how it crawls about so we are going to put windows in the bellies of cats and rabbits and then watch it.' He had also been working on permeability, by studying the diffusion of haemoglobin through gelatine. 'I think I have proved beyond doubt that diffusion occurs more rapidly into stretched than unstretched gelatine. This falls in with my theory of capillary permeability. When the capillary dilates, the endothelial cell is stretched and becomes more permeable, like a gel.'

He deplored the slowness of the mails in his correspondence with Ethel. 'Things requiring action cannot be discussed because an answer demands about three months; you have failed to grasp that I talk a lot of hot air and that I revolve schemes that never fruit.' Ethel had suggested deferring her departure from Adelaide for a further six months to save up more money. She thought she must be cold-blooded because she felt the financial aspect was so important. Howard wrote, 'I hope your sane attitude is not indicative of a general caution in all matters. Caution is all very well in the aged, but we're surely too young to be enclosed by it yet? I'm always jumping at the long chance.'

In July he wrote to say that he had heard from the Rockefeller Foundation that he had been awarded a Fellowship. 'As usual, there's a fly in it. They are reducing my stipend by £300 because I've got that from Cambridge. I shall leave England at the end of September. Where I'm going I've not yet fixed up. What I'm going to do, I've no notion. I'll have £400 clear of travelling expenses, laboratory fees, etc. If you think it can be done, I'll marry you as soon as you like in America.'

Howard had spent some weeks in Oxford working with Carleton. They were going to do a further week of experiments and then go to Cambridge to finish them. He asked Ethel about current teaching on the omentum and its functions. 'Be careful to notice if a great surgeon like Simpson used the word "chemotaxis".' He had given another paper to the Physiological Society, this time on capillary permeability. He was thinking of learning tissue-culture and micro-dissection techniques in the United States, but was going to 'mix his drinks' so that he 'sees as much of the country as possible'.

He made a new suggestion about their meeting and subsequent marriage. He might be able to visit Australia from San Francisco at the end of his Fellowship. They could then marry in Adelaide, and go back to England together. He had been talking to Lady Sherrington, who advised getting married and taking a chance. But Ethel continued to express grave doubts that they might be incompatible. Howard pointed out that they would not have changed so very much. They liked each other then, why not now?

In August, he wrote that he was 'very busy finishing off experiments before packing up. Have been putting in splenic windows in cats and rabbits. When cats are irritated, their spleens contract! I'm going to try the effect of massive doses of X-rays on various organs. The method is beginning to shape in an interesting way.' He had been to London to get his American visa and swear that he was not a polygamist, an anarchist, or wishing to overthrow the Government by force. 'They're a queer lot—a bit naive in these matters.'

In her next letter, Ethel told Howard that she had read of his Rockefeller Fellowship in the Adelaide paper. She would join him in America about six months after she finished her hospital job. (She had not had his letter about his plan to come to Australia.) He wrote again saying that he would arrange to travel from California to Sydney. He also told her about the X-ray experiments.

We have had consistently negative results which enable us to write a paper blowing some previous work out of existence. It was done by a man in the lab, who patronises me, so I have the greatest pleasure in telling him he's a fool. He overlooked a grave experimental error and all his results are due to sparks and not x-rays. I'm also trying to knock some stuff on capillaries into shape. I've got a method for looking at mammalian mesentery and am in a position to settle the histamine question, the Rouget-cell question and the action of pituitrin in shock. I'm trying to get all this finished to bring into the thesis for the Caius Fellowship. If I can get that I'm quite confident we wouldn't starve at any rate. It's a beastly bore writing these things up—I'd far rather do the experiments, let it go at that, and get someone else to write. Still, one must put one's wares on the market.

At the beginning of September he wrote: 'The suitcases have all been packed and I am a homeless wanderer once more.' His main activities had been writing his Caius thesis. Now he was going with J. G. Stephens, the Australian with whom he did the X-ray experiments, to Chamonix for a little fresh air. He looked forward to their meeting in Adelaide. 'Do you think we'll be able to walk about the hills? It's so much nicer to talk out of doors—or else over a good fire.' Apropos of fires, he told her he has burnt all her letters. 'It was this wise. I had either to leave them in the lab with the possibility of prying eyes or take them to America, so, in a firm moment, I did it. You won't think I don't value your letters? I think

I could answer all questions thereon by heart—I've read them so many times.'

Then he wrote from Chamonix.

Had a day in Paris. Went to the International Exhibition of Arts. Very fine, but the chap I'm with is more interested in machinery and women ... To pander to his taste we went to the Moulin Rouge ... Went on to Aix-les-Bains and Annecy. Beautiful weather. From there by mountain railway to Chamonix, 3,500 ft. Housed in quite a decent joint. From the bedroom one looks across the valley to Mont Blanc, with its snow and glaciers all complete. We climbed up a glacier and slid about on it yesterday in magnificent sunshine. In fact, it's not a bad place. But I'm always bickering with the chap I'm with. I trust I'm not getting into the bickering habit.

A week later, he was still in Chamonix.

My temper is returning towards normal. I don't get so annoyed over trivial details. I wonder if I'll ever be able to control my temper with you? Anyway, you've got one too, and we both think we're always right.

There is a woman who plays the violin in one of the cafes who is very like you—fair and has that semi-smile of confidence in the goodness of things. She plays the violin execrably and sings, if possible, worse.

The sun here is magnificent. Stephens and I spend a fair proportion of our time with most of our clothes off absorbing the ultra violet. Last night we climbed up to about 7,000 ft.—I was jolly frightened the last bit, but it was worth while, quite the best view we've had. We were there for the sunset on Mont Blanc. The snow changed from yellow to pink, then grey. Quite wonderful. I wonder why one can't appreciate these things alone, I'm always left with the feeling that I could get more out of the view if you could see it too. One of the worst things, if we don't get on, will be to face the fact that I see all I can see and that the other, in any situation, is gone.

So Howard's first year at Cambridge had ended, and including his last year as a Rhodes Scholar, he had now completed two years' research. What had he achieved? First, there was the work on the cerebral circulation, begun in Oxford at Sherrington's suggestion and under his personal supervision, and completed in Cambridge. This investigation posed a number of formidable technical problems: the difficulty of exposing the surface of the brain without damage to or even disturbance of the minute superficial vessels so that they could be studied in their natural state; the arrangement of the optical conditions of lighting and microscopical examination to give the greatest information; and, in the last few experiments, the insertion of glass windows that would allow long-term observation. On the experimental side, Howard studied a number of drugs that might be expected to cause contraction or dilation of these cerebral vessels. Some of the results were surprising and important. For example, the cerebral vessels contracted when stimulated mechanically, but adrenalin, which causes vessels to contract in most other parts of the body, had no effect on the cerebral vessels. He demonstrated the effect of heat and cold, and concluded that there is some justification for the old-fashioned ice-pack as a cerebral decongestant. He also tested nervous mechanisms known to affect vascular tone in other parts of the body, and found that the cerebral vessels did not respond. Finally, he investigated the response of these vessels to various sorts of local injury, chemical, thermal, electrical, and mechanical. In doing so he observed and described a phenomenon that has since become famous and is today being widely investigated. This is the formation of a 'white body' inside the vessel at the site of a slight injury. The mass grows until it almost blocks the flow, then breaks away, but reforms to repeat the process at more or less regular intervals. Florey showed that these 'white bodies' are largely composed of coherent blood platelets, the smallest formed elements of the blood, which in 1925 were still unfamiliar objects to many haematologists. Florey recognized the phenomenon as representing the early stages of thrombosis, and predicted that it would provide a method for studying the problem of thrombosis in general. Early thrombus formation had been observed microscopically by previous workers in other vessels, but Florey's description of the breakdown and rapid reformation of the mass, and the methods by which it can be produced, was a most important and fruitful contribution. These first experiments on the cerebral circulation formed the substance of the B.Sc. thesis which was so highly commended by his Oxford examiners. Unfortunately, no copy seems to have survived, and much detail of the work is therefore lost. But an abridged version appeared as an eighteen-page paper in the journal *Brain*, which embodies the main conclusions.[6]

Running parallel with this work were his experiments on thujone, which was known to cause convulsions like those of epilepsy. Howard investigated the possibility that thujone produced this effect by causing a contraction of the cerebral vessels. He found no such action, and though he did a good deal of work on other effects of the drug, and published the results in the *Journal of Pathology and Bacteriology* in 1925,[7] he was not pleased with this paper, which he later thought to contain fallacies. Also in Cambridge with J. G. Stephens he studied the work on the immediate effect of X-rays on living tissues, partly by his perfected technique of inserting transparent windows which needed meticulous surgery in those pre-antibiotic days. Direct observation failed to show the contraction of the spleen which Swann had reported to occur after exposure to X-rays. Florey and Stephens found that the changes in blood pressure, respiration, and so on described by Swann were due, not to the X-rays as such, but probably to high-tension current leaking through the animal's body. This, though largely a 'destructive' work, cleared away serious misapprehensions on the effects of X-rays. It was published in the *British Journal of Experimental Pathology*.[8] Then came his experiments with Harry Carleton, a brilliant histologist and an amusing person. Carleton and Florey became, in fact, quite close friends. Their collaboration on the movements of the omentum resulted in a positive and ingenious demonstration of the true behaviour of this structure which, sometimes called the 'abdominal policeman', had been credited with a mysterious active movement towards trouble spots within the abdominal cavity. Florey and Carleton showed that the omentum moves quite passively following changes in body posture and the movements of the intestines. Its ability to wrap itself round any area of inflammation is, they maintained, simply its ability to stick to such roughened areas, having been brought there by chance. Finally, there was their extensive and fundamental investigation of the blood and lymph vessels of the mesentery, the membrane through which vessels pass to and from the intestines. This work dealt with the whole problem of blood and lymph flow, vascular contraction and its mechanism, and the changes that occur in inflammation. It very largely disposed of the 'Rouget cell', a probably mythical structure which many histologists believed they had seen in the vicinity of minute vessels (capillaries) and which was supposed to regulate the bore of the

vessel by contracting or expanding. Florey and Carleton failed
to find evidence of any such activity, or indeed, for the existence of any such cell. They suggested that capillary blood-
flow was regulated by an activity of the endothelial cells that
formed the actual vessel wall. Forty years later, this suggestion
is now receiving considerable support, having been dis-
counted by other experts for most of Florey's working life.
Much of this work was evidently contained in his Caius College thesis which has not been found.

Florey's achievements during these two years are most
impressive for a young man doing his first research work. He
had given a paper to the Physiological Society, which was very
well received. He had submitted two full-length theses, full
of original observations, and covering different fields. He had
published four scientific papers. All this entailed a great deal
of careful work, many disappointments and setbacks, but rewarded by the steady development of advanced and original
techniques and real contributions to knowledge. His undoubted promise was recognized by his senior colleagues, not
only Sherrington (who clearly was very fond of him and therefore, perhaps, partisan), but by Dean, Haldane, Adrian, the
Master of Caius, and scientists in other fields who could
recognize the talents of a fellow-scientist when they saw them.
It was a genuinely brilliant start, all the more impressive when
seen against the background of self-doubt and emotional confusion revealed by his letters to Ethel.

5

AMERICA

Howard sailed for New York in the Cunard Liner *Caronia* on 19 September 1925. He enjoyed the voyage, his first in a ship with 'all home conveniences'. Among the passengers were 'buckets of delegates from the Inter-parliamentary Congress. It made us all weep to think that we are ruled by such a collection of nincompoops—however, I had the satisfaction of being very rude to one of them who pinched my chair.'

There was an immediate complication when he arrived. The plan had been for him to work in New York with Robert Chambers, Professor of Microscopic Anatomy at Cornell Medical School. Chambers was internationally famous for his dissection of living cells and tissues, using an almost incredibly delicate and precise micro-manipulator. But no micro-manipulator was immediately available for Howard's use. Howard, never prepared to await the convenience of even the eminent, immediately made other arrangements, persuading Professor A. N. Richards of Philadelphia to receive him in his laboratory at a week's notice.

Howard did not take kindly to New York. 'Truth to tell, I'm very glad I don't have to stay in this beastly town. It's noisy, and the part I'd have to live in is filthy as well. It costs a dollar every time one opens one's mouth—food is prohibitive. One small room—dirty—costs about 8 dollars a week; if clean, about 10–15. Food costs *at least* 2 dollars a day, so bang goes 24–30 dollars a week just to support life. Truly depressing.' 'Of all the towns I'd dislike most to live in, I think New York takes the prize. It's full of Jews and Italians. It's remarkable mainly for skyscrapers and dirt.' But he praised the pictures in the Metropolitan Museum, and the brilliance of Broadway at night. 'Fifth Avenue contains some very good shops, but they've nothing to show Paris or London, except the prices which are terrific. It would cost about £20 to get an ordinary suit made!' At least he found the railway station impressive when he left New York. 'Immense, beautifully clean and noiseless. The trains are all underground and electrically drawn until out of New York.'

On 7 October he wrote saying that he had started with Richards, a pharmacologist who had worked with Dale in England. 'He is a very blunt chap, but honest. He's well equipped, the University is a pleasant spot, and he's going to give me some work on kidneys to do which will need very considerable technical skill, which is what I want to acquire. Philadelphia is a moderate sized town. I knew a couple of chaps from here at Oxford. I'll beat them up—it's this being alone and mucking about in hotels which is so nerve wracking.' The indications were, he wrote, that he would be in America until the middle of August 1926. This resulted in yet another change of his plans to meet Ethel. If he came to Australia then to join her, they would not get back to England until December. So he asked her to come herself to New York in August so that they could travel together to England 'not necessarily married—surely we could travel on the same boat without a world-shaking scandal'.

During the next few weeks their correspondence was dominated by misunderstandings and cross-purposes. The most important factor in this situation was, of course, the delay of about three months between the dispatch of a letter and the receipt of its answer. The other factors were personal. Howard was persistent in his pleading that they should meet—and marry—as soon as possible. But, as he did in tackling laboratory problems, he produced a profusion of plans by which the objective could be achieved. He put forward a new suggestion in almost every letter, and by the time Ethel had given her views on it, it had been superseded by several others still in the mail. Nothing could have been more irritating than this enforced procrastination to a young man of Howard's temperament, or more unsettling to Ethel. She was cautious to an extreme, fearful of possible poverty, apprehensive about leaving Adelaide, and concerned for her own health. She was unwilling to give up medicine, and clearly was not swept off her feet by any romantic feelings for the man she had known briefly four years ago. The long-drawn vacillations finally caused a crisis in their relationship.

The beginning of this situation was a letter from Ethel, written at the beginning of September and received by Howard on 14 October. In this she announced her intention of coming to America, as he had suggested in letters written before he left England. She had not yet received his plan that he should, instead, join her in Australia; nor of course his latest idea that she should join him in New York in August. Ethel intended, she wrote, to arrive in America in April. Howard's reply conveyed not elation but evasion. Living in America was very expensive, he wrote, and Ethel would not be able to work as a doctor, because her qualifications would not be recognized. Howard himself had only £200 to last from March until September. Rather belatedly in his letter he assured her that he was not trying 'to put her off'. He had not yet written to her father for his formal consent to their marriage, but would do so.

But there was a more serious cause for concern in Ethel's letter. Howard wrote 'your deafness is pretty bad news. I had no idea you weren't joking, until you told me you were off to Melbourne about it. Otosclerosis, if my memory serves, is a nerve thing? Is there any treatment? Anyway, I'll divorce you if you can't hear what I say. But really, it's most distressing news.'

Howard was making the best of his time in Philadelphia. In Richards's laboratory he was learning to insert minute glass tubes (cannulae) into blood vessels about a quarter of a millimetre in diameter. 'The work progresses rather slowly, but still I'm learning quite a bit of technique and that's what I came here for. I don't know whether it'll gratify you, but I felt like a real scientist when a friend of mine from Harvard [John Fulton] wrote and told me that the great work on the cerebral circulation was treated in detail in lectures to third-year medical students. Also by the way, I'm going to spend Christmas with him in Boston, which ought to be very pleasant.'

Outside the laboratory Howard's main pleasure was the Philadelphia Symphony Orchestra, which he wrote 'is one of the world's best. It plays about three times a week. It really is magnificent, without exception the most delicate playing I've heard. I was thrilled to goose-flesh by it.' He was looking forward to hearing Galli-Curci, McCormack, Paderewski, and Rachmaninov during the next two or three weeks. He also went to dances occasionally to get to know people. 'A peculiar fact is that I saw more drunken young men than I've ever seen at a dance anywhere—a result of prohibition.'

But he still felt lonely and bored. At the beginning of November he wrote to Francis Wylie, the Secretary of the Rhodes Trust in Oxford: 'You were so kind as to mention that you would give me some letters of introduction to ex-Rhodes Scholars over here. I am finding it extremely dull and monotonous here, so would be glad if I could meet a few people.'[1] He had accepted the fact that Ethel would be coming in April and his letters were full of plans and instructions. 'The immigration authorities this end are both rude and cantankerous and make trouble on the slightest provocation. Take a N.Z. boat to Galveston, Texas, and thence by train. Its quite natural to be frightened travelling around. I was scared stiff in New York, but if one endeavours to be calm and asks questions, you generally manage to do the right thing.' He sent her a photograph of himself wearing glasses. 'If these make any difference to your matrimonial intentions, now is the time, or for ever hold your peace.' He calculated that Ethel would arrive at the end of April. She had a friend in New York (a nurse from Adelaide) and Howard suggested that she should stay with her. He would arrange to take a holiday in May and join her there. 'With you, perhaps I may be able to stifle my phobia for the great city. We could mess about together, just to tune in as it were, and then get married, go away somewhere for a fortnight and then down to Wood's Hole [the Marine Biological Station] till the end of August.' He thought that Maine would be suitable for a honeymoon.

But all these plans were dashed by the letter Howard received from Ethel in November, in which she wrote that she would not after all be joining him in America. Since this letter, like all her previous ones, was destroyed by Howard the reasons for this change can only be inferred from Howard's reply and from the probable circumstances. Of the latter, the most reasonable must be that Ethel had by then received his plan for his own visit to Australia in the autumn, a plan that

would appeal to her because it gave her another six months
in Adelaide and a better financial start to their married life—
if indeed they were to marry. But Howard, both furious and
bitterly disappointed, wrote by return on 4 November
upbraiding her for indecision and faint-heartedness.

I think this is the only letter I've ever written a second time to you. I
was really too angry when I wrote the last, and I've calmed down a slight
amount since. I'm most keenly disappointed that you're not coming. But,
far more important, you yourself seem disappointing too ... If you thought
it right to act in a certain way you should do so, and not be battered about
by every contrary opinion!

The contrary opinions to which Howard referred seem to have
been those of Ethel's parents and Dr. de Crespigny. The
former he could understand, though he did not agree. De
Crespigny's opposition, apparently on financial grounds,
he would not accept.

You apparently harbour the idea that in all matters outside my immediate
scientific work, I'm a complete fool. The thing I take so hard is that you
have for the second time been put off by de Crespigny, and that you haven't
mind enough of your own to know what you want to do. One thing is
eminently clear, and that is I will not marry you till I have £600 a year,
unless you specifically ask me. And perhaps it would be just as well to shelve
our engagement until we meet again.

He referred to the Sherringtons. 'They said if you were worth
your salt and were keen on the project, you wouldn't hesitate.'
    Within two days Howard was full of remorse. He wrote:

I wonder if you're ever going to forgive me for the last letter? I really am
most sorry for the beastly way I wrote. The fact is, I'm most unhappy
and lonely here. The people in the lab are about as cheery as clams. They're
quite decent, but its all so different from England. No-one has had the
decency to ask me to call on them. There's a Magdalen man here too, and
I might be a leper for all he cares. At the moment I have all meals except
lunch, alone in a restaurant. One finishes work, but there's only this one
room to come back to, or a walk in the streets. When you said you could
come over here, it fell on rather fertile ground ... and then to have it
crumple up 'made me mad' as they say here.

He went on to urge her to come to England when he returned
from America. He even suggested that she should persuade
her parents to come too, adding, 'I don't know why you think
I dislike them.'

With the idea that my outburst of temper was connected with dyspepsia, I have started to clean myself up in earnest. I've had my eyes refracted—there's more astigmatism and I've got new glasses, so I hope that'll remove headaches. I'm getting my teeth seen to and, what I'm most proud of, I can swallow a stomach tube almost like breakfast and have a jolly good time running bicarbonate in and out. I'm really afraid I've got a beastly mucous gastritis.

My friend Carleton, at Oxford, is married to a medical woman. He tells me she has been to Berlin for three months doing skins, and is going to put up a plate in Oxford. I wonder if you'd ever think of doing the same thing? We could go to Vienna in a long vacation, you could go to the skin clinics and I'd do something too. I think it would be satisfactory for you to practice ... I merely throw it out as a suggestion.

On 16 November Howard sent Ethel a cable to forestall 'that appalling letter' which would be arriving in Adelaide in about a month's time. He also wrote with important and exciting news, sending a copy of a letter he had just received from Dr. Philip Panton, Director of Hospital Laboratories, the London Hospital:

Dear Dr. Florey,

I am writing to you because a research appointment in these laboratories has just become available, and because several of our mutual friends at Oxford and Cambridge, including Sir Charles Sherrington and Professor Dean have warmly recommended me to approach you. The appointment is briefly this: whole time research in any branch of pathology which the holder wishes to interest himself in. Adequate laboratory accommodation and the free provision of apparatus and materials. Free access to all available clinical material. Freedom from all routine work and teaching. Salary £85o a year, with a guaranteed tenure on the part of the hospital of five years. I understand that Cairns is writing to you.

After giving details of laboratory facilities Panton went on:

I understand that your present intention is to stay in America for a year, but I thought it possible that you might curtail your time there and we are naturally anxious to get our research work started here as soon as possible. Perhaps you will write me definitely on this point. I shall be particularly obliged if, after consideration, you would cable me 'yes' or 'no' on the broad question of whether you would accept the appointment or not ... should you decide to accept ... I can assure you of a very warm welcome here and all the help and co-operation in your work of which we are capable.

This opportunity could not have come at a more apposite time for Howard. The terms of the appointment were, on paper, ideal for a young and energetic research worker and the salary was exceptionally good—well above the £600 a year that he had stipulated to Ethel as the minimum for marriage. The offer was also a flattering one. He wrote to Ethel that he was 'pretty pleased one way and another with the new job. There are very few so well paid. It means too that the older men consider I've got sufficient ability to do research, and they are willing to encourage me in it to the exclusion of all routine. So many men have just frittered away their time teaching dull medical students—till they're old. It's really a very golden opportunity.' But on the question of curtailing his time in America, he had this to say: 'Now I have different ideas on that subject. It would be a flat waste of 6 months to go back now, and I have much useful technique I can acquire here and not in England. I'm going to put it to them and feel moderately sure they'll agree.' He cabled 'yes' to Panton, but wrote that his acceptance was conditional on his spending the full year in America.

With this prospective change in his finances, Howard then assumed that Ethel would have no further objection to their early marriage and he urged her to come to America in February, bringing with her some of his things from Adelaide. They could go back to England at the end of the summer, via Canada, 'see a bit of the country, the woods, Niagara and take a boat from Quebec'. But his scientific self-doubts were still in evidence.

I've had some moments just lately when I've been paralysed with fear that I haven't the guts and the cortex to stick steadily to research, and yet, at my brighter moments, it seems possible . . . Things are moving a little more cheerfully here. I went out to lunch and stayed for tea and supper, to some very decent people. I also met a young English woman married to a chap who works here. She seemed a little vacuous but still a very pleasant change to the raucous American wenches one sees about. My professor [Richards] asked me out to Thanksgiving dinner, which was very decent of him. The more I see of him, the better I like him. He's a humorist and doesn't mind what one says to him.

Howard was, he tells Ethel, treating his dyspepsia by taking pepsin and hydrochloric acid. Then he told her of a momentous consequence of his trouble.

Incidentally, it has given me a new problem which I think could be worked out. Do you know anything about the secretion of mucus? How does it happen in the stomach, the large intestine, etc.? How do the goblet cells act? The literature is meagre, and I think the thing has never been tackled systematically—one might get some therapeutic procedure if one knew the physiology and pathology. You might let me know if you come across anything. The other secretions—gastric juice and so on, have been studied, but little attention is paid to mucus. Anyway, I've now got quite a number of things I could do and what I want is some enthusiastic student or graduate (an Australian come to London, perhaps) to do some of the hack work—I'm such a lazy beggar that way!

Thus, it was Howard's preoccupation with his own gastric function that triggered his interest in the whole subject of mucus and its secretion—a subject that he was to prove was important enough in its own right. And since it led him to the study of one of the properties of intestinal mucus—inhibition of bacterial growth—it was to have unexpected results of immense consequence fifteen years later.

Meanwhile he was still concerned with his other interests. 'I've just got some reprints of a paper of mine on thujone. I wish to heaven I'd never published it. It is probably true, but it can be shot to pieces by any competent person. Still, one lives and learns and I hope it will receive decent burial. I've a couple of others in print, one for December and another for January, which I think are better.' (These two latter papers were: 'An investigation of the immediate effects of X-rays on living animal tissues'[2] and 'The nature of the movements of the omentum'.[3])

About a week later Howard received a cable from Ethel saying that she would join him in England, not America. Howard, if he was again seriously disappointed, concealed it in his letter of reply though he speculated on her possible reasons. He had also received letters from her, written before his condemnatory outburst, and these letters are the first to survive.[4] They give a glimpse of her personality and her attitude to her work and to Howard. 'There is no startling news,' she wrote in October. 'The sun still shines by day and the stars by night. Alan Jones has just come back from England. His appearance has considerably improved. I must say, I do like people's clothes to fit them. I suppose you don't bother about that, but I shall be quite firm with you in future if I find you very slack. Please don't think I'm not interested in the omentum

and your experiments with it. So go on telling me about them
in as much detail as you wish.'

Two weeks later at the end of her house-physician
appointment she wrote: 'I took a tearful leave of Cresp the
other day, in which I think the regrets were mutual. Do you
mind if I'm awfully vain and tell you that he said he thought
I was the best house physician he'd had? Of course, I know
he thinks I'm a great deal better than I really am ... Still,
I couldn't help feeling rather pleased about it and hope you
won't turn up your nose at me for telling you.' She wished
she could come to America. 'It would be such fun pottering
about unexplored territory together ... I hope you won't get
cross with me much; I simply wilt when people do that—
or else I develop a burning fiery furnace of injured innocence
inside my chest. And please don't state opinions as if that was
the last word on a subject ... I'm afraid I find it very irritating
and may get quite cross with you, if you do it.' She asks
Howard if he has read *Martin Arrowsmith*. She liked 'old
Gottlieb' but felt that he would not 'be dreadfully interesting
to live with if he only talked about experiments and science.
I should rather have someone a little more human in the way
of a husband. I wonder how you are liking the Rockefeller
Laboratories. I suppose they are wonderful, with every
modern convenience and no bother about the means to pay
for necessities. I get so sick of putting my patients on an un-
palatable diet because the Government want to cut down
expenses before the election.'

On 15 November she wrote: 'I feel an awful cow after getting your last letter, in which you are rejoicing at my suggestion to come to America. I hope you won't be dreadfully
annoyed when you get my letter calling it all off again.' She
thanked him for the photograph of himself. 'I like it very
much—except for the glasses, which I hope you won't have
to wear perenially. I'm afraid I must be very earthy, as I find
attraction in things quite unspiritual or intellectual. For instance, I do hope your clothes will fit you ...'

Her next letter was written after she had received his cable,
in which he 'forestalled that beastly letter' and also told her
of his new opportunity in London. She has, she wrote, spent
a hectic night thinking about it. 'After talking it over with

Mother and Dad I've decided not to come. You see, you
haven't got the £850 yet, and we don't know how much debt
my coming to America would run us into. You may be quite
sure that there won't be the slightest difficulty in our disposing
of all your princely salary in the first year of married life ...
I don't mean to be horrid, but one of us must be sternly practical if we are to get along at all. Living with constant financial
worry is a frightful test of one's nerves and temper, and of
our companionability to one another.'

Then she received the letter which Howard had tried,
clearly unsuccessfully, to 'defuse' by his cable. She replied:

Dear Floss,
    I don't know exactly how to answer your last letter. It's all—or nearly
all—quite true, and I don't think it's worth while haggling over minor
points. I am dreadfully sorry I've disappointed you, but at least I've the
satisfaction of knowing that I told you, time and time again, that I was
not by any means the perfect person you imagined me ... And there are
lots of other flaws in my character as well as having 'no guts'. I suppose
I'm like Lot's wife and cannot keep my eyes on the things I'm coming
to without looking back at the things I'm leaving behind with considerable
sadness. It would be awkward to marry a pillar of salt—but probably it
would be a salt-lake as I shall certainly dissolve if you go on being cross
with me.
    I was thinking of trying to leave the children [her next house-
appointment] early and get to America for about two months. But then
you will only be still more angry with me for vacillating again. And I am
thinking, as there are still a good many disappointments in store for you,
that it would be better for me to wait and go to England where I would
have some means of supporting myself ... if you were too disgusted with
me altogether.

    Ethel explained her attitude to money. Although her father
had a good salary, and the Red House was large, comfortable
and in a fashionable part of Adelaide, domestic stringency was
the rule there. She wrote:

It seems funny, somehow, for you to be the one to talk about not caring
how little you have to live on, when I suppose all my life I've had to get
along on a great deal less than you have. Perhaps it is because I cannot
remember the time when I did not have to go without things I wanted
most dearly ... that I literally dreaded the thought of being poor all my
life. ... I see I must have made enemies of the Sherringtons already. That
doesn't sound very propitious does it—when he is your guiding star. I
should have to be conveniently sick, or have to stay at home to mind the
baby, when they wanted you to have dinner with them.

She often wondered, she wrote, why he wanted to marry her.
'But I suppose I must have forgotten that you endowed me
with all the virtues you thought a woman ought to possess,
and so fell in love with a false image. You must remember
that I'm me and if you don't love that very person, it isn't
the slightest use marrying ...' It was a tragically accurate
assessment.

Ethel followed up this letter with another a week later, in
which she justified her decision not to join him in America.
She compared his situation to her own. She had been advised
not to cut short her house-appointments and give up her
chance of a job at the Children's Hospital by coming to
America. He would not come back to the job in Adelaide in
order to marry her, and she did nothing to try to persuade
him. He had been advised (by Sherrington) not to jeopardize
his career. She had been dreadfully disappointed.

But at least I stuck to a resolution that I made when we first talked about
getting married. That was that no matter what I had to give up, I should
never voluntarily stand in the way of your career in the smallest possible
thing. Cannot you see how torn in two I am, between you on the one hand
and my profession which, after eight years of it, has filled my life and
thoughts and interest pretty fully? ... I know I wouldn't be a dud if I
stuck to medicine. But you know I am quite prepared to give the whole
thing up when I marry you. But, in return, I must feel absolutely certain
of your unreserved and entire love for the rest of my life. Your last letter
shows how little you know the real me. You still seem to think I ought
to be some sort of inhuman angel instead of a quite typical modern girl
with a lot of weaknesses and a fair amount of intelligence.

Shortly before Christmas, Howard set out for Boston. He
stopped in New York to discuss his work with Chambers, and
then went on to the Fultons.

They were very pleasant and did me very well. His in-laws are rotten with
coin. They threw a feed on Christmas Eve. Very pleasant company, with
some extremely attractive young females. Then all out, at about 10 o'clock
to see Beacon Hill. An old custom is that houses keep open house that night
and put lighted candles in the windows. A very brilliant sight. Streets full
of people singing carols and wandering about. Christmas day, a real Christmas card affair—snow, and all that. Went out to a country uncle and had
an immense meal, then on to tea with Harvey Cushing. Heard the Boston
Symphony Orchestra, a really fine performance. On Sunday pushed off
to Cleveland. Travelled with Cushing, talked a great deal, and was given
dinner by him. All very pleasant. The meeting in Cleveland [of the American Physiological Society] very amusing. Met lots of people and heard

numerous papers. Taken out to tea by Cushing to his sister-in-law. Met
people from Oxford, Lake Erie was frozen solid. A contingent from Philadelphia came back in the same train with me. In Cleveland met Carlson,
a big Swede, who has done a great deal on the stomach and he agreed to
have me out in Chicago. My present plans are to remain here [Philadelphia]
until the end of January, then go to Chicago for six weeks, and then back
to New York [to Chambers] for six weeks, then England some time about
the beginning of May.

Howard's conditions for accepting the research
appointment at the London Hospital were agreed but 'for
diplomatic reasons' he had offered to cut short his stay in
America by two months.

Howard found Ethel's letters waiting for him at Philadelphia. He was distressed that she was 'hurt with him', despite his cable. 'I was a maniac to fly off the handle at the
girl I want to marry because she won't come to a foreign
country and try the pleasures of financial embarrassment,' he
wrote in reply,

I'm a fool at throwing off plans and schemes. I've deluged you with at
least a dozen alternatives in the last week or two. Let's stop getting heated
over the fact that I can't have my own way.'

But, later in the same letter, his inveterate plan-making re-
asserted itself. Ethel had said she would come to England in
August.

I'll endeavour not to grumble and rumble if you don't turn up till then,
but if it can be managed earlier, so much the better, as you'll see England
at the best time of year. Also, there's an International meeting of physiologists in Stockholm on August 3rd to 6th. We could get married, go to
this show, perhaps, and honeymoon in Sweden which, from what I've seen
is a perfectly charming country.

When he received the letter in which she stated her own
case for not curtailing her medical training, he himself felt
hurt by her attitude.

Your most clear and logical exposition of events is so convincing that it
leaves me convicted of being a jabbering and insane baboon. God knows
what I've said—I only hope you won't pick up each phrase and award it
marks of approval or disapproval ... You've made me most terribly afraid
of you with your pounding logic and undeniable truths. I think that we
might defer further discussion till we meet, as it's so devastating to get an
answer three months after the propounding of a proposition. As a wavelet
on this stormy sea, I leave for Chicago in five days. My passage has been

booked leaving N.Y. on May 4th, so should be in England by May 12th. It will, perhaps, be as well to leave this country which has been somewhat ill-starred from our mutual point of view.

Before leaving Philadelphia finally, Howard went to the Johns Hopkins Hospital in Baltimore, where he met Abel and was shown the first preparation of crystalline insulin. Then over to Washington, which he thought 'really quite impressive. It must be very beautiful in summer, with its trees and flowers.' Then back to Philadelphia to hear for the last time the orchestra play, and to pack up for Chicago. There he went straight to Carlson's laboratory. 'He seems a very good fellow and introduced me to some of his chaps. I'm to join a "fraternity" for the time I'm here, which will be pleasant as I'll have someone to meal with and talk to.' Then he told Ethel of another offer of an appointment in England.

There seems to be quite a rush on to procure my invaluable services. I got a cable from Barcroft yesterday, offering me a job in the Cambridge Physiological Laboratory. I am, of course, flattered to death, as it is about the best in England. Langley died recently and Barcroft has been made professor and I expect that there is a general reshuffling. What is precisely offered, I don't know, but it would certainly be to teach part of the time— the salary not mentioned, but I'd bet my last cent it isn't better than £850. It would, of course, carry a fellowship with it—but I feel safe in judging that it does not offer better facilities than the London job, so I've cabled refusing. I trust I've done the right thing. I at least seem to have caught one bird.

At the end of January 1926, he wrote:

Chicago is enjoying a cold wave. I staggered to work this morning in the teeth of a high wind and the thermometer at $-7°F$, which is the coldest I've felt. My breath turned to ice on my spectacles, which was disconcerting. I'm beginning to dig in at the lab. Have got a contact with a very excellent person named Ivy, who will probably be one of America's foremost physiologists in the near future, and has already done some brilliant work on the stomach and pancreas. He's going to show me some useful operative procedures. . . . I talked over a problem I had in mind with Carlson. It is this. No-one has ever investigated systematically the production of mucus by the gastro-intestinal tract. No-one even knows its function for certain. Carlson says its an excellent thing to do—there's years of work in it and the results are bound to be interesting . . . There will be a lot of surgery attached to it, and it is the thing I wanted to know if you'd care to help in. If you'd rather not, I'll look about for someone as soon as I get back, so you won't mind letting me know.

I'm rather excited about the prospects of the thing. There is a possibility of finding out something about mucous gastritis and colitis—it was from introspection that the whole thing occurred to me. In the meantime, you'll be seeing cases of gastro-enteritis in children. Would you do this for me if not too much trouble?—Would you do fractional test meals on them and get data on their gastric juice? This may have been done, so would you look through the literature?

He went on to give precise instructions on the sort of stomach tube he wanted Ethel to use on her children, and the performance of the test, which required aspiration of the stomach every fifteen minutes for three hours. And she might, he suggested, get similar data on adult cases of mucous colitis. 'The big idea is this,' he went on,

There's an inhibition of gastric secretion by irritation of the large gut. What is required is a series of stomach contents analyses in a large number of patients with inflammation of the colon. If the patients come to P.M. make particular notice of histological and naked eye appearances of the stomach and colonic mucosae—collect slides and bring them with you—you might keep the 'great idea' quiet as I find this world full of plagiarists who are perfectly content to sneak round the corner with one's ideas.

It's amazing the vast numbers of problems one could tackle with a fair prospect of success if one has energy enough to do the experiments. I may be in a wave of enthusiasm, and reach the bottom again when my projected experiments all fail me—but then one always gets satisfaction in thinking that no other silly fool has done that, anyway, and it was amusing while it lasted.

In many of his previous letters to Ethel, Howard had, in his concern for her health, begged her not to overwork in her house-appointments. Now, in his enthusiasm for his 'great idea' this was forgotten, and he asked her to do exacting and extensive investigations in addition to her routine duties. Moreover, he asked her not to mention the purpose of tests which, since they were particularly unpleasant and difficult to carry out, would have been bound to cause objections and criticism.

Howard wrote that he was enjoying his fraternity house.

Most of the chaps are younger than I am. I like them quite a lot, on the whole. It may be the unction of flattery soothes my soul, as Oxford and England are held in a deal of respect. They come from all over the United States so that I'm really seeing some of the best youth of the country. Their system of medical education is one stage worse than Adelaide. The students are simply spoon fed all the time. The physiology lab is extremely crowded. The room in which I work houses other humans, cats, dogs, mice, rats and an owl. The mixture of faecal odours is truly astonishing, but the Lord has provided a nasal fatigue phenomenon.

He had been to concerts by the Chicago Symphony Orchestra, and had heard 'a violinist named Szigeti, a pianist named Gabrilowitsch, and Chaliapin'. He enclosed in his letter a newspaper cartoon that showed a wrecked living-room. A young man was saying 'I have a nasty temper, Ethel, but it's all over in a moment'.

Howard had been learning operative technique from Ivy. The object of these complicated operations was the study of mucus secretion in the intestine. They were designed to by-pass parts of the intestinal canal, the secretions in these isolated sections then being studied by making them open on to the abdominal surface. The experimental animals were dogs. 'I've done a few operations and am acquiring a certain degree of skill. I transplanted a piece of colon into the skin, leaving it attached to a pedicle for blood supply. I don't think its going to be of much use ... I have greatest faith in a Thiery fistula for a preparation to work on—if I can do this while I'm here the time won't be wasted.'

Two weeks later he reported that the operations were a success. 'My activities here are drawing to a close. I think I've got a method for tackling the mucus of the colon. I'm glad to say that the end-to-end anastomosis dogs and the Thiery fistula dogs have worked all right. The difficulty is to avoid infection.' He expected 'to push off from here for Toronto, have a peer at Niagara, thence to Ottawa, Montreal, Quebec, and down to New York, where I'll be till the boat sails'.

Howard spent two weeks on his Canadian tour. The Niagara Falls were frozen. 'Really extraordinarily beautiful, with ice stalactites everywhere.' Then on to Toronto.

The place has its merits, though not immensely impressive. From there to Ottawa, where the Parliament Buildings are nearly finished. Very impressive architecturally. The place was under 2–3 feet of snow. Beautifully clear sunshine, but too chilly for my taste ... McGill is less impressive than Toronto, but still quite a place. Montreal is a big city, largely French. The people drive about in sledges, bells ringing and so on. Then to Quebec, which is a most fascinating town. Almost entirely French, beautifully situated on a cliff above the St. Lawrence. Then back to N.Y. I have a room with a private bath—as Mr. Rockefeller's paying for it.

Meanwhile, Ethel's letters had been arriving. They described her clinical experiences and her quarrels with her colleagues. Dr. Alan Lamphee recalls that, as a house-officer, she was unpopular because she insisted on doing a nightly ward round, and would then take a prolonged bath after midnight in the residents' quarters, where the protracted noise of the plumbing kept her fellow residents awake.[5] The nursing staff found her autocratic but conceded that her insistence on the proper performance of every technical detail was probably to the benefit of her patients.[6] In reply to Howard's suggestion that she might help him with his animal experimental surgery she wrote, 'Don't depend on me ... surgery isn't my *pièce de résistance.*'

She wrote that Alan Lamphee was soon sailing to England. Another friend of hers, Aubrey Lewis [later Sir Aubrey Lewis], a psychologist, was also leaving Adelaide having, like Howard, been awarded a Rockefeller Fellowship. 'I do hope you like him, because I do very much, and you should find him a most interesting companion. It is such a relief to find a medico who has eyes for other things except just his profession.' She gave details of her own future plans. She would be doing an appointment at the Children's Hospital until the end of July, followed by two weeks at home, then leaving for England, so she would be arriving at the end of September.

Your suggestion that we should go to Sweden and have a honeymoon after the Physiological Congress sounds delightful. It seems to be my lot to throw cold water on all your schemes, but circumstances won't let me be in two places at one time. Please don't think I'm just a stay-at-home who doesn't care to try new adventures—I should just love to travel around any old how with you and poke our noses into all sorts of curious places, preferably where you haven't been before.

Soon after Ethel began her new appointment she wrote:

I had a horrible time here to begin with—3 deaths in the first 3 days and all lovely children—I felt I couldn't bear to stay any longer. But one of the delights of the place is Dawson Hanna, who was brought in just about moribund from diabetic coma four or five years ago. He is now a sturdy youngster with no ailments because he was lucky enough to get ill just when insulin was first heard about here.

Then, a few days later she developed a throat infection. This became a general sinus infection and laryngitis, a locum had to take over her work, and she was sent home to recover. 'Cresp comes and chatters to me for a short while each day, and Aubrey. Please don't stop making suggestions,' she wrote to Howard, 'they are all most entertaining even when they are not practicable.'

She thought that Howard's new job in London sounded 'simply wonderful, and especially to think that you got it through good men's recommendation of you. Will we have to get a flat? I don't like flats. I suppose it's quite out of the question to have a house and garden in London?' But when she heard that Howard had, almost simultaneously been offered a post at Cambridge, she wrote: 'It seems an awful pity to have to give up that position in Cambridge under Barcroft; and also the Fellowship. I don't think it matters at all that the income is less—£600 a year ought to suffice ...'

She told him that de Crespigny was shortly coming to England.

I suppose Cresp wants to see you in London. Please don't think of him in the light of the advice he gave me. You see, he knows me so well, and he doesn't know your point of view and he honestly tried to give an unbiased judgement. If he talks about my health and how you're to look after me, you needn't listen very carefully as I'm sure he is extremely over-anxious. He has been a most awfully good friend to me, so I do hope you will be nice to him.

In New York, Howard was again unlucky when he tried to start work in Chambers's laboratory at the end of March.

I arrived when the department was moving into new quarters. I had a trial today—nothing worked—all my apparatus will have to be altered. I think I told you I was going to try to insert pipettes into capillaries in a rat's mesentery and measure the pressure in them. I hope eventually to get into lymphatics. ... Spring is almost with us—the grass shows a sign of green. It seems rather astonishing that I'm only due to stay here for another five weeks. I suppose it must strike you as astonishing too that you've only got another four months. I don't mind how quickly they go, if only Old Time would throttle down for a bit when we do see each other. Spring is a very trying season. I'm consumed with envy watching the happy couples strolling.

Ethel's next letter told him that she had recovered and was back at work. She had quarrelled with the hospital pathologist, who was more interested in his private work than in doing investigations on her patients. 'I could willingly cut his throat,' she wrote. 'We've had an awful lot of dip [diphtheria] here lately, and some of the cases nearly break one's heart. Their parents don't bother until they have difficulty in breathing, and by the end of the next 24 hours, when the antitoxin should be beginning to work, the poor little things are in the mortuary. I think they are the most tragic of all the deaths because the children are often such bonnie little ones.'

She had just received his letter asking her to investigate cases of gastro-enteritis.

I am afraid you have set me a hard task if you want me to do test meals on people without letting anyone know what I am doing them for. As for material, we practically only diagnose gastro-enteritis as a last resort and, apart from food-poisoning, it is questionable if it exists as an entity ... I don't think it is, consequently, much use doing test-meals on babies with 'gastro-enteritis', do you? If we get any chronic dysenteries, I could try it on them—I will get Aubrey Lewis to look up the findings of test meals and correlate them with the disease at the P.M., though I must warn you that I don't think the results will be of the slightest interest to you.

What do you mean by saying that the teaching in the Chicago Medical School is only a stage worse than Adelaide—thereby insinuating that ours is not good? You went through the course a good many years ago, and probably with an unenthusiastic lot of fellow-students. There is very little spoon feeding now ...

Ethel expounded for two pages the merits of her medical school and concluded:

However, this is merely wasting breath as you aren't the slightest bit interested in medical schools or practice—Adelaide or elsewhere, but you will always make me bite if you run down the old place to me—I suppose because I love it and have put so much of my life and interest into it.

Howard, meanwhile, wrote of his own work.

My luck has been in during the last week. I think I've made an observation which may turn out to be of considerable interest if not importance. To my knowledge it hasn't been observed before. It's this—the lacteals of the mesentry are pulsatile. They pump their lymph quite vigorously. Of course, this opens up the whole question of contractility of all the body lymphatics—I've suspected this thing for nearly a year—so you can imagine my excitement on seeing the rats' lacteals nearly bursting themselves in an endeavour to show me they contracted. On the same kind of preparation I was fortunate enough to witness the resolution of capillary stasis. [In capillary stasis—a stage in inflammation—the red cells become packed into motionless masses.] As this escaped Krogh I'm also tickled to death about that too. Very pretty process—a pulsatile movement gradually loosens corpuscles from the compact mass in stasis and these are carried away by the rapidly moving stream in the adjoining vessel. In this way the capillary is gradually canalised, and once more carries a stream. It is immensely interesting to watch. Altogether, the rat's mesentry has proved a perfect gold mine.

I've only got 25 more days in this country. I suppose you're still counting in months. Be sure to let me know at the earliest possible moment the boat you're coming on.

He had been to a performance of Verdi's *Requiem* by the Metropolitan Opera, and heard Chaliapin in *Don Quixote*. 'The music of the opera is rather second-rate, but Chaliapin is a first-rate actor, and his voice of course, amazing. The setting was gorgeous—ballet and all complete.'

In his next letter he wrote:

I was somewhat taken aback today to find an article published [by A. Heller[7]] in 1869 describing contractions of the lacteals in guinea pigs in a very accurate way. However, I have added to those observations, which have been almost completely forgotten. But it is a disappointment. Nevertheless, it is a most fascinating sight to watch them—they are so transparent that under the high power one can see the nuclei of the endothelium comprising the valves—one can see these in action in the most beautiful way.

Of the Australian cricket team he said: 'I trust they beat the English—nothing gives me greater pleasure than to see the Englishman humiliated on the field of sport. They are so full of excuses.'

By the beginning of May, Howard was preparing to return to England.

I seem to be all set to sail tomorrow on the 'President Harding'. Precisely what the conditions will be when one reaches England it is hard to gauge. [It was the year of the General Strike.] Whatever the outcome, it seems that England must be a very unhappy place for years to come. I've been very busy wringing people's hands for the last few days—Americans are very keen on the process. Anyway, they've been extremely decent to me, and I only hope to be able to come back sometime. I have found that only the rats and guinea pigs have contractile lacteals of the animals I've studied. It becomes of interest to enquire why other animals don't. I've not any good notions on the subject at present, but there won't be much to do on the boat except cogitate.

6

LONDON

The North Atlantic was calm, and Howard found the voyage back to England dull. The passengers were not congenial and seemed, he wrote, to be mainly German Jews. 'I find it very annoying to have to make conversation for days with people at table with whom one hasn't a single idea in common.' Despite his complaints of loneliness in all the places in which he had worked he was usually sorry when the time came to move on. 'I've really had a most enjoyable and instructive time in America. I would like to go back some time. Everyone was most kind and English people—contrary to anything you hear—are very well treated. It was rather astonishing to see how English ideas, though outwardly sneered at, are the standard for the intellectual world.' 'The strike in England is very disquieting, but I trust it may be settled without stirring up class passions. The Government must win, it seems, unless England is to be ruled by the Trades Unions.'

The ship arrived in Plymouth on 13 May, the day after the General Strike ended. 'We must take off our hats to the British for their conduct of the whole show. They really are a remarkable people,' Howard wrote. He went on to answer Ethel's indignant defence of the Adelaide Medical School. 'I only know it as it was, of course, and I can't say too much bad about most of it. I'm very glad to hear it's all altered—I trust the spirit of it has too. Your beloved Cresp sinned, in my opinion. He once said to Souter (who, I confess, was a little stupid but had had an idea): "Now Mr. Souter, it is far better for you to learn what is in your books. It is not for you to question." As a point of fact, I take a bit of interest in medical education as it seems possible that I will have to teach sometime.'

He had been to the London Hospital, where he met Panton and some of his future colleagues. 'All very decent and willing to help. The only snag is the accommodation which is quite inadequate and inconvenient, but I hope in time to remedy that.' As the laboratory was not ready for him, he went to Cambridge for a week at Barcroft's invitation, to collaborate in a research project. The post that he had offered Howard had gone to Anrep, an already distinguished physiologist. 'It is extraordinary on what slender threads one's future hangs,' Howard wrote. 'It turns out that a letter sent by Barcroft never reached me in America and came back to him a couple of days ago. It now transpires that I was being offered a senior job in the physiology lab and probably about £850 or so. Very ironical. However, and this is very soothing—you know how vain I am—my reputation seems to have grown a bit, and I'm positively being sought after. It is quite astonishing. I was talking to the Master of Caius today. It seems almost certain that they're prepared to do quite a lot to get me back here. I really and honestly snigger to myself—and out loud sometimes to the astonishment of the onlookers—when I have people like the head of the Medical Unit paying respectful attention to my pronouncements. It's too damn funny, when one realises what a charlatan one is.'

'I've been to a dinner given by the Vice-Chancellor for the inauguration of the Scott Polar Institute in Cambridge. A large proportion of Scott's last expedition were there—two of whom I know fairly well. I sat between two F.R.S.'s and engaged in heavy conversation. Lady Scott was the only woman present. But the poor boob of a Colonial put his foot in it—as usual. Most of the gathering were in full evening dress, complete with decorations and academic robes. I arrived in a dinner jacket. I was so shaken at first that I very nearly walked past the Vice Chancellor who was receiving. However, there were some other ignorant youths too.'

He had been invited to spend the weekend with Lord Knutsford, the Chairman of the London Hospital. 'I had an extremely good time. He's a very affable old man and couldn't have been nicer. We played golf. He had 129 golf balls teed up 3″ apart and we alternately drove. He was so pleased with my progress that he gave me a club. I was also initiated into billiards, croquet, and clarinet playing.

'I've got a lot of irons in the work fire. There are so many things one can turn one's hand to that it's bewildering to know which to take up.' He asked Ethel what she knew about counter-irritation. Carleton had suggested to him that they might investigate its supposed action in reducing inflammation. He ended his letter, 'Just think. Possibly only a couple of more months of this interminable correspondence.'

Ethel's contributions to this correspondence were still arriving. In her next letter she put Howard in his place over his suggestion (already long superseded in his research programme) that she should pass stomach tubes on her child patients. 'Did you say kids swallowed stomach tubes easily?' she wrote. 'If so, my one experiment must have been a glorious exception to the rule.' After six attempts she had given up. (In reply to this letter, Howard admitted that he had only been *told* that it was easy to pass stomach tubes in children. Her experience showed, he said, how important it is not to believe everything one is told.)

'I wonder how you will like being back in England,' Ethel continued, 'I suppose it is almost like coming home to you. Is London always foggy? Or do you see the sun sometimes? We are having such heavenly weather that I want to revel in the sun all the time to make up for the future.' She writes of her quarrels with her medical colleagues, particularly her fellow house-officer. 'I disagree with nearly all his views, manners, and appearance, and the less I see of him the more harmoniously we work together. You may note that I always have to have somebody to complain about.'

'I have cold shivers up and down my spine every time I realise how soon I'll be leaving this little island for ever and ever. Sometimes I feel rather scared—the whole thing may turn out, not only a fiasco, but a tragedy for one of us ...' She expected to sail in the Commonwealth liner *Moreton Bay*, leaving Port Adelaide on 21 August, but not as ship's surgeon, as she had hoped.

'Do you realise, my intellectual young man, that you are preparing to marry a most horribly ignorant young woman? I have no French, my knowledge of English literature could be covered by a 6 inch square of paper. I cannot talk to you about motor cars, nor even drive one decently, and my geography is simply appalling. I shall never be able to live up to you.' She wrote of her work, and described how she helped to save the life of a child with both legs and an arm crushed by a railway engine. 'He looked as though he might die at any minute, he was so shocked, but he has improved wonderfully. I don't think there is any profession guaranteed to give one so much excitement and thrills as surgery does.

'Did I tell you Mother and Dad intend coming to England next year? I hope they will give us twelve months clear before they come, but there is no changing Dad's course of action if his mind is made up to come earlier. Is coal very expensive in London, or does one have central heating? I hope clothes are cheap, for I'm relying on that. I suppose eggs and milk are more of a luxury than here.'

Howard replied to some, at least, of these questions. Of her fears of intellectual inferiority he said, 'I don't want to marry you because you're a perfect mine of literary information or a travelling atlas.' Eggs, milk, and coal were, as far as he knew, as cheap and plentiful in London as in Adelaide. Central heating was rare. Where they would live must be decided when she arrived. At the moment he was living in a hotel at Chingford, close to Epping Forest, where he could get fresh air and exercise. 'Everyone remarks on my thinness, which is depressing; city life is very unhygienic. I've just come back from a walk in the forest. You can't help being charmed with the English country and I do assure you that, after a time, the odd spot of rain doesn't become so aggravating.' He was pleased that she had settled her date of sailing and was travelling on a Commonwealth boat. 'The food is quite good, I've been told, but the company will be the worst drawback, from which you will gather that I'm a snob ... I'll meet you at Marseilles, barring accidents.' He planned to spend two weeks in France and then join Ethel's ship for the rest of the voyage home. He went on to picture their meeting:

It will probably end by us shaking hands and being entirely speechless or mumbling incoherently. If the boat stops long enough we'll have a first class dinner, and wear evening dress—a woman always looks her best in it. I do hope the atmosphere will be right for us when we meet. If you've got a cold and I have a colonic spasm—what an initial blight! I have a hunch we won't be in London more than a year or two if my good fortune still holds. As soon as money appears I understand I'm to be offered a job in one of the Cambridge labs. They appear to be quite anxious to have me now that the letter (non-receipt thereof) has cleared up some doubts about my wanting to stay. The Master of Caius opened up a bit, in confidence. He wants to establish a chair of Experimental Medicine in Cambridge within the next few years, and if I do some decent work it seems as though he'd push me into that. In the meantime something in Pathology may turn up. I am applying for a Caius Fellowship again—the Master suggested it—I can hold it in London for one year and have the prestige—and about £250.

I had a most enjoyable week in Cambridge and did a few experiments which seem promising as offering evidence of contractibility of lymphatic glands on nerve stimulation. I've started on the tetanus work I told you about. It is difficult to do an intraspinal injection in a guinea pig without damaging the cord, but it can be done, I don't doubt.

I shot across to Oxford for the week end and stayed with Carleton. We are going to see what, if any, is the foundation for counter irritation, observe uterine changes in pregnancy without the presence of a foetus—that's got you, I trust!—and make some observations on fat absorption. It was very pleasant to be back and see some of the old people again. I had dinner with the Sherries. He was in good form.

My boom of popularity is quite appalling—I wonder when people are going to crack the bottle and see me deflated. It's rather an appalling thing to be expected to do something when one has the uneasy feeling that very little will eventuate.

Early in June Howard's laboratory at the London Hospital was ready for him. Almost at once he was confronted by the feuds between Panton's Clinical Pathology Department and the Medical College departments of Pathology and Bacteriology. Indeed, the various laboratories seemed to be united only by their common friction with the clinical staff of the hospital. This unhappy situation was not peculiar to the London Hospital. It existed to some extent in every medical school where the growing importance of science challenged the traditional domination of the clinician. Since this sort of conflict was to involve Florey for much of his career, its development at the London Hospital, where he was first brought face to face with it, might be relevant.

Dating from 1740, the London Hospital had grown steadily on its site in the Whitechapel Road as a result of successive benefactions, particularly from the City Companies, to become the largest of the eleven metropolitan teaching hospitals. Though the medical schools of these hospitals were affiliated to London University, they preserved a good deal of autonomy. For many years their clinical teaching was carried out by the honorary physicians and surgeons, eminent doctors who received no pay for their services to hospital patients or students and had no University appointments (or merely nominal ones). They had, however, lucrative private practices and they were both powerful and independent. Towards the end of the last century the University prevailed on some clinical schools to establish paid teaching appointments, thus gaining an academic footing there. Those appointed were not greatly respected by the honorary staff, who felt that any doctor who had to accept a salaried post must have failed in private practice.

The clinicians' domination of the hospital medical schools was also being undermined by the rapid development of pathology, bacteriology, and biochemistry—subjects in which they were largely ignorant, and which needed specialist teachers, laboratories, and equipment. And the clinicians, to their chagrin, found themselves increasingly dependent on the lowly laboratory worker for tests and investigations required for their patients and also for help in their own research. Thus there was a double load on the laboratories: teaching and research in the medical school and the ever-growing demand for hospital work. It was a load that was often too much for internal cohesion and the resultant separation of the academic from the clinical laboratories could have unfortunate consequences.

In 1897, the London Hospital Medical College appointed three full-time specialist teachers, and it is a comment on English medical education at that time that the men selected were all Scots: Arthur Keith (anatomy), William Bullock (bacteriology), and Leonard Hill (physiology). Before Bullock's arrival the rudiments of bacteriology had been taught by clinicians and his opinion of their efforts, expressed in uncompromising Aberdonian accents, quickly made him enemies. His expected appointment as a hospital consultant was blocked and he was thus deprived of access to cases of bacteriological interest. Bullock's reaction was to withdraw into his college department and have nothing to do with the hospital. This, in turn, deprived the hospital of a bacteriological service and the clinicians responded by creating a 'Clinical Pathology' department in 1909. P. N. Panton was appointed director and its essential function was to carry out laboratory work including bacteriology for the hospital staff. It was thus, in a sense, in rivalry with the college laboratories, though these were actually neither able nor willing to cope with the load of clinical work that they would otherwise have had to shoulder.

In its own sphere Bullock's department was highly productive, because it attracted young men of energy and talent who were to achieve lasting fame. The first of these was Paul Fildes, who became Bullock's (unpaid) assistant in 1907 while still a medical student. Bullock, more of an historian than a laboratory worker, was compiling a vast monograph on haemophilia—a strange diversion from bacteriology. Fildes helped him with this work and in 1911 the monograph by Bullock and Fildes was published, a classic and monumental record of every known haemophilic family in the world.[1]

Paul Fildes[2] was the son of Sir Luke Fildes, the Royal Academician whose famous picture 'The Doctor' had wrung the hearts of countless Victorians. He was educated at Winchester and Trinity College, Cambridge, and entered the London Hospital under the wing of its most eminent surgeon, Sir Frederick Treves, a friend of his father's. Treves wanted him to become a surgeon and was not pleased when he showed a preference for laboratory work and, in particular, for Bullock's laboratory. Paul Fildes had, however, found his niche there and he would have been out of place in the glamorized world of fashionable medical practice. Even as a boy he had been reserved, and later he developed a daunting manner in which dry, laconic cynicism was combined with a critical and acid humour. But behind this façade (which many failed to penetrate) there was generosity and kindness, as Howard Florey was to discover.

After he qualified, Fildes became Bullock's paid assistant, at a salary of £25 per annum, in succession to F. W. Twort, who left to become Superintendent of the Brown Institute and there helped to discover bacteriophage, one of the greatest advances in microbiology. Meanwhile, James McIntosh, another Aberdonian, had arrived in the department. He and Fildes were about the same age, and they set out on a joint research into syphilis, for the treatment of which Ehrlich had just produced his synthetic drug Salvarsan. Bullock was a personal friend of Ehrlich, who sent him a supply of his new drug for clinical trial. Bullock handed it on to McIntosh and Fildes, believing that they would make a better job of organizing a controlled trial than any of the hospital clinicians. They were, indeed, successful and published eighteen papers on their researches during the next four years which put them among the leading authorities of their day. On the outbreak of the war in 1914, they turned their attention to the anaerobic organisms causing gas gangrene and tetanus. These can be cultured only in the absence of oxygen and they devised a method, soon in universal use, for removing the last traces of oxygen from a sealed jar containing the culture plates. They were thus able to discover the natural history of these organisms, and to show that they were normally present in cultivated soil. Fildes had set up a bacteriology laboratory at the Naval Hospital at Haslar, with the support of the Medical Research Committee (later to become the Medical Research Council), the secretary of which was Walter Morley Fletcher who had been his tutor at Trinity College, Cambridge. In 1919 Fildes returned to Bullock's department, still at his salary of £25 p.a. but now also paid part-time by the Medical Research Council. Bullock had become Goldsmith Professor of Bacteriology, but had not improved his relations with the hospital staff. In 1920 McIntosh left to take the Chair of Pathology at the Middlesex Hospital. But he and Fildes made another important contribution of a different kind. In that year they founded, with one or two colleagues, the *British Journal of Experimental Pathology*, which was to become an outlet for the best research in this new subject, including many contributions by Howard Florey.

The Freedom Research Fellowship to which Howard had been appointed was created in Panton's department in 1925 by an anonymous benefactor.[3] Neither Panton nor his staff had much inclination or opportunity for research. Their task was to carry out the laboratory tests demanded by the clinical staff for their hospital and their private patients. This latter occupation was a lucrative one since the clinical pathologist charged a fee for his services. A research appointment in such an atmosphere was not entirely a congenial one, particularly to Florey. His opinion of clinicians and of clinical research was already low, as he had often made clear to Ethel. Panton, an Old Harrovian and also (like Fildes) a Trinity College, Cambridge, man, was the antithesis of Fildes. Though a good hospital pathologist, he was primarily an administrator and a politician. His polished, courteous manner pleased his clinical colleagues and concealed a strong personality and an ability to gain ends by farsighted manœuvring. For example, in 1933, when Bullock retired, Panton became Director of the College laboratory departments as well as his own, causing Fildes to leave. And during the 1939–45 war Panton became the Ministry of Health Adviser in Pathology, with the virtual control of every civilian hospital pathology department in the country.

Howard, when he finally started work in London in June 1926, took up the experiments begun in Carleton's laboratory on mucus secretion, and also his idea for treating tetanus. The obvious person to be interested in this latter project was Paul Fildes, already one of the world's experts on tetanus. Howard approached him though, since Fildes was working in Bullock's laboratory, such a move would not be appreciated by Panton. On 7 June Howard wrote to Ethel. 'I'm getting more or less into shape at the hospital. The work on tetanus has begun. I'm working with Fildes, who is an extremely good fellow, though he at first impresses one as being very bad tempered. We've got a technique now for intrathecal injections in g. pigs, and today I've started a series.' 'I'm continuing my work on the lymphatic system, and have managed to expose the thoracic duct in the guinea pig, the first time anyone has done it, as far as I know.

'Last Friday I went down to Panton's at Bourne End, which is on the River. I had a miserable time. Mrs. Panton is rather queer, and they had a domestic upheaval, though there seemed to be loads of servants about. We were to have gone to the Tattoo at Aldershot. Mrs. P. became temperamental, and two of us arrived $1\frac{1}{2}$ hours late. Next day there was wrangling over the most futile points—not in an antagonistic way, because they seem fond of each other—but oh, such futility! Anyway, I pushed off then, as I couldn't stand any more. My dear, if we ever have people to stay with us, or have them about the place at all, we must treat them well ... I don't think there's anything more disagreeable and annoying than to be made to feel uncomfortable in someone else's house.'

He took up various points in Ethel's letters. He disliked being called 'young man' by her—he pictured 'an acidulous and disillusioned female pointing a somewhat bony finger. I can't agree to sit with my collar on all the evening. You mustn't bite my head off if I'm late for meals occasionally. Sometimes an experiment takes much longer than one anticipates and it's a great ruiner of a day's work to have the thought hanging over one that I must go at 5 o'clock or my wife will be bad-tempered. I think we shall probably have rather different views on medicine which will produce, I trust, some fruity arguments. I've not the slightest doubt you can present me with a whole hat full of ideas just screaming for someone to work on them, but you, as a clinician, are doubtless looking at them from a different point of view. Like all clinicians, you are not in a position to prove your ideas.' He had been, with Carleton, 'to an excellent play "They knew what they wanted" ... Talullah Bankhead, a young actress of 24, was first class.'

Ethel's letters had been arriving with their inevitable time-lapse. 'What a disappointment', she wrote, 'to find that some-one else had discovered those pulsations in the lacteals in 1869! I suppose you'll get over it, though you sounded hor-ribly disappointed in your letter. How depressing it must be to think that you may discover something wonderful and important and have for its reward the pleasure of knowing that people will entirely forget about it.' She took up a suggestion in one of Howard's letters that her remarks on the masculine selfishness of her colleagues were innuendoes aimed at him-self. 'Now, young man,' she wrote, 'you don't know me very well, but you may rest assured that if I have any fault to find with you I shall let you know direct, or else hold my tongue. I do not resort to innuendoes.' She had now definitely booked her passage in *Moreton Bay*, leaving on 21 August and arriving in either Plymouth or Southampton about 24 September.

I am busy turning out drawers at home and collecting all my worldly goods I cannot bear to part with. I have even thought of making a will in case I am drowned at sea or somebody strangles me with a deck quoit. I could leave you all the letters you've written to me, or, if you don't want them, direct them to be burnt. As a matter of fact if you don't keep a diary, they would be very useful to you and form interesting reading matter. I am keeping them so that, in the event of your death at a timely and revered age, I shall be able to publish them as 'The Love Letters of Lord Reading' or something equivalent, and make a pile of money.

Do you realise I shall arrive in England on your birthday? I shall wrap myself up in tissue paper tied with pink ribbon ... will you accept the present, or throw it into the sea?

I have become so used to thinking of you as my prospective husband that I am quite sure I could never face the ordeal with anyone else. So, from my point of view we might just as well get married on the wharf and leave it till later to decide whether we can put up with living together. If we don't actively want to be with one another, then I'm sure it would be much better to separate than to drag on in a humdrum existence together.

Ethel had been to see Howard's mother, and gave her a newspaper cutting about his appointment in London.

But she knew all about it already. She knows ever so much more about you and your doings than I do, I feel quite ashamed of myself when I go there. I never can understand why you don't care for her. She seems such a fine type ... and I imagine you have inherited quite a lot of your temperament and intelligence from her.

Your plan of joining the boat at Marseilles is out of the question, unless you hire an aeroplane and drop into her—she calls at no ports between Port Said and Southampton. I shall come armed with about £180 to stand between me and starvation, or to assist in buying the table linen if I marry you ...'

Howard, disappointed in his plans to join Ethel at Marseilles, wrote that he will meet her at Plymouth or Southampton. She had suggested that her friends will bore him. 'If they haunt the doorstep there might come a time to protest, but otherwise I see no objection, so long as you don't insist on playing cards. I think that's the last resort of the conversationally destitute.' He regretted that she did not appreciate music and liked going to the cinema.

He went on to warn her (yet again) that he was bad-tempered.

I had a frightful row today and kicked a man out of the lab. He was the hospital photographer, a chronic grumbler who has been there a long while. I wanted a photo taken of a dog. In the rudest way he raised difficulties, and I swallowed my annoyance as I wanted the photo. However, when he came he was ruder still, so I shouted at him and told him to take his bloody camera and get out.

I'm getting a few experiments done now under constant, slight irritation. I have to work with builders looking in at one window, patients and every passer-by at others. It nearly drives me to fury at times. I've screamed myself hoarse trying to get the windows whitened, and it wasn't until I swore today and told them to be down in half an hour that anything happened. It's that kind of place. To cap it the head of the lab suggested I was trying to do too much—the silly idiot. He's too damn lazy to do a thing himself—an armchair critic who looks wise and mutters. He thinks I've never done an experiment before, apparently, and rebuked me for using 'American' methods. I pay the very greatest respect to anything Sherrie or Barcroft would say but I'm hanged if I'm to be harangued by this man who has done practically nothing himself. It's a curious fact that the harshest and most persistent critics are the people who've never tried to do anything themselves.

If Panton had hoped that a vigorous young research worker would add to the prestige of his otherwise routine department, he had failed to foresee the strains that must develop. And in Florey he had someone with a temperament least likely to submit quietly to the bureaucratic authority which Panton habitually exerted. In fact, he was faced with open rebellion—even, on one occasion, with the threat of a punch on the nose. For once, he was defeated. He went to Fildes, asking him to supervise Florey's activities, as he himself was unable to control him.

On 20 July Howard wrote from Cambridge sending his letter to Ethel to Fremantle, where *Moreton Bay* would arrive about 26 August. 'I'm feeling very restless so have migrated here for a few days to work in Barcroft's lab. I'm rather fed up with the London Hospital. They annoy me—but I dare say when I have you to soothe me I will be able to see things more clearly.'

At Cambridge he worked with Stephens on the photography of the microcirculation in the cerebral cortex, using the transparent windows he had perfected and special cameras. The results, if the apparatus could be made to work would, he said, 'astonish the natives'. He felt, however, that descriptions of his work were 'unlikely to interest or cheer someone feeling very homesick and horribly doubtful about the future. Don't fail to make the best of your time at the few ports you do call at—don't miss anything for the sake of a pound or two—you'll always regret it if you do.'

Ethel was still immersed in her own work and wrote of her harrowing experiences.

I had such hopes of one little fellow, who had a septic arthritis of his ankle. He developed a spreading infection of the skin, and septicaemia, and was given up as hopeless. However, he rallied wonderfully after an intraperitoneal transfusion of blood; his temperature was coming down steadily and his colour improving when, all of a sudden he began Jacksonian epileptic fits. So I suppose he's got meningitis and everything is up ... Another boy did the same thing a month ago. It takes all the heart out of one—if they'd only get better, it wouldn't matter how much one did for them.

What a nice, warm comfortable feeling it must be for you to be wanted by lots of people. You seem quite sanguine about leaving London in a year or two for Cambridge, but can you really break your contract as easily as that? Wood Jones tells us that there is a poverty of young English research workers, and America is the place where everything is being discovered. Did you find that so? English scientists are much too conservative, in his opinion.

I wonder if you have seen Cresp? I'm afraid he won't give you a favourable impression of me. But you must tell him that I'm heaps better now. I've only three more weeks here—the time is getting appallingly close, isn't it? Don't you feel a bit frightened? I'm sure our first meeting will be most embarrassing.

Howard wrote, at the beginning of August, to Suez. He had seen de Crespigny.

He paints rather a dismal picture—tells me you are exceedingly subject to infection—in fact I gather that you are liable to dissolve at any moment. And what infuriated me, he told me in what an utterly insensate manner you worked. He's given me the name of a man in Glasgow to see about your ears. He also thinks you must live in the South. What is certain, however, is that you are going to live out of London. I went with the de Crespignys to a meeting in Oxford of the British Association, and showed them round the University, complete with historical quips. You may gather my amiable temper from the fact that I did not ask Madame to shut up in the name of the deity. She is one constant succession of banal remarks—I could cheerfully have wrung her neck. And she has a refained accent which is intolerable.

Ethel had still been trying to comply with Howard's request for information on patients with gastro-enteritis.

I've been looking through reams of test meals but after copying several dozen, I've come to the conclusion that they will be useless to you, because hardly any of them have mucus charted. However, I'll bring them along.... Do you know anything about intraperitoneal transfusion of blood? If so, are the cells absorbed unchanged? I have given quite a number here, and have found the improvement in the child to be noticeable after the second day.

She had been down to see her ship, *Moreton Bay*, which had called at Adelaide.

She seems to be a good boat, with a fair amount of deck space to sun round on. My cabin is a four-berth one, which I am to have to myself. But I believe Dad will get a deck cabin for me, which is heaps to be preferred. I don't know anyone travelling on the boat, which may be a mercy or the reverse. There were 700 immigrants on it this time, and I cannot say I was very impressed with their appearance. However, there will only be 300 going back to England, and I ought to find a few congenial ones. My aunt has come out on it, and says the people are very free and easy, and one has to keep one's distance from them. Ship's life is most demoralising, I'm told, so you must be prepared to meet a completely demoralised woman. However, I've had a fairly good training after a year's residence at the Adelaide.... I'm going to see Pavlova—10/6, mind you, but not at my own expense. The prospect thrills me, as I love dancing and have never seen anyone like this before.

In her next letters she wrote:

I have one blow which is going to make things a bit more hurried for me. I thought I'd have a clear 3 weeks at home, but now Scottie [M. L. Scott] expects me to spend a week in the Memorial after he has dug out some cervical glands. It is a nuisance, but I suppose it's safer than risking the effect of a cold, damp climate on them. I have been to see Pavlova, and am going again tonight. It was absolutely delightful. I hope we will go to lots of Russian ballet when we are in London. Dad has been generous and given me a deck cabin on the boat, it is a great relief.

Howard wrote what he said would be his last letter to her: 'In a sense it is a relief—but perhaps the next I write will be to "Dear Ethel" or even "Dear Dr. Reed". In fact, I'm coming to the conclusion that I can't possibly compete in your affections with medicine.' His own work was progressing slowly.

As far as I can see the lymphatics and lymph glands in the cat are under sympathetic control—for what reason I am at present at a loss to explain. They contract to adrenalin, and so on. I'm examining a series of animals, including a pig, hedgehogs, etc, but in only 2,—rat and guinea pig—have I seen rhythmic contractions. The mouse is a very pretty sight, as one can see right through the gut wall, it is so transparent. In a month I hope to get the thing in shape for a bit of a paper to pacify the Lords of the Hospital.'

He is also working on his Caius Fellowship thesis.

I do hope I can get it, if only for its monetary value—about £250 for the first year anyway. I have to overcome the fact that I'm not a Cambridge graduate, which may wreck me, but as the Master advised me to put in,

I'm making a bit of an effort to doll out my mediocre stuff to look like the scintillations of a brilliant mind. It is a mercy that people are easily deceived in these matters!

'The test match finished today with a rather severe defeat for Australia, which is depressing. I hate to think of the deluge of bilge in the papers about the old country not done yet and so on. I'm going down to Devon about September 14th to get a little less peevish. I can't find out if your ship will call at Plymouth or Southampton, so I've decided to find a golf course on a bit of sea midway between, so that I can go either way as occasion demands.

I've asked several people about giving children intraperitoneal injections of blood. They'd none of them heard of it. A priori, I should say the whole idea was absurd, and that typing was a sheer waste of time. The way such stuff is removed is by phagocytosis and thence into the lymphatics—did you ever try it compared with saline, and serum and, say, gum saline? If not, it is obvious one can arrive at no conclusions.

Despite these dismissive comments, Howard was himself to show a few years later with L. J. Witts that red blood cells injected into the peritoneal cavity *are* absorbed into the bloodstream.[4]

Though Howard had written his last letters to Ethel, there were still some of hers on their way to him, and they were disturbing.

I am afraid this letter is going to be about myself, but I had better tell you everything quite plainly. I came in here [the Memorial Hospital] because I have had some enlarged cervical glands for about 9 years, but which have been increasing in size lately, and rather rapidly. Scottie thought it would be as well to take them out. I am afraid I haven't got any satisfaction out of him yet, but I found out from Dick Grant, who gave the anaesthetic, that it was a large mass, very difficult to remove, and that a good part of it was caseous. The point I want you to understand is this, that a good many doctors, besides Cresp, think I am very foolish to go to England, and since they know the nature of the gland, they are still more insistent about it. Dick Grant and Harry Wunderly have each spent about half an hour trying to persuade me to give up my plans. Helen Mayo and Scottie have also spoken to Mother and Father, and there are a lot of others. Some of them know that I am not going to work very hard and that I could get out of London if necessary during the coldest months. If you really are prepared to marry me, and I'm afraid it's too late now for me to wait for you to change your mind, it means that you will have to be prepared to run the risk of possible heavy expenses added to an invalid wife. Of course, it's only a risk, and I wouldn't come to England if I didn't think that, after all, it is a fair exchange—I mean, if you want me to give up living in a comparatively safe, warm, dry climate for a cold damp one, it's up to you to look after me if I crock up. You can't blame me altogether

if I don't view the future with perfect equanimity. I suppose it was partly
this that made me rather regret your missing the Cambridge appointment
and taking the London one instead.

In her next letter she apologized for her depression:

The chloroform was still in my bones. Added to that someone had just
been haranguing me about what a little fool I was and making me imagine
I was walking straight into the Jaws of Death ... please forgive me.

This is the last letter, I think, so the next time I address you it will probably be as "Dr. Florey". Still, four hours in a train may work wonders—
we may be quite good friends by the end of two, have quarrelled by three
and made it up and resolved to live happily ever after by the fourth.

She did, in fact, write once more, and this time from her
ship.

I've got something dreadful to ask you, so I'd better get it over first. The
sad truth is that I've run through all my money. I thought £10 would have
seen me through. Do you think you could bring a little extra down with
you to meet me—say about £5? I've a draft on the Bank of Australia, which
will enable me to meet my liabilities as soon as I get to London.

The trip so far has been very good, though I am steadily degenerating
into an inveterate gossip. I hope you have forgiven me for that letter I wrote
from the Memorial. I'm really not bothering a bit about such things now.
I just eat and sleep and chatter. I'm afraid what I have seen on board does
not make me exactly enamoured of married life. There is one couple, however, who seem to be perfectly happy and contented with one another.

There is still some doubt as to whether we are going to Plymouth or
Southampton, so I do hope you will not go to the wrong place. I should
feel so lost if you weren't there to meet me. I'm afraid I've got a bit
spoilt on the boat, so you will have to consider it your duty to unspoil me.

So the long correspondence between these two young
people came to an end. At its beginning, they scarcely knew
one another. In the five years that followed they had built up
pictures of each other and a tenuous relationship which had
little foundation in reality. Ethel recognized this more clearly
than Howard, who had fallen in love with his idea of her. It
was a dangerous basis for the marriage to which both were
now committed.

With the ending of their correspondence there is an end
of the only source of first-hand information on Howard's
thoughts and feelings. For no one else did he record these
as he had done for Ethel during the previous five years. He
did not keep a personal diary. His letters to other correspon-
dents are business-like and brief. He had no close friends in

whom he confided. Thus, with the exception of one document, it is only from outside evidence that the emotional stresses and strains of the next few years can be inferred.

The bare facts are that Ethel's ship called at Plymouth on 24 September, Howard's birthday. He went out in the tender to meet her, and they then travelled to London by train, where Ethel stayed with a Mrs. Bathgate at 119 Westbourne Terrace. Howard moved from Chingford to be with her in London, and found lodgings at 37 Gloucester Gardens. They made arrangements for their wedding, at Holy Trinity Church, Paddington, to take place as soon as possible. In the interval they went together to Oxford where they spent two weeks with the Carletons. Within a day or so Ethel retired to bed with a bad cold and stayed there for a week. She seemed to Dr. Alice Carleton remarkably unexcited about her impending marriage, even allowing for her obviously poor health.[5]

The wedding of Mary Ethel Hayter Reed and Howard Walter Florey took place on 19 October with the minimum of fuss. Harry Carleton was to have been Howard's best man, but he could not attend, and Noel Wigg, an Adelaide contemporary of Howard's who happened to be in London, took his place. Dr. de Crespigny gave the bride away—one may guess with reluctance. The few guests included Howard's half-sister Anne, and an uncle by marriage, James Burrin. Uncle James Florey had died in 1923. Susan Florey, his sister, had married James Burrin, but died in childbirth and her baby lived for only a few weeks. The bereaved 'Uncle James II'—as he signed himself in his frequent letters to his favourite niece Anne—was a forlorn figure who clung to the Florey family in Islington while they survived. His attendance at the wedding was an expression of this attachment, but it seems to have been his last appearance in Howard's life.[6]

Dr. Alice Carleton believes that Ethel became ill with a throat infection immediately after the wedding. She writes: 'I think this may explain the absence of a honeymoon. Later, they went on holiday to Rome. Ethel insisted on climbing the big modern Vittorio Emanuele monument. A gendarme went up to see what was going on and Floss, demented, rushed after him shouting "E mia moglie, mia moglie!" '[7]

They returned to England in November, scarcely an attractive time of year for someone who had spent her life in Adelaide. Howard kept to his promise that they would live outside London, and they settled down in a rented house, Heather Lodge, near Chobham, Surrey, though the daily journey to the London Hospital was tiresome and frustrating. It was not a happy time for either. Ethel's fears and doubts seemed to be fully justified. Howard's romantic fantasies were shattered. The real Ethel, though she had undoubted charm, was not the submissive, adoring companion and playmate of his dreams. She was a critical, unromantic, and determined woman whose affection for her husband never allowed her to overlook his shortcomings.

Howard's reactions to the reality of this longed-for reunion with Ethel are available in his own words. Eight years after their marriage they exchanged memoranda in which they set out their mutual grievances and the conditions under which they could continue to live together. Only Howard's memorandum (a reply to Ethel's) has survived.[8] In it he recalls his feelings before and after their marriage:

You complain now of my not writing to you when you go away. I suffered the same bitterness before you came to this country. It may seem a strange thing, but I looked forward to the days when letters arrived—how often was I disappointed and that for some frivolous reason too often. Your arrival in this country and your treatment of me are etched on my mind and I see the things crystal clear even now. Before the arrival of your ship I spent a week at the sea trying to get fit and presentable for you ... I can remember my straining to catch a glimpse of you from the tender, but it was not until I had run about all over the ship that I found you. This I dismissed as maidenly embarrassment at the time, but the journey to London was as if you'd hit me with a wet towel. You yawned in my face the whole way up to London. Imagine my astonishment when you were naïve enough to tell me that it was because you'd been up late the previous night rejecting an importunate suitor. During your stay at Mrs. Bathgate's there were episodes you forget but which cut me to the quick. Do you recollect the night you cheerfully left me, to go to the theatre with two men from the boat? Mrs. Bathgate was most astonished at your conduct and I may say she fully sympathised with me while I aimlessly spent a little time wandering about her house. Do you remember going down [to Tilbury] to see the people on the boat? You were so little pleased with me that you wouldn't take me the whole way, and then during the part of the journey down which I did make, I sat a miserable supplicant ignored in one corner. It astonishes you, perhaps, that I have feelings. Nevertheless, I thought all would be well, and that these things were not so serious as I thought. At great inconvenience we lived in Surrey, where I was struck with horror when I realised how sick you were. You came at the time you did after insisting on doing not only your general hospital work but also the Children's. It is all very well to tell me now of the brilliant career you had in front of you, but you have evidently forgotten that you told me in those far-off days that you thought your health would have precluded a very active medical life ...

Soon after we were married, you blandly informed me that if you didn't like married life you were going back to Australia. A nice conversational opening. The bickerings I used to have with you trying to keep a regime for the benefit of your health are fixed in my mind. As a result of these things and many trivialities I unfortunately did not realise you had any real affection for me. My disappointment was, and is, and always will be immense that you are even now not strong enough to share my pleasures ... you must realise that you are not a physically normal woman—I refer to your deafness. It is a tragedy for you—I am fully aware of that, but what you don't realise is that this tragedy extends in ever widening circles. You must accept my statement that to talk to a deaf person for long periods of time is very exhausting and it may be exasperating ...

That the memory of these early events should have persisted for those eight years is a measure of Howard's profound disappointment. Ethel's initial coolness to him must have seemed heartless, but it is perhaps understandable: she hardly knew him, and ship-board friendships have a magnetic, if temporary, glamour. And she had not, as he had done, fallen in love with an imaginary paragon. That she did come to love him is probable, but not to the point of exercising sympathy or self-sacrifice where her personal interests clashed with his. Her ill-health was a constant source of anxiety and frustration to both of them. The glands removed from her neck had proved to be tuberculous. Her repeated respiratory infections must have seemed ominous; and the threat of consumption was constantly with her. But it was her deafness that was the main barrier. Howard was not the most patient of men. He disliked having to raise his voice and repeat remarks in the home, he was selfconscious about doing so in public places and, as he remarked to a friend, it is impossible to shout endearments. Deafness is, in fact, one of the most tragic forms of isolation. And Ethel, as is so often the case, tended to become suspicious, misinterpreting half-heard remarks and believing that she was the target for unheard criticism.

It is easy to understand, now, why Howard married Ethel. He was irrevocably committed by five years of fantasy. As he remembered her she was gay, pretty, intelligent, and charming. He was desperately lonely in England and apparently unable to make close friends of either sex and of his own age. His image of Ethel, the perfect companion, was his emotional outlet. It is not easy to understand why Ethel married Howard. She tried on more than one occasion to extricate herself from the undeclared commitment that grew steadily during their correspondence. She made excuses not to come to England, and delayed doing so as long as possible. It seems clear that Howard had no serious rivals. Despite her charm, she was too critical, determined, and domineering for masculine comfort. Perhaps she did have some affection for Howard and felt that she should not 'let him down'. She was proud of his achievements, though her letters reveal a tinge of the jealousy which later became more evident. Howard may have been right in his accusation; she may have realized in Adelaide that her health would not allow her to make a career in medicine on her own, and that marriage to a successful research worker would give security and also professional opportunities. She had no great hopes of her marriage: in fact she had predicted disaster. In the event it was not much worse than she had feared and indeed her affection for her husband seems to have grown. But for Howard it was the shattering of every romantic fantasy. His sole emotional outlet from that point and for many years was to be his work.

7

LONDON AND CAMBRIDGE

Florey's appointment at the London Hospital lasted for little more than a year. He found the physical and mental environment at the Clinical Laboratories unsatisfactory and, whenever possible, he slipped away to work for a few days at Cambridge or Oxford. He found himself feeling at home in Cambridge, and he could not forget that the mere chance of an undelivered letter had robbed him of the post with Barcroft. His working life in London was complicated too by the long and awkward daily journey from Chobham. He fretted at the waste of time and, in particular, at the constraints of the railway timetable which might curtail an experiment. During the previous two years he had done an astonishing amount of work, published eight papers, written two theses, and travelled widely. But the year following his marriage was less productive.

In 1926 two of his papers had appeared in the *Proceedings of the Royal Society*—communicated by Sherrington. The first, 'Rouget cells and their function', described his work with Harry Carleton on capillaries.[1] They concluded that these minute vessels could contract or expand and that 'the motor activity of the capillaries resides in the endothelial cells'. They decided that 'Rouget cells' played no significant role and probably did not even exist. This work therefore demolished a well-established myth. It might be mentioned here that these conclusions on capillary contraction were attacked some years later by American workers who, largely because they could see no contractile elements in their walls, decided that capillaries were inert tubes. But recently electron microscopy has revealed fibrils in capillary endothelial cells that resemble contractile elements, and Florey and Carleton's conclusion of fifty years ago seems likely to be vindicated.

The second paper described Florey's original observations on the mechanism by which a static mass of red cells choking an inflamed capillary becomes mobilized by progressive loosening at the outlet.[2] Perhaps city authorities might learn something from the natural resolution of traffic problems in the microcirculation, which Florey some years later illustrated so beautifully by cinematography. A third paper that appeared in 1926 was on his studies with Carleton of the movements of the omentum.[3] Once more, it was the elegance and ingenuity of the approach that had given the direct answer.

In London he was writing up work done during his few weeks in New York. This had been on the contraction of the lacteals and other lymphatic vessels, which was published in the *Journal of Physiology* in 1927.[4] He showed that a rhythmic contraction combined with the presence of valves resulted in the pumping of lymph from the tissues to the thoracic duct and thence into the blood-stream. The work required the highest degree of surgical skill combined with the ability to operate on a minute scale.

Florey's other literary activity in London was the writing of his thesis for a Fellowship at Caius College. Though he suspected that not being a 'Cambridge man' would tell against him, his standing with the leaders of scientific research there was extraordinarily high. Dean had a profound admiration for Florey who, he had said, had the greatest technical skill of any young pathologist he had met.[5] Sherrington, in a letter to Florey after his appointment at the London Hospital, wrote: 'Dean has asked me to try to find him another Florey.' Dean's opinion was shared by Barcroft, Adrian, J. B. S. Haldane, and Anderson, the Master of Caius. And, of course, Florey had equally powerful admirers in Oxford. A reputation of this sort, so quickly gained, prompts some examination. His excellent academic record was not the main issue, since many of his contemporaries who had done equally well at school and university had subsequently done little else of note.

Florey's most striking characteristics were his energy and enthusiasm for research and his complete scientific honesty. He was a prodigious worker, full of ideas for the practical solution to some immediate problem, impatient of delay, and with an infectious vitality that was to attract a succession of collaborators who often found themselves doing the best work of their lives under his influence. He had already done more sound research in two years than most promising young men had done in ten, and it seemed likely that he would set this pace for those who would work with him. It was the recognition of this peculiar research vitality that had so impressed his seniors. His personal relations with them, too, were friendly—provided (unlike Panton) they had earned his respect.

With contemporaries—other than collaborators—relations were less friendly. Florey disliked the social life that most young people enjoy. His supposed aloofness was a positive cause of unpopularity in some fields. His refusal to join the drinking bouts that followed Cambridge tennis matches made him disliked by members of the college team, though one of their best players. The dedicated ferocity with which he would pulverize a lesser opponent to the last point was also thought rather excessive. His remark to Ethel, 'Nothing gives me greater pleasure than to see the Englishman humiliated on the field of sport', partly explains this ruthless determination. But it was also an expression of a fundamental attitude. 'In everything he undertook,' said a lifelong colleague, 'Florey went all out to win.' [6]

In addition to his outstanding talent for research, other factors were working in his favour. Nearly a whole generation of Englishmen of Florey's age had been wiped out during the war, and between the old men who staffed the medical schools and the flood of new students there were many vacancies in the middle ranks. Another, though less important, factor was that overseas students (particularly Rhodes Scholars) received a special welcome in Britain. Much is forgiven in an Australian or Canadian that would be less tolerated in a native Englishman, and genuine excellence is the more remarked and rewarded. It was as an Australian that Florey was first invited into the Sherrington household, but it was his own qualities that kept him there.

Florey's laboratory notebooks for 1926[7] show two main interests—the circulation of the lymph and the secretion of mucus—that had begun in America. Both required long and careful operations in a variety of animals and the facilities for these at the London Hospital were unsatisfactory, including objections to his work with dogs. It was for this reason that he carried out experiments in Oxford and Cambridge on cats, dogs, rats, mice, guinea-pigs, rabbits, hedgehogs, squirrels, bats, and pigs. The London Hospital at least provided him with some human material since he was able to study, in the mesentery of patients undergoing abdominal surgery, the special lymphatics (lacteals) that drain the intestinal wall.

These experiments led to a second paper on 'The contractility of the lacteals' published in the *Journal of Physiology* in 1927,[8] which emphasized the strange species differences in lymphatic vessel activity. Only the guinea-pig and rat showed spontaneous lacteal contractions. In all other species studied, the lacteals contracted only when stimulated directly or via the nervous system. Florey also observed that the lymph nodes contracted on stimulation and increased their output of lymphocytes.

In the second line of research, the study of mucus secretion, he continued to develop the methods of direct observation he had begun in Chicago. These included bringing isolated segments of intestine to the surface of the abdominal wall where their lining membrane, still functioning normally, could be kept in view. He also investigated human cases of mucous colitis—the sort of work he had asked Ethel to do in Adelaide. But none of this work was published for several years.

Concurrently with these two main interests Florey made some branch-line excursions. One of these was with Paul Fildes on his idea for the treatment of tetanus, about which he had written to Ethel. The spasms of tetanus are due to the action of the bacterial toxin on the nerve control of muscular movement. They can be prevented if antitoxin is given before the toxin reaches the nerves, but not if it is given later. At the time of Florey's interest it was supposed that antitoxin in the blood could not reach the nervous tissue, and thus could not neutralize toxin present there. It was therefore suggested that antitoxin injected into the cerebrospinal fluid might be more effective, and Sherrington had done experiments in monkeys that seemed to support this view. But the evidence was inconclusive. Florey hoped that the shrinking of the cerebrospinal nervous tissue caused by an intravenous injection of salt solution would draw antitoxin into the affected area.

His work with Fildes began in November 1926. The first problem was technical, the difficulty of injecting antitoxin into the spinal fluid of guinea-pigs, the animals on which much previous work had been done. Usually the needle entered a blood vessel so that the supposed intrathecal injection was actually into the blood-stream. This finding at once cast doubts on the validity of previous work. A long series of experiments then started on rabbits, in which intraspinal injection was easier, and the results were published in 1927 in the *British Journal of Experimental Pathology*.[9] From the point of view of Florey's idea they were disappointing. The new method had no advantage over simple intravenous injection of antitoxin. Such negative results are not valueless. As was so often the case with Florey's experimental work his critical approach and meticulous technique had exposed errors that might have continued to mislead.

The thesis for the Fellowship at Caius achieved its object. Florey became an Unofficial Fellow in 1926, an appointment he could hold while working in London and which gave him most of the privileges of the full Fellowship, except rooms in College. Meanwhile he had completed yet another thesis, one for the degree of Ph.D. in the University of Cambridge. The regulations allowed the inclusion of published work and his thesis, which ran to seventy-five pages, was entitled 'The physiology and pathology of the circulation of the blood and lymph'—a comprehensive title.[10] It contained, in fact, most of his observations on the capillary circulation, inflammation, the resolution of stasis, and his work on lymph flow and the contractility of lacteals and lymphatic glands. It was profusely illustrated with photographs and drawings. E. P. Abraham, in his admirable *Biographical Memoir* for the Royal Society, wrote: 'The account of those experiments, in straightforward sentences, illustrates the type of research that he liked and was so well-fitted to do—direct observations on animal tissues of the effects of various stimuli, which were made possible by first rate techniques and established a factual basis that no future investigators in the field could ignore.'[11] The result was never in doubt, and the degree was awarded in 1927.

In the summer London became less of a health risk for Ethel and the Floreys left Surrey for a flat in No. 42 Belsize

Square, Hampstead. Relieved of the interminable commuting between Chobham and the London Hospital, Howard was able to devote himself more to his laboratory work. But his remaining time at the London Hospital was to be short. In January 1927 the Huddersfield Lectureship in Special Pathology in Cambridge became vacant, following the death of Strangeways, and Howard was offered the post. Not unnaturally he accepted with enthusiasm. It renewed the opportunity of the previous year. Cambridge had every advantage over London: splendid facilities for research, stimulating companions, a college fellowship, and a healthier environment for Ethel. His departure from the London Hospital caused little regret on either side, despite the fact that he had not completed the term of his appointment. Only Fildes had been a congenial colleague and Fildes was to pay many a working visit to him in Cambridge and later in Sheffield. And he was to spend the many years of his retirement working in Florey's department in Oxford.

Howard took up his post in Dean's department in the autumn of 1927. The room in which he had worked during his previous Cambridge appointment had been used by Neil Goldsworthy, a fellow Australian, but Howard was able to re-occupy it. He did not regain his previous technician, Sivell, who flatly refused to work for him again on the grounds that he had been driven too hard. Into this vacuum walked a small fourteen-year-old boy named Jim Kent whose job it was to clean the room for Goldsworthy. Howard's greeting was to ask if he was interested in animal work. The boy had hopes of becoming a vet, and said 'yes'. 'Would you like to work for me?' said Howard. 'You'll get plenty of animal work if you do.'[12] The Florey reputation for long hours, hard work, and exacting standards was already notorious among the technical staff, but there was no hesitation on Kent's part. In those first few minutes was decided a lifetime's partnership as if each had recognized the compatibility that made Jim Kent Florey's devoted and indispensable assistant for the next forty years. His rapidly acquired skill, not only technical but in the understanding of Florey's methods—and temper— made a major contribution to the vast output of research that was to follow.

Florey's immediate decision to make Kent his personal assistant caused the first row to mark his return to Cambridge. The Superintendent of the Laboratory staff was W. A. Mitchell, who had come with Strangeways to Cambridge in 1897. He was a formidable person of the sergeant-major type, but he had the welfare of technicians at heart and had been largely responsible for founding the Institute of Medical Laboratory Technology. He had also introduced a rotational training scheme by which each technician served a period in each department of the laboratory: bacteriology, histology, chemical pathology, photography, and so on. The full rota took several years, by which time Mitchell could place a fully trained man in some senior post elsewhere. This system, admirable for the trainee, was anathema to the research worker. The latter needed the exclusive services of an assistant trained in a limited but advanced field, and to have him replaced every few months by a raw recruit was exasperating. It is an indication of Mitchell's personality that he had imposed his system on all the graduate staff of the Pathology Department—except Florey.

On the morning that he met Jim Kent, Florey went to Mitchell and told him that provided Kent suited him, he would remain as his permanent technician. Mitchell was furious at this flouting of his authority, particularly by a young man who had joined the staff only that day. But somehow Florey got his way. He was, in Kent's words, 'a firebrand in those days', and Mitchell realizing, perhaps, that a refusal would provoke a storm that might reach unknown heights, agreed with a poor grace.

Florey, as usual, started his experiments on the day of his arrival. This was not too difficult since much of his previous experimental work had been done in these laboratories. Now, as a more senior member of the staff, he shared a responsibility for the teaching of pathology. And there was a new circle of colleagues to be met on an equal footing. Among them he was to find research collaborators and, as far as his detached nature would allow, friends. The closest of these, perhaps, was the Demonstrator in Bacteriology, R. A. Webb, an American. Webb had graduated at Johns Hopkins University, Baltimore, and during the First World War enlisted in the British Royal

Army Medical Corps. After being demobilized he studied with Sir Almroth Wright at St. Mary's Hospital, London, then with McIntosh at the Middlesex Hospital, supporting himself financially. His first appointment was as a Lecturer at Manchester, where H. R. Dean was then Professor of Pathology. When Dean was appointed to his Cambridge chair, Webb went with him and became his 'right-hand man'. He had known and liked Howard Florey when the latter was Walker Student, and they used to sit together at the 'research table' in Hall at Caius.[13]

Another member of the department was Alan Drury, who had the research rooms next to Florey's on the top floor of the old Pathology Building. Drury, fundamentally a physiologist, had developed pulmonary tuberculosis while serving in India during the war. When he recovered, he found it difficult to restart a career until Dean gave him a teaching post and obtained a grant from the Medical Research Council for him to work on gastric function. Florey, with his interest in mucus secretion, found in Drury an obvious collaborator. Drury described himself as being 'rather at a loose end' and Florey provided exactly the mental tonic he needed. During the next two years they worked together on gastric and intestinal secretion and published three joint papers. But Drury began to have reservations about their collaboration. He himself was a gifted scientist, later to become a Fellow and Vice-President of the Royal Society and Director of the Lister Institute. He was, too, Florey's senior by four years and the younger man's natural assumption of leadership began to be irksome. Drury, again in his own words, felt that he was becoming 'a dogsbody for Florey', who would plan an experiment, allow Drury to carry out all the routine preparations, and then step in to finish the job himself. 'He was always a great finisher,' said Drury—a penetrating comment on the key to Florey's success. Behind the established achievements of science are the forgotten volumes of unfinished experiments. It was Florey's flair that could recognize those unfinished experiments that were promising in his own field, and it was his skill and determination that allowed him to finish the work successfully. He was the antithesis of that too common character, the scientific dilettante.

Other people in the department with whom Florey worked were A.Q. Wells, the bacteriologist, who carried his 6 feet 5 inches and striking good looks with the bearing of a Guards officer; N. E. Goldsworthy, and L. J. Witts from Manchester, the current Walker Student. Presiding over this spirited and generally harmonious group was the Professor, whose bland, relaxed manner did not conceal hidden depths of scientific activity. But Dean had two great attributes. He had the ability to pick men who were potential scientific winners; he then backed them heavily. His second great attribute was influence. For two decades Dean was the most influential man in British academic pathology, a self-supporting phenomenon since his ability to place his pupils in high positions naturally strengthened his own. It was said to be his ambition to fill every chair of pathology in the British Commonwealth with his own men and he did not fall far short of his mark.

On the whole, Dean's influence was beneficent. Pathology had for many years been dominated by the morbid anatomists. Dean encouraged the experimental approach in all branches of pathology, a refreshing departure from the purely descriptive methods imposed by classical morbid anatomy. The forum for research activities in his field was the Society of Pathology and Bacteriology. Dean was one of its elder statesmen and usually sat in the front row of the lecture theatres in which the Society met. Young men nervously delivering their first papers were only too aware of the senior critics on the front bench. It was said that if, on their way back from the rostrum, they received a word of praise or—supreme award—a pat on the shoulder from Professor Dean, their future was assured. Dean was a personal friend of Sherrington's and one can discern Sherrington's influence extending through physiology into this vitalized pathology. It was natural that Sherrington should send Howard Florey to Dean when he decided that 'experimental pathology' was a subject that could both develop and be developed by the talents of his young Australian pupil.

It was not until December 1927 that the Floreys found a house in Cambridge. Then they bought, in their joint names, No. 75 Cavendish Avenue. The price was £1200, rather a high one in those days for a four-bedroom semidetached suburban villa, with a small front and back garden. But the quiet neighbourhood is pleasant and the two miles separating it from the centre of Cambridge are over flat roads that are no problem for a cyclist. Howard and Ethel were, as she had feared, still relatively poor. So £900 of the purchase price for the house had to be raised on mortgage. The remaining £300 was paid by them both and it is probable that Ethel's parents made a contribution.

Howard cycled to work, arriving punctually at 10 a.m. each day, except on class days when he had to be an hour or more earlier to supervise the preparations. It was his habit to do an experiment almost every day, including Sundays, starting precisely at 10 a.m. Most experiments involved hours of delicate surgery and Kent was invariably his assistant. Kent soon became a skilled animal anaesthetist and able to anticipate every stage of the operation. Sometimes Ethel would assist Howard in his experimental surgery. It was the situation to which he had often looked forward in his letters to her, but the reality was far short of his hopes. Ethel had some surgical skill, but Howard required an assistant able to respond immediately and her deafness was a fatal handicap. Howard, intent on some minute manœuvre, would demand a particular instrument. Ethel, mishearing, would hand him the wrong one, or not hearing at all, do nothing. In operations in which speed and precision are vital such failures could and did ruin an experiment. Howard on such occasions was reduced to unrestrained exasperation and Ethel to tears. Nevertheless, she did collaborate in work with Howard and Drury on colonic reactions, and with Howard and Szent-Györgyi on suprarenal activity, and she was a co-author in two papers on these subjects that appeared in 1929. But her initial desire to share in Howard's work gradually evaporated and, during her first pregnancy, she ceased to come to the laboratory.

A common interest in research was not, therefore, to be the bond between them that Howard had hoped for. Perhaps this is not surprising. The research was essentially Howard's, and Ethel's part was to be, in Kent's words, 'just a pair of hands'. Yet she was an intelligent, critical, and ambitious young woman with a good medical training. She had wanted a career of her own and was convinced that she had sacrificed it to Howard. And Howard soon ceased to make the effort to include her at all in his research enthusiasms. The eager description of his ideas and experiments that he had poured out to her in his letters no longer had this emotional outlet.

His work, in fact, became a barrier rather than a bond. Since Ethel could not share it she came to resent it. She grew jealous of his absorption and even, it seems, of his success. The outward sign of this resentment was her constant complaint that she was expected to run the home with little appreciation or help and to prepare meals for which Howard was often late. It was precisely the situation which Howard had jokingly anticipated in his letters, but it now assumed the proportions that are symptomatic of a deeper malaise. Few of Howard's colleagues were invited to his house. Webb was among those few, and he found the emotional atmosphere there oppressive. On such occasions, when Howard had made an effort to be on time for dinner, Ethel, not the best of domestic managers, might be an hour late in preparing it. Mutual recriminations, uninhibited by the presence of a guest, invariably followed.

But Webb succeeded in maintaining good relations with both Ethel and Howard. He lent them a car—an Austin Seven—for a motoring holiday to Devonshire. Ethel was a bad driver, specializing in minor collisions—usually with stationary objects such as walls or gate-posts. Webb's car suffered, as every car the Floreys subsequently owned was to suffer. Howard was himself a careful driver and his wife's apparent carelessness infuriated him. He finally refused to pay any repair bills incurred by Ethel and insisted that she should meet them herself out of her dress allowance.

Another source of domestic friction was Howard's gramophone. His love of music had deepened and by then he was particularly enamoured of Mozart, not at that time the vastly popular composer that he is today. Ethel did not share his enjoyment and, perhaps like many deaf people, she found the recorded sound unpleasant. But Howard's sessions with his gramophone remained his main relaxation for many years. He extended its power by building, with Webb's help, a large exponential horn. Its construction involved the use of glass wool and Webb recalls that they both suffered from tracheobronchitis in consequence, and that Howard was convinced that they would develop silicosis.

An added strain was the long-threatened visit of Ethel's parents. They arrived during the summer of 1928 and Howard, who had never concealed his feelings for them, found their presence a considerable tax on his temper. Her father was described by an Adelaide contemporary of Howard's as 'thin, dry and domineering' and her mother, to whom Ethel obviously had a strong emotional attachment, was conventional to excess. The inevitable unburdening to her mother of Ethel's marital troubles took place, and in the thin-walled little house the raised voices dictated by Ethel's deafness were all too audible to Howard. It was a humiliating experience that he did not forget.

The Reeds bought a Morris car and took Howard and Ethel on a tour of Britain and the Continent. No record of this tour seems to have survived except some photographs taken in Scotland and a reference to it in a letter written by Howard on 30 September 1928 to Wylie at Rhodes House: 'Very many thanks for your letter on my birthday. I only read it today as we returned from the Continent this morning. My father and mother-in-law are staying with us, and we all went touring in a car. It is a very strenuous business, but all told we enjoyed it quite a lot.'[14] The touring party may have returned together, but it seems that Howard did not remain a member of it for the whole time. According to Ethel's sister 'he walked out half way through' and apparently went off 'to visit a laboratory'. There was one material advantage to the Reeds' visit to Cambridge. When they returned to Adelaide they left their car with the Floreys. Kent remembers that the name 'Hayter Reed' was engraved on a brass plate on the dashboard of the Morris that Howard and Ethel used for the next six years.

Life in Cambridge had some compensations for Ethel. She was a keen gardener and a good dressmaker, making most of her own clothes from *Vogue* patterns. She was, if not a good housekeeper, active about the house, making all the curtains and doing a good deal of the decorating. She and Howard joined the Cock and Hen Tennis Club which brought them into contact with many of the young married couples of the

University. Howard, of course, was a good player, and though he was openly intolerant of her mistakes, Ethel was herself above the average. Howard played with a ferocity that genuinely intimidated his female opponents to the extent that at least one of them (Alan Drury's wife) refused to play against him. This reputation was recalled thirty-five years later when the Public Orator, presenting Sir Howard Florey to the Chancellor of Cambridge University for the conferment of an Honorary Degree, declared (in Latin): 'he shows no diminution in that energy and brain-power, that agility combined with tenacity, with which, as a tennis player while he was here among us as a young man, he stood alert at the net like some fire-breathing Chimaera and struck terror into his opponents' and he ended: 'Perge modo, precamur, et ut floruisti flore' (only go on, we beg, and flourish as you have flourished).[15]

Ethel had friends in Cambridge, particularly Mrs. (later Lady) Drury, Mrs. Thomson (who, as Nan Mitchell, had been a friend in Adelaide), and Lady Anderson, the wife of the Master of Caius. The obvious bickering between the Floreys naturally led to speculations on their marriage and it was natural too that sides should be taken. Ethel's friends took the view that Howard was a selfish, rough-mannered husband who treated her badly, while Howard's thought that he had to put up with that classic irritant, a nagging wife, and that when it came to an argument Ethel gave as good as she got. It was agreed, however, that the marriage was not likely to survive for long, on the grounds of simple incompatibility. What both sides failed to recognize, it seems, was that the Floreys were Australians—who are seldom restrained in the expression of their feelings—and that they were pretty outspoken even by Australian standards. The fact that Howard and Ethel *were* Australians was, in reality, an important bond between them. They were compatriots in a foreign country and if they criticized each other, they found much mutual enjoyment in criticizing (and ridiculing) the behaviour of the English. The marriage was always further from the rocks than it seemed to the onlookers.

From the point of view of his career, the first year at Cambridge was a satisfactory one for Florey. His research was yielding definite results. He wrote up and published, with

M. N. Marvin, work they had begun at the London Hospital on the relationship between the cerebral and the general blood pressure. At Sherrington's suggestion Florey had looked critically at the conclusion by the German physiologist H. E. Hering that pressure changes in the cerebral vessels are important in regulating the systemic blood pressure. Florey and Marvin soon showed that most of Hering's results could be due to the effect of the urethane he had used as an anaesthetic, rather than to the pressure changes he had induced in the carotid arteries. Florey completed this demolition work in a second paper with Marvin and Drury, in which he showed that the conclusions of Anrep (then in Barcroft's laboratory), Starling, and Hering on the existence of pressure-sensitive receptors in cerebral vessels were not generally valid. Both papers were published in the *Journal of Physiology*.[16] They describe direct physiological experiments made possible by Florey's operative skill and experience of the cerebral circulation. Once more the apparently destructive is really constructive if it reveals fallacies.

This criticism of Anrep's work made no difference to Florey's welcome in the Physiology Department, and Barcroft's suggestion of a collaboration became a reality. Barcroft was famous for his high-altitude experiments and for his work on the functions of the spleen, which he had studied by perfecting the operation of 'exteriorization' by which the spleen, still functioning normally, becomes an external organ. Florey's studies with Ivy in America had perfected the similar operation of exteriorizing a segment of the colon, and Barcroft's suggestion was the two techniques should be combined in the same animal. But their first collaboration concerned the curious ability of the spleen to concentrate the blood within it, that is, to increase the proportion of cells to fluid. Barcroft and Florey were able to show that a reflex contraction of the spleen forces fluid plasma into the lymphatic vessels, thus raising the concentration of cells remaining in its blood vessels. They published these findings in the *Journal of Physiology* in November 1928.[17]

Florey's experimental notebooks reveal the true extent of the work behind his published papers. For example, his work with Drury which began early in 1928, was based on a series of over 130 long and difficult animal operations. And they also reveal his ability to diversify without losing continuity. On each of three successive days in October 1927 he began a new line of investigation. On the 19th he started experiments with L. J. Witts on the fate of blood cells injected into the peritoneal cavity; on the 20th he began his work with Drury on the control of the cerebral blood pressure; and on the 21st he did his first experiments with Barcroft on splenic blood concentration.

The work with Witts had been prompted by Ethel's description of intraperitoneal infusions of blood given to her young patients in Adelaide, which Howard had thought useless. But his distrust of theoretical pronouncements extended to his own, and he decided to 'do the experiment'. Witts was interested in haematology (he was later to become a leading authority) and Florey asked him to collaborate in work designed to trace the fate of red cells injected into the peritoneal cavity. They found that, in the dog, red cells pass through the walls of the lymphatics and thence via the thoracic duct into the blood-stream. But the process is slow and incomplete and thus a poor substitute for intravenous transfusion. Nevertheless, Ethel's interest was justified, though it was not mentioned in the paper published by Florey and Witts in 1928.[18]

Meanwhile, Florey had been studying the reactions of the exteriorized segments of the colon in the dog with Drury and Ethel. Their most interesting finding was the blanching that occurred when the dog was frightened by, for example, the banging of a door. It was due to a contraction of the colonic blood vessels and muscles, beginning four or five seconds after the stimulus and lasting for about half a minute. Most of the subsequent work sought the mechanism involved, but though it was clearly nervous in origin, the nerve pathways could not be found. On the significance of the phenomenon they had this to say in their paper: 'These observations have a considerable bearing on the James–Lange theory of the emotions ... Though the physiological evidence against the theory is very considerable, the psychologists have been slow to recognise this, and these observations may be of service in providing one more piece of evidence against it.' They then went on to speculate (a rare exercise for Florey) that 'nervous dyspepsia' might be due to changes in the blood supply to the colon.[19] Following this demonstration of the value of Florey's method, Barcroft made his suggestion that they should combine their two operations so that changes in the spleen and colon could be observed together. This was done on a dog named Betsy, who was none the worse for the surgery. One of the factors studied was exercise, and this was shared by Kent, since he had to pace Betsy round the park on a bicycle. At a meeting of the Physiological Society, Betsy demonstrated the effects of exercise by running cheerfully up and down the corridor. As Barcroft and Florey described in their 1929 paper, the spleen contracted and the colon blanched, but the former effect lasted for twenty minutes whereas the latter was more transient. They concluded that the spleen reacted to exercise and the colon to excitement.[20]

Like most of Florey's work these experiments were designed to answer straightforward questions by direct observation. If he wished to learn about the reactions of the colonic mucosa he looked at it and measured its secretion. If such an approach involved great technical difficulty, this did not daunt him—in fact it presented a challenge that he gladly accepted. He was at his best devising and mastering difficult methods that would give immediate answers to simple questions. He had little time or sympathy for those who avoided technical difficulties by making a roundabout approach through layers of inference. 'What exactly have you discovered, and what does it actually prove?' was often his only comment at the end of some involved exposition—a question that the deflated speaker usually found himself unable to answer.

A new line, but one based on an old interest, had begun in January 1928. Florey was aware of the dangers of overpopulation, and had spoken on this subject in a school debate in 1915. He had read a good deal of what had been published on methods of birth control, and it occurred to him that he might apply his own surgical techniques and experience to a more scientific study of the physiology of mammalian fertilization and of devices designed to prevent it. He started this project with Carleton, an extension perhaps of the uterine fistula work with him that Howard mentioned to Ethel in a letter in June 1926. During the next three years Carleton and Florey paid working visits to each other's laboratories and developed a surgical technique for establishing vaginal fistulae in dogs, cats, and rabbits. This method allowed a direct study of the fate of injected spermatozoa, the efficiency of various spermicides, and their possibly harmful effect on the vaginal mucous membrane. Their papers on this subject appeared in 1931.[21]

One of Florey's research interests, the physiological control of the blood pressure, led to collaboration with a young Hungarian biochemist, A. Szent-Györgyi, who had been working with Drury on respiration. Szent-Györgyi was one of Gowland Hopkins's pupils in the field of vitamin research, and within a few years he was to become world-famous as the discoverer of the chemical structure of vitamin C, for which he received the 1937 Nobel Prize. Florey's interest converged with that of Szent-Györgyi's on the question of the measurement of adrenal activity, one of the most potent factors in controlling blood pressure. There were two methods in current use, one involving an effect on respiration, the other on muscle fatigue. Florey suggested putting both to the experimental test and he, Ethel, and Szent-Györgyi proceeded to do this. In their joint paper, published in 1929, they showed that neither was, in fact, reliable.[22]

By 1928, Florey had become firmly established in Cambridge. His Unofficial Fellowship at Caius had become an official one, with all its privileges. He was a College lecturer and as Director of its Medical Studies was responsible for the work of its medical students. His researches had been sound, technically excellent, and important to physiology. They were rewarded in this same year by two prizes, the Rolleston Memorial Prize from Oxford University and the Chapman Memorial Prize from Magdalen College, Oxford. This quite notable recognition drew a letter of congratulation from Wylie, Secretary of the Rhodes Trust in Oxford. Florey replied: 'Very many thanks for your letter of congratulation. My luck was certainly in this term. It is very hard to convince one's wife that one is not a congenital idiot, however.'[14]

This period, too, was important for Dean and his department. Oxford and Cambridge had an approach to the teaching of pathology that was unique among British medical schools. In all other schools, pathology was taught during the student's clinical studies. But in Oxford and Cambridge there had been a strong move to create schools of pre-clinical medical science which widened, as knowledge grew, from anatomy and chemistry to include physiology, pathology, bacteriology, pharmacology, and biochemistry. And in both, pathology had developed an experimental approach, whereas the clinical schools favoured systematic routine teaching. The professors of pathology in Cambridge from C. S. Roy onwards had been experimentalists and their teaching had been neither systematic nor comprehensive. In 1910, when pathology became an examination subject in the Part I Tripos (together with bacteriology and pharmacology) the course, which had to be completed by the students before their clinical work could begin, was rather scathingly known as 'bugs and drugs' and the examination regarded as a nuisance.[23]

Dean could not be satisfied with this low academic status. He engaged in a battle to improve it by making pathology a subject for the Part II Tripos—the equivalent of a Final Honour School in Oxford. Dreyer had fought a similar battle there, and lost. But Dean was a more powerful man, and the Cambridge opposition weaker. He gained his object, and he gained too a new building for pathology. This, completed in 1929, is almost a replica of the Sir William Dunn School of Pathology in Oxford, on which Dreyer had lavished his energies and talent for laboratory design three years before. In the ensuing reconstruction of pathology teaching, Dean delegated much of the work to Drury and Florey. Florey enjoyed this. He had strong views on teaching and here was an opportunity to put them into practice. The result of their efforts was the creation of an excellent course of general pathology and a most valuable introduction to scientific medicine.

In 1929 came the move from the old Pathology building to the new one. In the case of Florey's laboratory equipment, it was made by horse and cart and, of course, by Kent. Florey now had a suite of rooms on the top floor of the building, and all the facilities that he could wish for. His tempo of work increased and he demanded even more of his young assistant. On the rare occasions when Kent was not in his room when needed, Florey wasted no time on searches, messengers, or the telephone. The department would be shaken by Kent's name, bellowed from the top floor, and echoing down the stair well. Experiments might begin at 8.30 a.m. and continue until late at night. Kent was perfectly willing to work in this way for Florey on weekdays and Sundays alike. 'I would have jumped into the river for him if he had told me to,' he said. When, after some years of this devoted service, a colleague suggested that such long and irregular hours were rather hard on the boy, Florey was surprised. 'Kent?' he said, 'Kent's got nothing to worry about—he's not married!'[24]

Professor Dean had, as might have been expected, the best laboratories in the new building for his own use. In fact, he used them very seldom and their gleaming equipment remained idle. He had many outside commitments—the consequence of his wide involvement in the affairs rather than the practice of pathology—and he was an important figure in the University, becoming Master of Trinity Hall in 1929. In consequence he seldom appeared in his department except when he was due to lecture or attend a meeting. Kent's only contact with Dean was through the Professor's bulldog, also named Jim. Kent, being the junior technician in the department, and having some experience in the exercising of dogs, received a shilling a week for walking Jim through the streets of Cambridge during the lunch hour. Unfortunately, Jim the bulldog was roughly the same age as Jim Kent, and thus in his canine dotage. On their last walk together he collapsed in the street with a heart attack. Kent had to fetch a barrow from a near-by garden and wheel him home to Trinity Hall amid dark suspicions of vivisection and scientific malpractice.

Every University department has its own inner social life more or less stratified in terms of rank and seniority. In the Pathology Department the daily occasion for the graduate staff to meet was at tea-time. Here the departmental affairs would be aired, students and experiments discussed, anecdotes retold, and views on the world in general pronounced. Most people contributed to and enjoyed these communal conversations, and Witts remembers in particular Dixon and Cobbet (a previous Walker Student and Professor of Pathology at Sheffield) as 'marvellous talkers'. But he also remembers that Florey never came to these tea-parties, nor indeed to any of the social functions of the laboratory. It was an attitude that naturally suggested aloofness, though Florey was probably avoiding a waste of useful time rather than the company of his colleagues. Previous biographers have noted that, even with such close companions as Robert Webb and Harry Carleton, Florey never used their Christian names. Too much should not be made of this. In the middle classes, men in those days seldom used Christian names outside their own family circle; never at school, and almost never at university. For example, it would not occur to the readers of Conan Doyle to be surprised that, after all they had been through together, the Baker Street immortals always addressed each other as Holmes and Watson.

Florey was not anti-social in the ordinary sense, though he found English polite society rather ridiculous. As a young man he avoided social occasions because he found light or aimless conversation boring. For him, talk should be productive, and thus he was only interested in discussions likely to lead to concrete results in the laboratory, the classroom, or the practical affairs of life. He was not, therefore, much happier among the University dons at high table, whose conversation on most subjects (except science) is supposed to sparkle with an unmatched brilliance. But the usual topics were scarcely to Florey's liking. He had little interest in the classics, and though he derived real pleasure from painting, music, and architecture he derived none from the abstract discussions about these arts to which the non-creative academic is sadly prone. The least popular subject with Florey was philosophy, that engulfing whirlpool into which almost every late-night collection of convivial dons is finally drawn. 'Talk about talk' could hardly be further from Florey's taste, but it was inescapable in Oxford and Cambridge college gatherings.

Philosophy was, at that time, going through a phase of more than usual obscurity in England. It was still shaken by the effects of Bertrand Russell's discovery that 'the class of all classes that are not members of themselves must be, and at the same time cannot be, a member of itself'—a discovery that, in Frege's opinion, undermined the whole foundation of mathematics.[25] It was also preoccupied with the precise logical significance of such statements as 'the present King of France is bald'. It is difficult to imagine anyone less likely to appreciate the contortions of logical positivism (or atomism) that would develop from such a conversational gambit than Howard Florey. Curiously, there was at least one eminent philosopher who had the same sort of aversion to conversation for its own sake. Wittgenstein, Russell's star pupil, was in Cambridge at that time, and became a Fellow of Trinity College in 1930. Pitcher writes:

The one thing he could not stand in anyone was affectation or insincerity—in short, dishonesty. His belief that the academic life is afflicted with this sin doubtless accounts in large part for his intense dislike of it. He had, he said, only once been to High Table at Trinity and the clever conversation of the dons had so horrified him that he had come out with both hands over his ears. The dons talked like that only to score; they did not even enjoy doing it.[26]

Florey's reactions were less extreme. Like Wittgenstein he was an 'outsider', but unlike him he did not become a recluse. Though Florey seemed unwilling to join his contemporaries in their play he could work with them. His magnetic energy drew people into his orbit so that he was always surrounded by willing collaborators. Barcroft had much the same power; Cambridge referred to his satellites as 'Barcroft's white slaves'. But Barcroft had the most open and friendly of natures, with a delight in the company of others, particularly the young. Barcroft used to describe himself as 'Britain's senior medical student'—an apt description for he had, like Sherrington, that quality of humility and a perpetual youth of mind and spirit that distinguishes so many great men.

At the beginning of 1929 Ethel became pregnant. This created, of course, a new emotional situation for Howard and herself. They had both wanted children and were pleased at this fulfilment of their marriage. But it was not an easy time for Ethel. She suffered severely from sickness and stopped coming to the laboratory. Howard seems to have been less sympathetic to this new manifestation of her ill-health than he might have been. His own health was not good. He suffered from hay fever, asthma, and, like Ethel, from respiratory infections. Always, in the background, was his chronic digestive trouble which was sometimes bad enough to keep him away from the laboratory. He experimented on himself, partly from scientific interest, partly in a search for relief. He swallowed a stomach tube from time to time to see if the natural secretion of acid was improving, and he continued to take dilute hydrochloric acid, which he drank through a glass tube to avoid damage to his teeth. But the acid did discolour them, and the rather tight-lipped smile that became a characteristic was a self-conscious reaction.

In the laboratory, Florey's work on mucus secretion had reached the stage at which he needed to know more about its possible nervous control. This could be studied by the use of pharmacological inhibitors or stimulants, but he also wished, if there were nerves to the glands, to see them directly. The greatest advances in neurohistology had been made in Madrid by Professor Ramón y Cajal, who had shared the Nobel Prize with Golgi in 1906 for work on the micro-anatomy of nerve cells. In 1921, at the age of seventy, he retired from his university appointment, and the Spanish Government, recognizing his great contribution to science, founded the Cajal Institute in Madrid, of which he remained director until his death in 1935.

Sherrington, as the greatest neurophysiologist of his day, was naturally in close contact with Cajal and had often drawn Florey's attention to the aspects of Cajal's work that concerned the vascular responses and the nervous control of secretion. Florey, always anxious to learn new techniques from their inventors, felt that a visit to Cajal's Institute was highly desirable. Sherrington wrote on his behalf to Madrid, and it was arranged that Florey should go there for a stay of six weeks in June 1929.

Ethel's condition posed problems. She was by that time six months pregnant, easily tired, often sick, and generally uncomfortable and unwell. Her friends felt that she should not go to Spain with Howard, and they considered him unfeeling and overbearing when he insisted that she should. Ethel herself seems to have gone willingly enough. It is a long and tedious journey to Madrid by train and steamer, and it was the height of the summer. But Ethel had been used to higher temperatures in Adelaide and she craved for the sun. In fact the dry heat of Madrid on its high arid plateau was not unpleasant. Both Howard and Ethel enjoyed their Spanish visit and the happy memory of it was often recalled by both. With his usual thoroughness in the preparations for any such venture, and his interest in languages, Howard had taken a Linguaphone course in Spanish to such effect that he was able to give a lecture, more or less in their native tongue, to his Madrid colleagues.

At the outset, however, the visit was almost a disaster. Howard was quite unprepared for that timeless Spanish way of life that is summed up in the word *mañana*. Having arrived at the Cajal Institute and having been received with much courtesy by the great man himself, he proposed to follow his usual practice and begin work at once. 'Tomorrow perhaps' was the reply. But tomorrow is apt to be inexplicably delayed in Spain. When, after two or three days, it had still not arrived Florey's never-elastic patience snapped. He walked out of the Institute saying that he was going back to his hotel to pack and return to England immediately. It was Cajal himself who persuaded him to stay. One version of the story maintains that this world-famous scientist, then aged seventy-eight, ran out into the street after his impetuous young visitor and begged him to change his mind. 'If you will come back,' he pleaded, 'I will give you a bottle of the original methylene blue made by Ehrlich himself.' Whether the pursuer was really Cajal or a messenger, the result was that Florey did return, and to his own great advantage, since he learned what he had come to learn, and saw much of another ancient European culture. But, as usual, he was not given to expressing his real appreciation of such things with much eloquence.

'Dear Mr. Wylie,' he wrote on 29 September, in reply to the birthday letter which Rhodes Scholars usually received from the Secretary, 'I've had a good series of changes this year. In June my wife and I went to Spain and I did some work in a laboratory in Madrid. Most interesting place, and I thoroughly enjoyed it. I then went to the International Physiological Conference in Boston. Most enjoyable and, on the whole, worthwhile.'

International conferences were much rarer and more significant events in those days than now. Travel was relatively slow and expensive, and only people with important things to say, hear, or see could justify the time and money needed to attend meetings abroad. Howard, with his rapidly developing lines of research, needed to take the opportunity of meeting the leaders in his various fields who would be together for this occasion. Moreover, the European Physiological Societies had chartered the liner *Minnekahda* to transport delegates and their guests to America, so that the voyage itself became a social preliminary to the Congress. Florey went to Boston, and apparently at his own expense, because he later had to meet the accusation of extravagance from Ethel on the cost of his fare, which was £35.

The four letters that he wrote to her while away are more revealing of their relationship than of the Congress and his part in it. The first, dated 10 August, written on board the liner: 'All told the ship is considerably better than the second class Nelson Line, but the stewards tend to be as rude as most Americans of that type. Be good, and gestate carefully.' On the 18th he wrote: 'My love; we get in tomorrow. Very few people ill. There have been lots of activities on board and I wish you had been here—it would have been much more pleasant. None of the women have caused me sexual excitement.' Next day he wrote from Boston: 'My love: I'm posting you another letter, but perhaps you'd like this. I wish you were here, the arrangements are admirable. I have a room in a very elegant place, like a Cambridge College—two rooms, in fact, plus a bathroom. Most of the English people in this place I don't care for, but perhaps they will brighten up later in the week.' His final letter, dated 26 August, was from New York.

My love; very disappointed at getting no letter from you at all. Finished conference, and by bus to Wood's Hole. Bathed and messed about, and then at night caught a boat to New York, where we arrived at two next day. It is very hot and everyone is bad tempered—in fact I shall be very glad to be home again as I'm thoroughly fed up with going about in droves. The *Olympic* leaves on Friday night, so we should be in Southampton the following Friday. I don't think I'll stop for the Schneider Cup, but come straight on. I should be home Friday night. I hope you are getting on alright, it's rather worrying to have heard nothing. I am missing you all the time and am just longing to get back to have you around again. Perhaps it is a good thing to go away like this occasionally as the little things seem to assume much less importance.

This was, for Howard, a deeply affectionate letter, and the reality of his feeling is confirmed by the fact that, in order to return without delay, he was missing the seaplane race due to start on the day after his ship docked at Southampton. The 1929 Schneider Trophy race, flown over the Solent, was one of the most spectacular events of the decade. Thousands of people had travelled or slept out all night to get a view of it from ships, boats, or the shores of the Isle of Wight and Southampton Water. It was won by the British Supermarine Napier. Howard must indeed have been anxious to get home. It was not the sort of occasion he would otherwise have missed.

Ethel's baby girl was born—with no complications—on 26 September, two days after Howard's thirty-first birthday. With their Spanish memories still fresh, they called her Paquita—soon abbreviated to 'Paq'—Mary Joanna. From the beginning Ethel took not only a maternal but a clinical interest in her baby. Everything was weighed and measured and she kept elaborate charts. She was more contented than at any time since her marriage. Howard, too, enjoyed and was proud of his little daughter, and his home life gained in tranquillity and importance. There was a significant decrease in the volume of entries in his experimental notebooks during the next two terms, which probably reflects the pull of domesticity.

8

THE DISCOVERIES OF ALEXANDER FLEMING

Neither Florey's published papers nor his experimental notebooks give direct evidence of the line of thought that led him to work on lysozyme, but his decision to do so was, eventually, to have tremendous consequences. The first mention occurs in the notebooks on a page dated 17 January 1929.[1] The entry refers, without details, to rats reared on vitamin-deficient diets. The animals had been killed and it was recorded that various organs and tissues were then 'sent to Fleming'. The purpose is not stated, but since the page is headed 'Lysozyme' and Fleming was the discoverer of that strange antibacterial enzyme, one might suppose that Florey had sent these specimens to him for an assay of their lysozyme content.

But why should Florey have done this? He had shown no previous interest in vitamin deficiency, nor in lysozyme. There is a clue in the paper by Goldsworthy and Florey published in 1930.[2] In this they refer to an observation by Cramer and Kingsbury that vitamin A deficiency abolishes the secretion of mucus in the colon and causes an invasion of its wall by bacteria. This would have a double interest for Florey. He had been studying the whole mechanism of mucus secretion, and the possible involvement of vitamin A was therefore relevant. And he was also interested in the function of mucus, so that the association of a failure of mucus secretion with a local bacterial invasion was suggestive. One can think of several hypotheses that might explain this association. The one that would have attracted Florey is the supposition that mucus normally prevents the entry of bacteria. Protection by mucus might be due simply to its existence as a viscous barrier, but it was also possible that mucus might itself have antibacterial activity. It was almost certainly such a line of thought that led to his experiments on lysozyme. It also led to the first contact with the interests of Alexander Fleming, then

Professor of Bacteriology in Sir Almroth Wright's department at St. Mary's Hospital, London. Fleming's work and Wright's influence on it are so closely bound up with Florey's future researches that they must now be the subject of a digression.

Almroth Wright,[3] who was born in Yorkshire in 1861, had an Irish father, a Swedish mother, and a remarkably cosmopolitan upbringing. He was educated in Dublin and studied science and medicine in Leipzig, Strasbourg, and Marburg. He was the first demonstrator appointed to the new Pathology Department in Cambridge, and then became Lecturer in Physiology in Sydney. He returned to Britain in 1892 as a professor in the Army Medical School at Netley, Southampton, and it was here that he developed his lifelong interest in active immunization by vaccination—the logical extension of Pasteur's work. He produced a mixed vaccine (TAB) against typhoid and the paratyphoid fevers, which was used, though not very wholeheartedly, by the Army during the South African war. When properly controlled the results indicated a dramatic protection by the vaccine, and it was later agreed that his work was a triumph in the fight against infectious disease.

The cult of vaccination for every sort of bacterial disease, from boils to rheumatism, flourished under Wright's leadership and with considerable profit to its practitioners. He himself was elected a Fellow of the Royal Society and knighted in 1906. In 1908 he founded the Inoculation Department at St. Mary's Hospital. This, though nominally part of the hospital and of London University, was virtually Wright's private research institute. It was financially self-supporting, since it made vaccines for sale through a commercial firm and the considerable private fees earned by its members paid their salaries and research expenses. Wright was Director in name and in fact. He had the final word on all appointments to the staff and on all the activities of the department. He was a man of great—even flamboyant—personality, of wide scientific experience, a classical scholar, and a dazzling conversationalist on almost any theme. He collected a band of the most talented disciples—Leonard Colebrook, John Freeman, Alexander Fleming, Bernard Spilsbury, John Wells, and many others, most of whom remained devoted to him throughout his long life. Outside this charmed circle there were those who considered that, at best, his scientific views were controversial and, at worst, unsound. But to many influential people (including George Bernard Shaw) Almroth Wright was the heir to the genius of Pasteur.

This was an overestimate. Wright followed lines established by Pasteur; he did not (as Pasteur himself would have done) discover new ones. The main objectives of his department were the extension of Pasteur's work on immunology and of Lister's on antiseptics. Lister had searched, in vain, for the perfect antiseptic, some chemical that would kill bacteria but not the living cells of the body. Wright was not enthusiastic about this approach: all antiseptics, he believed, were poisons for bacteria and cells alike. But he failed to see the glimmer of light that should have shown that this line was not a blind alley. When Ehrlich sent him samples of his drug Salvarsan, Wright soon confirmed that it did indeed kill the spirochaetes of syphilis without doing serious harm to the patient. It was a 'magic bullet' with a very restricted target; it was also a pointer to a new principle, later called chemotherapy. Ehrlich had tried 605 synthetic compounds before finding '606'—Salvarsan. The increasing power of synthetic chemistry, properly directed, would therefore be likely to create other and more widely useful antibacterial compounds. But Salvarsan was to remain the unique application of a general principle for many years, as did vaccination after Jenner. Pasteur saw the principle behind vaccination; Wright saw nothing behind Salvarsan.

Wright's main interest was in the natural defences of the body, and the ways in which these could be stimulated. Immunology had defined these defences, though not the nature of their mechanisms. One of these was the formation of 'antibodies' or 'antitoxins' in the blood, proteins specially fashioned to fit—like a key for a lock—a chemical structure (antigen) of organisms such as bacteria, or foreign proteins such as toxins, which had gained access (or been injected) into the body. Antibodies could coat bacteria and cause their clumping and removal from the blood-stream. Or they could cause bacteria to be dissolved (lysed) by an enzyme system in the blood called 'complement'. Or they could render the bacteria more appetizing for the blood leucocytes (phagocytes) which would then swallow and digest them. It was this last process that Wright thought particularly important. 'Stimulate the phagocytes' became a departmental motto. Wright believed that there was a special sort of antibody involved which he called 'opsonin', and he devised a method for measuring the 'opsonic index' of a patient's blood which for many years had to be tediously carried out by obedient pathologists in most clinical laboratories. Those who know Shaw's play *The Doctor's Dilemma* will recognize the source of its inspiration.

In 1906, Almroth Wright needed an additional research worker. It happened that the hospital authorities (who took a very serious view of sport) were looking for a way to keep one of their junior doctors, since he was a fine rifle shot and, if he left, the hospital team might not win at Bisley. So it was suggested to Almroth Wright that he might offer his new post to young Alexander Fleming, and to Fleming that he might give up his idea of becoming a surgeon (he had already passed his F.R.C.S. examinations) to join Wright. As it happened, Fleming had developed a great admiration for Wright during his student years and he was pleased to accept the offer. Thus Fleming joined the staff of the Inoculation Department, which was to become, in 1947, the Wright–Fleming Institute. He worked there until his death in 1955, first as a junior assistant, then as Almroth Wright's lieutenant until, in 1946, Wright retired (at the age of eighty-five) and Fleming himself became Director.

Alexander Fleming[4] was born in 1881 at Lochfield, a lonely but substantial farm in the Ayrshire hills about fifteen miles east of Kilmarnock. Like Howard Florey, he was the youngest of the family and the child of a second marriage. When he was old enough to walk the four miles to Darvel, he went to school there, as his brothers and sisters had done, and received the Scottish primary education that was probably better than its equivalent in England. At eleven, he was promoted to Kilmarnock Academy; and when he was fourteen he was sent down to London to live with a brother in Baker Street and to take a job as an office-boy. He combined this with evening classes at the Regent Street Polytechnic, and became a junior clerk in a shipping office in Leadenhall Street, which he found unbearably dull. One of his brothers was an ophthalmologist, and the idea of medicine as a career, and as an escape from the trap of office life, probably came from him. At all events, when Alexander received a small legacy at the age of twenty, he gave up his job and entered St. Mary's Hospital as a first-year medical student. His choice of St. Mary's was determined by his interest in sport. Though physically small (he was 5 feet 5 inches tall) he was good at boxing, football, shooting, and water-polo; and having played in a water-polo match against St. Mary's, he decided that it was a congenial institution. He competed for, and won, an entrance scholarship in 1901.

At St. Mary's it soon became evident that he had found his proper niche in life. He worked hard and showed that he had an excellent brain and an original, inquiring mind. He won prize after prize, and was in the honours list in almost every examination. He qualified in 1906, and then proceeded to take his F.R.C.S. with the idea of becoming a surgeon. It was then that his importance as a member of the hospital shooting team determined his change of course, and a career in laboratory medicine.

Wright and Fleming were unlike in almost every respect and their complementary qualities formed a perfect partnership. Wright was a large, ungainly bear of a man, with an astonishing range of knowledge and imagination, willing and able to fascinate his listeners. Fleming was small, neat and, with his invariable bow ties, rather dapper. He had, as Ronald Hare writes,[5] 'a bent nose and a broad Ayrshire accent' and none of Wright's cultural background. Like Florey, he had little time for aimless conversation or for ideas that could not be put to practical use. Leonard Colebrook, one of the most distinguished members of Wright's staff, described his attitude: '"Pain in the mind" was not the spur that drove him to do research, as it was with Wright, but rather an urge to do a job better than the next man. Competition was the breath of life to him. Wright and he made a fine team. Wright supplied the ideas, which Fleming usually received in silence, and then went away and devised some neat trick for working them out.'[6]

Fleming, like Florey, loved the actual performance of his

experiments and delighted in devising the techniques for doing them. Wright, with his admiration for the Greek philosophers, had acquired their veneration for pure reason. 'He looked on research as an intellectual exercise and used experiments to prove to lesser men, conclusions he had reached as a result of sheer cerebration' (Hare). It was Fleming, then, who was so often responsible for translating Wright's airy logic into factual observations.

In 1910 Wright received his supply of the new Salvarsan. He passed it on to Fleming, who had to devise improved methods for giving it. Salvarsan had to be injected intravenously—almost an unknown route in those days—and any leakage outside the vein could cause local tissue destruction. Wright's own researches involved the use of human blood, which was easily available only in minute quantities, and he had invented a series of micro-methods using capillary glass tubes instead of the conventional test-tube. Fleming took a delight in these methods, and invented several more of his own.

Then came the First World War. Wright, with the rank of Colonel, was put in charge of a special hospital for the treatment of infected war wounds, which was housed in the Casino at Boulogne. Fleming was also there, with the acting rank of Captain. Most of his work was on the antiseptics then available and their use in treating wound infection. His findings were a great disappointment to the surgeons who retained their reverence for Lister. Fleming found that, first, the antiseptic was unable to penetrate into the depths of bullet or shell-splinter wounds. Second, in the usual concentration employed, almost all antiseptics actually favoured the growth of bacteria in the tissues, because they damaged the white cells.

After the war Almroth Wright returned to the Inoculation Department, but the years had reduced his energy. Hare writes: 'Most of the team he had ruled with so firm a hand remained with him ... but they had grown too old to dance to his tune and were going their own ways. The one-time dictator retreated into his shell, taking very little interest in what the old hands or even new recruits like myself were doing.' One custom survived—the daily departmental tea-party which everyone attended and at which Wright presided. No one spoke until he had started the conversation, which then went on for an hour at least. Wright had known practically all the pioneers. Science, scientists, the classics, the arts, politics, women, architecture, gardening—Wright could hold his audience on every such subject. He had become more broadminded. 'One could even question the validity of some of his techniques without risking expulsion, a fate that would have been almost inevitable before the war.'[5]

Fleming, too, had returned. At the Inoculation Department he arrived in his small laboratory punctually at nine and left, equally punctually, at five. Between these hours he pursued his enjoyable invention of new techniques for studying his favourite bacteria. He had plenty of time to do so despite the fact that, since Wright was temperamentally incapable of dealing with the details of administration, he had become the mainstay of the department. One day, in 1922, when he was suffering from a head cold he plated out his own nasal secretions to see what would grow. The result was the expected mixture of organisms, but Fleming had a genius for noticing anything unusual, even when it was half-buried by the expected. Among the colonies were some that were apparently dissolving. He isolated a colony, subcultured it, and then tested a suspension of the newly grown bacteria with a drop of nasal secretion. The turbid suspension cleared completely within a few minutes. He had, in fact, made two discoveries. The first was an organism that could be dissolved in the most extraordinary way. The second was that nasal secretions contain something that can bring about this solution.

Fleming settled down to work out his discovery. The organism so prone to dissolve was one that had never been previously described. Wright had no difficulty in coining a name for it, and it became *Micrococcus lysodeikticus*. (Wright liked inventing classical names. His book on philosophy, *Alethetropic Logic*,[7] is so full of neologisms as to be almost unreadable.) The agent responsible for dissolving the organism Fleming found not only in nasal secretions but also in human tears, saliva, and bronchial mucus and also in white of egg and various animal and plant tissues. Though Fleming was never able to separate the active agent from these sources he concluded that it must be an enzyme capable of destroying some part of the bacterial wall. He called this (still hypothetical) enzyme 'lysozyme', again with Wright's blessing. He then set out to discover what sorts of bacteria lysozyme could attack. *M. lysodeikticus* was by far the most sensitive, suspensions of it being cleared by the addition of lysozyme literally within a few minutes. It formed, therefore, an admirable test system for the presence of lysozyme. But many other organisms were also dissolved, though much more slowly; or killed without being dissolved; or prevented from growing without actually being killed. There were thus three grades of antibacterial action.

Fleming, of course, was elated at this discovery of an entirely new natural defence against bacterial attack. The presence of lysozyme in the secretions of the eyes, nose, mouth, and respiratory tract must, he felt, be important in their protection from airborne bacteria. He found that about three-quarters of the many sorts of those airborne bacteria were sensitive to lysozyme. Unfortunately, he also found that the common pathogenic bacteria (the streptococci, staphylococci, pneumococci, and tubercle bacilli responsible for serious human infections) were resistant to lysozyme. Clinical bacteriologists were not impressed by Fleming's discovery, since lysozyme seemed to attack only harmless bacteria, and it came to be regarded merely as an interesting oddity. Although Fleming, during the next few years, published four papers on the subject, no one else had paid any serious attention to it.

Yet there were two obvious lines opened up by lysozyme which might have been followed with advantage. First, there was the possibility that lysozyme attacks only 'harmless' bacteria because it is the very fact of their susceptibility to lysozyme that renders them harmless. Conversely, the pathogenic bacteria might be harmful because they are immune to it. This perfectly logical argument was one that Fleming put forward but was unable to substantiate. It would require a wide study of lysozyme in different animals in relation to their natural diseases to show that it had an important protective function. The second line would be the purification of lysozyme itself, the definition of its mode of action and, if it proved to be an enzyme, the identification of the specific chemical structure (substrate) which it attacked. Since such a substrate would be part of the bacterial wall, its identification might provide new ways of attacking this by other enzymes. But Fleming was no chemist, and he had insufficient confidence in his own ideas to pursue the matter.

Neither did Fleming have the ability to fire other people with his own enthusiasms. His writings are prosaic to the point of dullness and, what is far worse, they are obscure and imprecise. He was equally uninformative in his speeches. Hare writes: 'I well remember an occasion when he spoke on lysozyme to an audience of medical men who knew very little bacteriology. He not only failed to tell them what the substance is, but where it comes from and its importance. About the only thing they gathered after half an hour was that lysozyme could evidently kill an organism of which they had never heard.'[5]

Fleming, to the end of his life, regarded his discovery of lysozyme as his best contribution to bacteriology. It was entirely original, a glimpse of a whole new mechanism that seemed to open vistas for exploration. It probably moved him to greater efforts than anything else in his research career, but these efforts were simply not effective. Discovery is exciting. Working out the implications of a discovery may need unexpected skills, patience, and determination. Persuading other people to accept a new discovery and to act on it needs what Fleming lacked, the ability to convince. So lysozyme research lapsed from inanition.

Lysozyme did, eventually, attract the attention of a man who was temperamentally the antithesis of Fleming. Florey had no proprietorial pride in lysozyme. He saw it as a possible lead in his own systematic study of mucus, an exploration that proceeded step by step, like a mountain expedition, each new difficulty overcome by patient technical improvisation and invention. The amount of scientific literature in those days was only a fraction of that which inundates research workers today, and Florey had not much difficulty in reading, or at least skimming, all that was published in his own field. He was certainly aware of Fleming's papers on lysozyme in the *Proceedings of the Royal Society* and the *British Journal of Experimental Pathology.* A feature of lysozyme was its occurrence in secretions containing mucus, and he probably wondered if it were a property of mucus itself. The paper by Cramer and Anderson in 1925 describing the failure of mucus secretion in vitamin A deficiency coupled with an invasion by bacteria strengthened the case for supposing that mucus had an antibacterial function. But it was not until 1929 that he actually carried out an experiment on the effect of vitamin A deficiency, and submitted specimens to Fleming probably for an assay of their lysozyme content. There is no available record of the results. One can only suppose that they were interesting, because Florey recruited a collaborator and embarked on a series of experiments in which he himself mastered the technique of lysozyme assay. These experiments lasted for many years, and it can be said here that Florey's determination finally produced the complete answer, resulting in the chemical identification of lysozyme itself and of its substrate. Fleming's unfinished work was thus brilliantly completed, and then Florey turned, astonishingly enough, to another of Fleming's neglected discoveries—penicillin.

Florey's first collaborator in his work on lysozyme was Neil Goldsworthy. Goldsworthy had followed L. J. Witts as John Lucas Walker Student, and was thus free to choose his own research. Florey seems to have chosen it for him, and they set out to determine the lysozyme content of a whole range of animal secretions. The purpose of the investigation was to examine the hypothesis that mucus has an antibacterial action, and that this action (if it exists) is related to its lysozyme content.

The first requirement was the establishment of techniques for measuring lysozyme activity. The most sensitive indicator was a suspension of *M. lysodeikticus*, and Fleming provided them with suitable cultures. Inhibition of growth was demonstrated by Fleming's 'well-plate' method, in which agar culture plates are prepared with a cavity or 'well' formed in the agar. These cavities are then filled with melted agar containing the secretion or tissue preparation to be tested. The whole plate is then 'sown' with the bacterial culture and incubated. Failure of the bacteria to grow around the well shows that its contents are inhibitory, and the width of the

sterile zone indicates the strength of the inhibitor. Besides
M. lysodeikticus Goldsworthy and Florey used two other
sensitive strains of bacteria isolated from the air of their own
laboratory.

They next tested their various animal preparations for
lysozyme activity. They used saliva, nasal secretions, tears,
stomach washings, colonic washings, and tissues from the
stomach, small intestine, and colon. The animal species examined were the rabbit, guinea-pig, dog, cat, and goat. It will
be realized that the testing of eight different preparations,
each from six different animals, against three different organisms, adds up to a formidable number of observations. It is
a tribute to their industry that the authors covered most of
the possibilities.

The results were, from the point of view of Fleming's
thesis, disappointing. Goldsworthy and Florey found that
there was a great variation in the occurrence of lysozyme in
different animal species. The cat had little or none in its tears
or gastro-intestinal tract, but a considerable amount in the
saliva. Dogs, rabbits, and guinea-pigs had lysozyme in all the
secretions tested, but the goat had none, except in tears. The
point is that all these animals normally lead healthy uninfected
lives despite the presence or absence of lysozyme. The conclusion is that lysozyme—as tested—has little influence on
natural immunity. The authors wrote:

The search for definite antibacterial qualities in mucus has given us discordant results, e.g. the mucus in the colon of the rabbit is rich in lysozyme,
whereas that of the cat seems to be devoid of detectable quantities. It
appears, therefore, that any protective action of mucus, common to all
species, against the entry of bacteria from the lumen of the gut into the
tissues must be sought rather in its mechanical properties than in its ability
to destroy bacteria.

They then proceeded to show, by experiment, that mobile
organisms were unable to penetrate a film of mucus.

Their paper, published in the *British Journal of Experimental Pathology* in 1930,[2] has a very significant appendix,
headed 'inhibition of one bacterium by another'. It describes
the inhibition of the growth of one of their lysozyme test
organisms on a culture plate in the vicinity of a colony of *B.
coli* (a normal inhabitant of the colon) that had appeared in

a sample of cat mucus. 'It was clear', they wrote, 'that bacteria existing in the colon were able to inhibit the growth of test organisms in a way indistinguishable from that of lysozyme.' This antagonism between one micro-organism and another was, as they pointed out, 'a very well known phenomenon' and they cited a review of the subject published in 1928 by Papacostas and Gaté. They did not, however, refer to a paper by Fleming, which had appeared in 1929 and in the same journal as their own, in which he described the striking inhibition of bacterial growth by a *Penicillium* mould.[8]

The story of Fleming's historic (and quite accidental) discovery has been told and retold in books, articles, lectures, and documentary films so often that it might be thought that nothing more need be related here. But, unfortunately for the historian, these accounts differ in important ways and a good deal of highly coloured invention has been added to an already complicated picture. The credit for establishing the probable facts by careful research, backed by personal knowledge of the people involved, must go to Ronald Hare. His book *The Birth of Penicillin*[5] published in 1970 describes a fascinating piece of detective work and disposes of many myths deeply embedded in the literature. The brief account given here owes much, therefore, to Hare's work.

In 1928 Fleming was studying staphylococci. These small, round bacteria tend to grow in clumps, like bunches of grapes; they form opaque colonies on culture plates and may vary in colour from whitish grey to golden yellow. Staphylococci, as local invaders of the human skin, cause septic spots, boils, and carbuncles. They can also cause serious and sometimes fatal infections of the deeper tissues, with invasion of the lungs, bones, and blood-stream. Most clinical bacteriologists during this post-war decade were mainly preoccupied with the highly virulent streptococcus, the main cause of septicaemia, wound infection, puerperal fever, and bronchopneumonia. The less lethal staphylococcus was therefore relatively neglected. Fleming, however, had organized a clinic for sufferers from recurrent boils and he was pursuing a leisurely study of the habits of the staphylococcus. He had an idea that the virulence of different strains could be related to their colour. He collected samples from hospital patients and cultured them on agar plates. These he incubated for twenty-four hours and then kept them at room temperature for longer periods to see what colour the colonies developed.

Fleming was not a particularly tidy worker. His room was small, measuring only 12 feet by 10; it had also an odd shape, since it was in one of the turrets on the frontage of the hospital facing Praed Street. He had little useful space and his bench was cluttered with culture plates, racks of tubes, an incubator, water-baths, a microscope, and odds and ends of apparatus. During this particular period—July 1928—the plates were mostly cultures of staphylococci in various stages of their growth. When, at the end of July, he prepared to go off for a summer holiday at his country cottage near Newmarket, he simply collected these plates into a pile at one end of the bench to leave space for S. R. Craddock, a recently recruited Research Scholar, to work there during his absence. Then, as Pasteur had done in 1880, he set off on vacation and left his cultures to their own devices.

On his return in September, Fleming had to deal with this pile of plates, which had, of course, remained at room temperature. He examined each one, subculturing any colony of interest, and then discarded them. The process is best described in Hare's words: 'In a properly organised laboratory this would have involved their being plunged into buckets of antiseptic such as lysol, in which they would have been totally submerged. The organisms would have been killed at once and so made the glassware safe enough for the technicians to wash and prepare for use again. But at that time the Inoculation Department, probably because of its love for micro-methods, did not employ anything so big or so efficient as buckets. It made use of shallow enamel trays containing so little antiseptic that it was only possible to submerge about half a dozen petri dishes at a time. Any more to be discarded had to be piled up on top.'

So Fleming piled up his discarded plates, high and dry above the lysol, and went on with his bench work. Presently a colleague, D. M. Pryce, who had helped Fleming with his staphylococcal studies in the past, dropped in to see him. Fleming complained of the amount of work he had to do since Pryce had left him, to become a morbid anatomist, and pointed to the pile of plates. He then started showing Pryce what was on them—a variety of staphylococcal colonies and many contaminating moulds and yeasts, which was not surprising since the plates had been lying about for weeks. But one plate arrested Fleming in his casual demonstration. 'That's funny,' he said. One of his gifts was an acute perception of the unusual, the ability to see, amid a jumble of irrelevant appearances, anything surprising and significant. What he was looking at was a plate liberally dotted with colonies of staphylococci. But near the edge there was a blob of mould and round this was a zone, about three-quarters of an inch wide, in which there were either no staphylococci, or almost transparent 'ghost' colonies. Mouldy plates are common enough and seldom get a second glance in the laboratory. But it was the appearance of the colonies in the vicinity of this particular mould that had so impressed Fleming. Once more it was chance—or rather, as it will be seen, a fantastic sequence of chances—that had favoured a mind prepared by previous experience of lysozyme. For there is little doubt that Fleming himself, and the various people to whom he showed the plate during that day, supposed that they were seeing a sort of lysozyme, and the general lack of interest was due to the fact that Fleming had become rather a bore on this hobby-horse. What they did not appreciate was the all-important fact that this 'lysozyme' was evidently attacking a virulent and pathogenic organism.

Fleming showed his plate to anyone handy during that day, including E. W. Todd who had just returned from the Rockefeller Institute; Hurst Brown, a Canadian Rhodes Scholar, and C. J. La Touche, a mycologist from the laboratory on the floor below his own. He then took it upstairs to show to the young men (including Ronald Hare) working in the 'big laboratory' and finally to Almroth Wright himself when he arrived in the afternoon. None of these people showed any great interest, though they were to remember the occasion for the rest of their lives. Fleming, however, kept his interest. He had the plate photographed, and he subcultured the mould into a tube of broth. During the next few days the mould grew on the surface of the broth, forming a thick pellicle. Fleming took some of the fluid from below this, mixed it with melted agar, and ran it into a 'gutter' cut into the solid agar of a culture plate. This was the technique that he used to study lysozyme, and a variant of the well-plated method. When the agar in the gutter had set, he made streaks, with a platinum loop, of cultures of several different organisms across the plate and the gutter. After incubating the plate he found that some types of organism had failed to grow in the vicinity of the gutter; others had grown normally. Clearly, something in what he called the 'mould juice' had diffused through the agar and inhibited the growth of these organisms. What was highly exciting was that these sensitive organisms included the streptococci, staphylococci, pneumococci, meningococci, and gonococci—germs that between them accounted for most cases of acute bacterial infection. The enteric organisms were, however, not sensitive.

During the next few months Fleming pursued his new discovery with enthusiasm but without much interest or support from his colleagues. He kept his mould cultures going, and asked La Touche to identify the strain. La Touche came to the conclusion that it was *Penicillium rubrum*—incorrectly, as it later transpired. Fleming harvested 'mould juice' and did a number of fairly obvious experiments. He determined the optimum conditions for producing antibacterial activity. He found that the crude juice would inhibit pathogenic organisms when diluted up to 1 in 800, that is, it was four times stronger as an antiseptic than carbolic acid, but not so strong as the newer antiseptics, such as flavine. But unlike these antiseptics, his mould juice did not impair the activity of living leucocytes. Further, when he injected it intravenously into a rabbit, and intraperitoneally into a mouse, he found that it was no more toxic than the broth in which the mould had been grown. It seems clear that, at this stage, Fleming thought of his mould juice as a possible antiseptic, something that could be applied to wounds or infected surfaces without the risk of damage to the tissues. And it was along these lines that he proceeded to make some rather haphazard clinical tests.

S. R. Craddock, the Research Scholar working with Fleming, must have been the first human penicillin guinea-pig. He grew Fleming's mould in milk, and then ate some of the solid scum which, he told Ronald Hare, tasted rather like Stilton and seems to have been no more toxic. Craddock was also suffering from nasal sinusitis; so he irrigated his nose with mould juice twice a day, but without either benefit or ill-effects. Then Rogers, one of the laboratory assistants, developed a pneumococcal conjunctivitis. What made this important to Fleming was the Rogers was in the shooting team and due to take part in a match. His eye was irrigated with the mould juice and the infection cleared up immediately. It was his consequent participation in the match rather than a historic contribution to medical science that seems to have impressed those involved. Mould juice was then used as a local application in some hospital cases. Fleming later (1941) wrote: 'About 1930 it was used as a dressing on a few septic wounds with favourable results, but as in peace time septic wounds are uncommon in hospital, and as the potency of penicillin rapidly disappears on keeping, the therapeutic aspect of this substance was dropped.'[9] And in his Harben Lecture (1945) he wrote: 'We tried it tentatively on a few old sinuses in the hospital, and although the reports were favourable there was nothing miraculous. When we asked the surgeons if they had any septic cases they never had any, and then perhaps when they asked us if we had any penicillin our whole supply had become inert.'[10] Even allowing for Fleming's peculiar style there is no sense of urgency or energy to be gained from these descriptions of his clinical trials.

Fleming soon gave up his hopes that his mould juice might provide something approaching Lister's perfect antiseptic, but two young men, Craddock and Ridley, embarked on a project in his department. This was an attempt to purify the active antibacterial substance. Though the mould juice was called 'penicillin' (a word not, for once, invented by Wright but by Fleming) it was, of course, a mixture of innumerable components from which the active material must be separated by chemical processes. Fleming had no more knowledge of chemistry than he had acquired as a student and no chemical techniques beyond those he used as a bacteriologist. Craddock also had no chemical expertise, and recognizing this he enlisted the help of Frederick Ridley, a young ophthalmologist who had worked with Fleming on the lysozyme of tears. Ridley had taken a B.Sc. course in biochemistry and was therefore better qualified than anyone else in the department to tackle the problem of purifying penicillin.

The two started work in the passage outside the main laboratory, the only space that their project was thought to merit. Craddock grew Fleming's mould in large flasks of broth and passed the fluid over to Ridley. Ridley treated this fluid in many different ways, and returned samples to Craddock, who tested them for activity. In this way they discovered that, by careful control of the acidity of the fluid and by extraction with alcohol or acetone followed by concentration in a vacuum, much of the inactive material could be removed. This was a considerable advance. Fleming was provided with much more active 'penicillin' for his further bacteriological work, and a good deal had been learned about the nature of penicillin itself. It was not a protein, for example; it had a relatively small molecule but it was, under these conditions of preparation, highly unstable. There was no sure way of preserving its activity for more than a few days.

Fleming's attitude to this work needs some clarification. Craddock and Ridley were his pupils and he must have approved their project. But he clearly gave them very little material help, or even proper laboratory facilities or space. Though they told him of their experiments he seems to have been little interested, but he did, of course, use the penicillin samples they provided. Hare suggests that the fact that Ridley and Craddock were working upstairs and some distance from Fleming's room accounts for his lack of interest. 'It is of course possible that Fleming was not fully aware of all that had been achieved . . .' he writes, 'And certainly, there is a remarkable series of omissions in the account he gave in the original paper [in 1929]. It is a single paragraph that reads as follows:

Solubility. It is freely soluble in water and weak saline solutions. My colleague, Mr. Ridley, has found that if penicillin is evaporated at a low temperature to a sticky mass the active principle can be completely extracted by absolute alcohol. It is insoluble in ether or chloroform.

'This does, it is true, give the facts, or more correctly, some of them. Quite apart from omitting all mention of the solubility of penicillin in acetone, he says virtually nothing about the procedures employed so that there is no mention of the necessity for a vacuum, the temperature employed is not given, there is no reference to the importance of the pH [acidity–alkalinity] and that it had to be periodically adjusted to keep it low, there is no indication of the titres of the penicillin in the masses, or whether a watery solution could be obtained from the alcoholic extracts.

'A second reason for suspecting that he knew very little about the work is that he never referred to it in after years in papers, lectures, or even in private conversations. In fact, he seems to have forgotten all about it.'

Since Ridley and Craddock themselves published nothing, their experiments went unrecorded. In any case, they were soon to cease. Ridley left the department in 1930 to take up his proper career in ophthalmology. Craddock obtained an appointment (with Fleming's help) at the Wellcome Physiological Research Laboratories as a bacteriologist. But he did nothing more on penicillin, and in 1932 became a general practitioner.

As far as penicillin is concerned, this is virtually the end of the story of its discovery and development at St. Mary's Hospital. Fleming had some hopes of its therapeutic use as a local 'antiseptic' during this first period, but these hopes were never high, and even they were not fulfilled. The systemic use of penicillin—with its immense therapeutic possibilities—seems never to have occurred to him. If it had, surely he would have carried out the obvious and crucial experiment—the injection of penicillin into mice together with a lethal dose of streptococci or pneumococci to see if it would protect them. He had indeed injected penicillin, but only to prove that it was harmless, that is to say, usable as a local antiseptic. Neither he nor anyone else at the Inoculation Department at that time seems even to have considered the possibility that penicillin might be used like Salvarsan: given by injection so that it could reach the site of infection wherever it might be in the body. Unlike Salvarsan, it would have a wide range of bacterial targets and yet be almost totally harmless to living tissues. These ideas, which seem so obvious from our present viewpoint, simply did not arise at that time. This was largely due to the climate of thought in the Inoculation

Department. The main preoccupation there was with the mechanism of natural immunity and the means by which it could be stimulated. A lesser preoccupation was the study of antiseptics, one that was mainly destructive, and aimed at devising methods for showing that they were of little value. If penicillin came anywhere in this line of thought, it was as a sort of antiseptic, and a messy and unreliable one at that. Chemotherapy, as we now know it, did not develop until 1935. 'Antibiosis'—the inhibition of one living organism by another—had been tried therapeutically by various workers during the previous half-century (including Lister), but if Fleming knew of this earlier work there was little in it to encourage him. Penicillin had to wait for a change in mental climate induced by chemotherapy, and for a man with the will to overcome all obstacles.

By the summer of 1929, Fleming had published his observations on penicillin and then, for all practical purposes, lost interest in it. He described his findings and experiments, listed the organisms which he had found to be inhibited by it, and suggested that it might be useful as a local application to infected areas. But the main use he had found for it, and which he stressed as an important technical advance, was in suppressing the growth of unwanted bacteria in a mixed culture so that other organisms, not sensitive to it, could be more readily cultivated. Penicillin thus found a use as a sort of bacteriological selective weed-killer. For example the 'influenza bacillus' (which does not, by the way, cause influenza) is very easily overgrown by other organisms but can be readily grown in pure culture if penicillin is added to the medium. And this was the most practical use that Fleming could suggest for penicillin, which many bacteriological laboratories adopted.

As to the origin of the mould itself, it has remained a mystery. It seems that the penicillin-producing strain that happened to land on Fleming's culture plate is a rare one. There are thousands of different moulds, some with antibacterial properties, but few in nature have bettered the activity of Fleming's original strain. Its identification as *Penicillium rubrum* by La Touche was later (1932) shown to be incorrect by the American mycologist G. Thom, who classed it as

Penicillium notatum. It has often been said that a spore happened to float in from Praed Street through the open window of Fleming's laboratory, and settled on the plate just when it was exposed to the air while being sown with staphylococci, or later when being examined. But Hare has pointed out that Fleming did not work in front of an open window (no bacteriologist would be likely to do so) and that it was physically impossible, in fact, to open or close the window without climbing on a bench loaded with tubes, dishes, and apparatus. But La Touche, working in the laboratory below Fleming's, was studying moulds that might cause asthma. His room did not have the fume cupboard which mycologists normally use to ensure that spores do not escape. It is more likely that mould spores would be wafted from this room up the staircase and through Fleming's perpetually open door than through his closed window.

There is a final and quite extraordinary point that must be made about Fleming's original observation. In the opening paragraph of his 1929 paper he described it as follows:

While working with staphylococcal variants, a number of culture plates were set aside on the laboratory bench and examined from time to time. In the examinations, these plates were necessarily exposed to the air and they became contaminated with various micro-organisms. It was noticed that around a large colony of a contaminating mould, the staphylococcal colonies became transparent and were obviously undergoing lysis.

The inference is that the plate had been sown with staphylococci which had grown into numerous colonies and that a mould spore had then or later fallen on the surface. This spore then grew into a large colony, and produced penicillin which diffused out into the medium and lysed the already fully grown colonies of staphylococci. This would have been expected if pencillin were, like lysozyme, capable of dissolving bacterial cell walls. But it is not. As Dr. Margaret Jennings (later to become Lady Florey) was to show in 1944, penicillin has no lytic effect on mature organisms, staphylococci included.[11] Lysis only takes place if penicillin acts on bacteria during their earliest stages of division and growth. At this stage, penicillin interferes with the metabolic process that forms the cell wall. The bacterium grows, but its wall gives way and the cell contents escape to form a transparent gelatinous mass. Thus, in order to produce 'ghost' colonies, the penicillin must have been present in the medium *before the staphylococci had formed colonies*. The mould colony, in other words, must have been well grown by the time the first staphylococcal colonies appeared. But how could this have been the case? No bacteriologist in his senses would inoculate a plate already contaminated by a mould unless he were studying the mould itself, which Fleming at that time was not. And if the mould spore and the staphylococci had started life together on the plate, and Fleming had followed the usual practice of then incubating it for twenty-four hours, the staphylococci would have grown into colonies and the slow-growing mould would not. After several days or weeks on the bench the mould colony might then have appeared, but the surrounding colonies of staphylococci would have remained intact. At that stage they would have been unchanged by the penicillin and there would have been nothing to attract Fleming's attention. But what circumstances could have conspired to allow the mould to grow and to produce its penicillin, while the staphylococci, though also present, were kept in check for at least several days? Ronald Hare suggests that these circumstances were the sequential temperature changes to which the plate happened to be exposed. He argues, convincingly, that Fleming did not incubate this particular plate, either by design or (more probably) by accident. The plate, bearing its mould spore and its staphylococci, was simply left on the bench for several weeks. During the first few days of August 1928 the weather was unusually cold, with temperatures not exceeding 60 °F. *Penicillium* grows well at this temperature, but staphylococci do not. The mould spore grew and multiplied into a large colony and released its penicillin. Then, about 6 August, there was a minor heat-wave with temperatures nearing 80 °F. This suited the staphylococci, which now also grew and multiplied—except those in the penicillin zone which grew, multiplied, and then burst to form the transparent ghost colonies that Fleming saw.

If this most ingenious explanation is the true one, which seems likely, the feature that strikes one is the astounding series of accidents involved, a sequence against which the odds must be incalculably large. First, Fleming's penicillin-producing strain of mould was a rare contaminant of bacteriological plates. Second, having contaminated his staphylococcal culture, this unusual mould could have produced its visible effect only (a) if Fleming failed to incubate the plate, (b) left it lying on a bench for a long time, and (c) if the weather during this time were at first cold and then hot. Third, Fleming had discarded this plate, but so carelessly that it happened to escape immersion in lysol. Fourth, Fleming would not have given it a second glance if Pryce had not happened to visit him and to turn his thoughts to the work that had been forced on him.

Every event is, in the last analysis, the result of a myriad antecedents. It is a favourite occupation for historians to identify trivialities that may have influenced some momentous happening. Tolstoy, in *War and Peace*, argues that if Napoleon failed at Borodino because of the heavy cold from which he was suffering at the time, then the valet who had forgotten to bring him his waterproof boots two days before might be regarded as the true saviour of Russia. The argument deliberately ridicules the following of lines of cause and effect into smaller and smaller channels, but chance often does play a decisive role. It is the business of a scientist, like that of a general, to anticipate and forestall unfavourable chances and to seize those that are advantageous. This is the strategy of planned research or a military campaign. Nothing in the *discovery* of penicillin was either planned or foreseen; it was the result of almost incredible luck and Fleming's genius for perceiving the unusual. In fact, in a modern, well-ordered, properly equipped, and hygienic laboratory it would probably never have happened. But it was only under these latter conditions, combined with the drive and organization of a military exercise, that planned research was able to bring the full potential of penicillin to realization twelve years later.

The first attempt by a professional biochemist to purify penicillin was made by Harold Raistrick and his colleagues in 1932.[12] Raistrick was interested in the chemistry of moulds—an interest stimulated by Gowland Hopkins, with whom he had worked—and he soon found himself a pioneer in a field that had been neglected since the days of Pasteur.

Within a few years he had isolated sixteen previously unknown organic compounds produced by moulds (some with
commercial value) and accepted the offer of a post at the Nobel
laboratories in Ayrshire. Here he discovered biological
methods for making the glycerine used in the explosives industry, and other products of industrial importance. But he
was more concerned with chemistry than its commercial
exploitation, and he was glad to return, in 1929, to academic
work as the first Professor of Biochemistry at the London
School of Hygiene and Tropical Medicine.

The chemistry of penicillin was an obvious target for Raistrick's attention, and he turned to it, as part of his general
research, in 1931. Ridley and Craddock's work had not been
published, and Raistrick knew nothing of it, nor was he told
about it by Fleming either then or later. Fleming did supply
a culture of his mould when Raistrick asked for it, but he
seems to have had no part and little interest in the research
that followed.[5]

Having decided that penicillin was a worthwhile project,
Raistrick handed it over to one of his assistants, Dr. P. W.
Clutterbuck. Since the assay of activity—the measure of progress towards purification—depended on a bacteriological
test, they needed a bacteriologist. Raistrick consulted Professor Topley, Head of the Bacteriology Department, who
collaborated to the extent of lending one of his own assistants,
Dr. R. Lovell. Lovell was already working with pneumococci,
which were sensitive to penicillin, and he therefore used them
in his assays. So Clutterbuck and Lovell started with their
mould cultures and soon confirmed most of the observations
published by Fleming.

They then, though quite unknowingly, proceeded along the
path already followed by Ridley and Craddock, and the
ground gained was really very little more. One positive advance was the use of a synthetic growth medium (Czapek Dox)
in place of the chemically messy meat-digest broth. They
found that penicillin is more stable in acid solutions and that
a good deal of impurity, including the yellow pigment, could
be precipitated by sulphuric acid, leaving the penicillin in
solution. They used low-temperature vacuum distillation to
concentrate this solution and found that the active material

could be extracted by alcohol. They also found that it was soluble in ether, if acidified, a property that Ridley had missed. But from this point they could make no further progress. Penicillin dissolved in ether is useless for biological purposes, and they could find no way of removing the solvent without destroying the penicillin. It was a chemical puzzle without precedent. Raistrick realized that it could not be solved without a full-scale attack along unfamiliar lines and he was not prepared to undertake a long and probably unrewarding diversion. There was no urgent academic reason for doing so, and there was at that time no suggestion that penicillin might be of any clinical value.[12]

The penicillin work lapsed, therefore, after a few months, and Lovell and Clutterbuck returned to their previous occupations. They had made little progress, and they were saddened by the death in a street accident of J. M. V. Charles, the mycologist who had helped them. The results of their work were described in a paper in the *Biochemical Journal* in 1932, but only three pages concerned penicillin.[13]

This chemical foray by Raistrick has been misrepresented by some writers on penicillin, including (as Hare has pointed out) Fleming himself. It was inferred that Fleming asked Raistrick to produce purified penicillin for clinical trials, or that the Medical Research Council did so. There is no evidence of either approach. And Raistrick's failure to purify penicillin was blamed for the subsequent general lack of interest. It was suggested that other workers committed to the problem were so discouraged by the defeat of a top-class biochemist that they themselves gave up the problem as hopeless. But there is no evidence that any other workers *were* committed to the problem. In later years Fleming felt it necessary to explain his inactivity in developing his discovery. He blamed lack of biochemical collaboration for his own lack of progress, and lack of bacteriological collaboration for Raistrick's.[10] The simple-minded observer might then wonder why these two experts could not have got together and supplied each other's deficiencies. But such a collaboration would have needed a common purpose, a conviction of the importance of penicillin, which at that time did not exist.

Looking back to that period when penicillin might have

been developed instead of lying dormant for another ten years, it seems that two men who had the authority, and the means, to promote it failed to take the opportunity. One is Almroth Wright, a man dedicated to the battle against bacterial disease. Why did he not insist that every available resource should be put at Fleming's disposal? An answer (and almost certainly the wrong one) is given by André Maurois, Fleming's biographer.[4] He suggests that Wright actively discouraged work on penicillin because it conflicted with his belief that the study of natural immunity would provide the key to success. It is far more likely that Wright neither knew nor cared very much about Fleming's temporary enthusiasm for a strange mould. The idea of its potential value had not occurred to him, for the same reason that it had not occurred to Fleming. The essential experiment that would have shown this value had not been done—an omission that now seems astonishing. If Fleming had shown the protective power of penicillin by injecting it together with pathogenic bacteria into mice he could have convinced himself and also Wright, who would then, surely, have thrown his considerable weight into a major research project.

The second man who might have acted is Topley. Topley was the leading bacteriologist in Britain and Raistrick's colleague at the London School of Hygiene and Tropical Medicine, and one might have expected him to see that Fleming had omitted the crucial experiment, and to have insisted that Lovell, who was working in his department, should do it. Though Topley was austerely academic and had rather a contempt for practical objectives, he could hardly have failed to respond to the magnitude of penicillin, had he recognized its potential value. And, like Wright, he had influence that could have mounted a large-scale project. But in both cases the opportunity passed unused. One should not criticize Fleming for what he did not do. His genius for discovery had unearthed penicillin as it had turned up other buried treasures. Having made a discovery he liked to go on to make another—this was his main enjoyment and his supreme talent. He had neither the wish, nor the sort of character needed, to stop this exploration in order to exploit his own discoveries.

There were two other minor attempts to purify penicillin

before the Oxford onslaught. The first was by Dr. R. D. Reid, the mycologist in Pennsylvania, whose attempts to extract the active principle went no further than the previous ones. But he did examine a large number of different varieties of *Penicillium* mould and found that none had the activity of Fleming's original strain, a point that emphasizes its rarity and Fleming's good fortune.[14] The other effort was made by Dr. Lewis Holt, in Fleming's laboratory in 1934. Holt was a chemist by training and it is not clear if Fleming suggested penicillin as a project, or if he thought of it himself. At all events Fleming gave him little help, and told him nothing of the previous work of Ridley and Craddock. So Holt started *de novo* and made two important discoveries. One was that penicillin could be extracted by amyl acetate from an acid watery solution. The second was that the penicillin could then be extracted from the solvent back into an alkaline watery solution. As will be seen later, this process, rediscovered in Oxford, became the keystone for large-scale production there. But Holt stopped his work on penicillin after a few weeks, because he could not stabilize his products. His work was never published, nor mentioned by Fleming, and it lay buried in his notebooks until disinterred by Hare in 1970.

In 1935 any latent interest in penicillin was submerged by the most exciting development of chemotherapy. This took place in the I.G. Farbenindustrie laboratories near Düsseldorf, when G. Domagk reverted to Ehrlich's method for testing antibacterial activity. In these laboratories enormous numbers of new compounds were synthesized, and a team of experts in different fields then tested them for properties that might be useful. A useful property, for instance, would be antiseptic activity, which was measured by the effect on bacteria in culture. More powerful and less toxic antiseptics had emerged from this quite empirical process. Domagk then added a test *in vivo*, to observe the ability of a substance to protect animals from lethal organisms—as Ehrlich had done many years before. The results were surprising. Some chemicals that were poor antiseptics had much greater antibacterial power in the body, and from this beginning Domagk worked through the likely compounds, recording any improvement resulting from changes in molecular structure.

It was a laborious, hit-or-miss type of research, which Almroth Wright saw in action when he visited the laboratories. Hare describes his opinion of it when he returned to St. Mary's:

He regaled us at tea with the story that he had been shown enormous laboratories in which they did nothing but take compound after compound and test its ability to deal with infections in animals caused by a variety of organisms. Blind groping in the dark in this way was so utterly foreign to someone of Wright's temperament that he looked on it as a form of sacrilege. To him, the only method was to think the thing out in the privacy of one's study, and then do the experiment to prove the theory.

Blind groping it might have been, but the light of reason alone would not have predicted Prontosil.

Prontosil emerged when Domagk happened to include the azo dyes in his programme. He then discovered that those with a sulphonamide group had the exciting ability to protect mice from infection by streptococci. The most active was sulphamido-chrysoidin, a red dye with the trade name of *Prontosil*. Domagk published these observations in February 1935, but in fact the dye had been given secret clinical trials in animals and human patients over the previous two years. An earlier announcement had been avoided, to allow the completion of patent formalities. And even after publication it was difficult for other workers to obtain Prontosil, so that they had to synthesize similar compounds for themselves. In fact this had the effect of stimulating independent research that was unfortunate for the manufacturers of Prontosil, since it was soon superseded by far more effective and less toxic compounds.

Prontosil had no antibacterial effect on microbes in the test-tube, but became lethal for them in the body. This effect was, for a time, quite unexplained, and Domagk suggested that Prontosil stimulated the natural defences of the body, which would have pleased Almroth Wright. It was soon in wide use in human cases of streptococcal infection, including puerperal fever, and proved effective, though it had unpleasant side-effects, the least of which was the bright red staining of the patients' skin. Within a few weeks of its introduction workers at the Pasteur Institute in Paris discovered that it was the sulphonamide part of the molecule that was active against bacteria, not the red azo dye as supposed. In fact, Prontosil was broken down in the body so that sulphonamide was released, and this could kill bacteria either in the blood-stream or in the test-tube. The way was then open for the synthesis of a whole range of sulphonamide compounds, of which sulphanilamide and sulphapyridine (better known as M and B 693) were the most effective.

One of the first doctors in Britain to use the new sulphonamides was Leonard Colebrook. While Almroth Wright had been lost in his abstractions, and Fleming in his laboratory discoveries, Colebrook had been concerned for sick people, and was doing his utmost to defeat puerperal sepsis at Queen Charlotte's Hospital. When Prontosil appeared he began using it in otherwise almost hopeless cases. In the first six months of 1936 he gave the drug to 38 patients and only 3 died. In the following six months 26 cases were treated, with no deaths. It seemed a clinical triumph, but it did not convince Professor Topley, because Colebrook had not done the statistically necessary controls. Yet with a disease as lethal as puerperal sepsis Colebrook did not feel justified in allowing half his patients to risk death untreated in order to satisfy the statisticians. Common humanity must, on such occasions, override cold logic; but Topley, the leading authority of his day, maintained that the case for Prontosil had not been proved. 'If he had had his way', writes Hare, 'it is improbable that anything further would have been heard of it. For he was a great believer in statistics.'[5]

Fortunately, the sulphonamides were so effective that they needed no statistical underpinning. But they were not the final answer. Though they produced a dramatic reduction in mortality their value was limited to certain types of infection, and they could have dangerous side-effects. They made a less tangible contribution, however—they raised the hopes that this method of selective chemical attack on bacteria within the body was the true path to success.

9

FROM CAMBRIDGE TO
SHEFFIELD

One might suppose that life in Cambridge was giving Florey all that he had hoped for. He had a secure appointment in a progressive department among congenial colleagues. He had his own laboratory in a brand-new building, a personal assistant, ample time for research, and almost all the facilities he needed. He was expected to teach, but since he and Drury had largely designed the new Tripos course, he had the satisfaction of putting his own views on scientific medical education into effect. Outside his department most of the advantages of university life were open to him. He had a Fellowship at a college in which scientists were actually respected. The Master, Sir Hugh Anderson—himself a physiologist and a friend of Sherrington—took a personal interest in Florey and his future. At Caius also there were teaching duties. Florey had been appointed Director of Medical Studies, and was much involved in the tutorial system. But routine teaching was not Florey's vocation, and he was not by temperament a 'college man'. His drive was too personal and dynamic to be sunk in an institution, as was evident to all who knew his restless, irrepressible urge to be up and doing.

Though Florey's decision to leave Cambridge came as a surprise to his colleagues it was one in line with his own ambitions. If individual research had been his sole interest, then he might well have chosen to stay where he was, since he could have pursued it to better advantage there than almost anywhere in Britain. Research *was* his vocation. He probably enjoyed working in his laboratory (which he usually called 'messing about') more than anything else. But he had another objective which, as will be seen, was related to his research ideal. He was determined to achieve professional promotion both for financial reasons and for the influence that comes with it. With his usual forethought he had planned the timetable for his ascent of the academic ladder. In January 1925 he had written to Ethel: 'I feel confident that I can pick up a lectureship within two years and a professorship 2 or 3 years after that.' His lectureship came, in fact, just over two years after writing that letter, and now, two years later, he was on the lookout for the next step up.

In 1929 there was little chance of promotion for Florey in Cambridge. Dean was firmly in the Chair of Pathology (and remained there until 1955 and his seventy-sixth year). Barcroft was equally in possession of the Chair of Physiology; Florey might, perhaps, have taken the Lectureship in his department which fell vacant when Anrep became Professor of Physiology in Cairo. But this would have been a step sideways and not upwards, and Florey was by now committed to experimental pathology. The Master of Caius was still urging the creation of a Chair of Experimental Medicine, but the years had gone by and nothing had happened. Florey was not given to patient inactivity sustained by vague promises, and his judgement was vindicated, since the Chair of Experimental Medicine was not created until 1945.

There were two factors that determined Florey's quite natural academic ambition. The first was financial. His Lectureship and his Fellowship together brought in about £900 per annum, quite a comfortable income at that time for a young couple with one small child. They owned a house in a pleasant neighbourhood and a car, and could have afforded to travel abroad and to take part in the social life of the University, had they wished to do so. But Ethel was perpetually worried about money. She had retained her almost pathological dread of poverty, the fear that some disaster would leave them destitute. She longed for greater financial security, and resented every penny that Howard might spend on the things he enjoyed and which she regarded as extravagances: his gramophone and records, his photography and the cost of a ciné camera, and the cost of his visit to America. Since she had abandoned the idea of a career for herself, she worked hard in the house and garden, dressmaking for herself and Paquita; making curtains and chair-covers; and doing most of the interior decorating, at a time when this was not usual. But she was a poor housekeeper and found it difficult to cater within a fairly generous budget. As a car driver she was accident-prone to an almost legendary degree and she had to pay for repairs herself. Ethel, for one reason or another, was always short of money and, with a feminine ambidexterity, she blamed Howard for being both mean and extravagant.

The second factor to be considered is the matter of power. Though professors in Britain were seldom the autocrats that they tended to be in many continental universities (Almroth Wright was one of the exceptions), they had a certain amount of power fifty years ago. They could spend their departmental funds as they chose, contrive to get the men they wanted, and to get rid of those they did not. They carried weight in their universities and, in their own subjects, on an international level. There were, at that time, relatively few professors (there were only four in scientific medicine in Cambridge, for example) and their recent proliferation has proportionately lowered their status.

It was not power for its own sake that Florey wanted. He detested autocrats. He disliked their pretentious assumption of superiority and deadening influence in subjects in which the recognition of ignorance is the greatest spur to progress. He also disliked them for personal reasons. Like many Australians he objected to being given orders. A few people had tried to be high-handed with him—for example the Captain of *Otira*, Philip Panton, and Mitchell—and none with impunity. Yet, when he recognized and respected the need for discipline—as on his Arctic expedition—he could comply loyally. It was a matter of respect for ability, rather than authority. And he did not like giving orders himself. Kent insists that Florey never gave him a direct command. It was always a request, beginning, 'Do you mind doing ...' or 'I think we'll start with ...' which, needless to say, Kent obeyed with the alacrity of an Able Seaman on the quarter-deck. And with his collaborators Florey adopted the same attitude. He discussed proposed experiments with them, and they felt that he paid careful attention to their opinions. But there was seldom doubt about who was the leader.

It was this question of collaboration, particularly where it concerned his new interest in lysozyme, that brought home to Florey the advantages of being in charge of his own department.

Even then it was becoming difficult for one man to follow a line of research to any depth. Florey was better equipped than most. He had worked in several of the best laboratories in the world, where his natural technical ability had ensured the best use of the experience. He had become one of the finest animal surgeons in the country, a skilled microscopist, and an expert histologist and photographer. But he recognized the limits of his own knowledge and deliberately chose as collaborators people who could supply what he lacked. In his published work he had as partners a cytologist, a radiologist, a bacteriologist, a haematologist, a biochemist, and several specialized physiologists. When he decided to study lysozyme in depth he realized that he would need long-term collaboration in fields—particularly chemistry—of which he had little practical experience. He needed, in fact, a team. As a lecturer in Dean's department, it would have been hardly possible for him to lead a research project involving several well-qualified people. He had no control over appointments, nor funds to meet any expenses but his own.

Florey made attempts to start his lysozyme project in Cambridge. He applied to the Medical Research Council for a grant in 1929, and was awarded £50 per annum for a period of three years. His first collaborator, Neil Goldsworthy, was more of a bacteriologist than he was, but even less of a chemist, and their research association did not last more than a few months. After their study, published in April 1930, Florey worked alone on the distribution of lysozyme in different animal tissues, and published the results later in the same year.[1] He asked A. A. Miles (later Sir Ashley Miles and Director of the Lister Institute) to study with him the antibacterial action of various tissues. Florey supplied extracts, and Miles tested them for activity against a range of bacteria, but the work proved unrewarding and the collaboration lapsed.[2]

Florey then approached E. G. D. Murray, also a bacteriologist, with a proposal that they should make a joint attack on the lysozyme project, but Murray declined, on the grounds that he had insufficient chemical knowledge. Finally Florey turned to Marjory Stephenson, the biochemist who later became one of the first two women to be elected to the Royal Society. She was working in Gowland Hopkins's laboratory, and could spare little time for collaboration with a lecturer from another department. Their investigations are recorded in Florey's experimental notebooks, but relatively little progress was made and the results were never published.

Florey thus had material reasons for seeking promotion, and only the inertia induced by the pleasant security of Cambridge to restrain him. But he was not subject to inertia. Change and the challenge of new difficulties were, for him, an enjoyable stimulant. He was not the sort of research worker who needs a settled location and seclusion to pursue an uninterrupted line. Florey was quite happy to set up his experiments in any laboratory. Some of his best work entailed repeated moves between London, Cambridge, and Oxford, and on arriving he could be hard at work within a day. But, having planned an experiment, he would allow no distraction or interruption. In this sense he took his seclusion with him. A move to a new appointment would be unlikely to disrupt his research.

Dean, in any case, encouraged his young men to try for promotion. They had the advantage of their training in a department in which the experimental approach to both teaching and research was unhampered by routine hospital work, and the advantage of Dean's great influence in academic pathology. There were Chairs of Pathology in most of the British medical schools and one or two vacancies occurred almost every year. The newer and smaller universities were proving-grounds for new professors. The best men would not stay long, but while they were there they would contribute ideas and energy and in return learn to look after a department with tact and sensitivity, and how to fight for it in the perpetual jungle skirmishing of university politics. A young professor who has successfully come through these initiation ordeals is in a good position to move on to a larger, older university.

The first evidence of Florey's intention to leave Cambridge actually antedates his interest in lysozyme. It is a letter from Sherrington, written in August 1928,[3] in which he gives his views on the Chair of Pathology in Liverpool, which had become vacant and which he had once held himself. It is evident that Florey had asked for advice on applying for it, and kept the idea in mind for some time, because in October 1928

the Master of Caius writes: 'I am relieved that you have decided not to apply for the Chair of Pathology in Liverpool.'[4]
He goes on to say that the hospital work would be bound to
distract him from research, and argues that Florey would be
better off, financially, in Cambridge. He would have to live
outside Liverpool, and the daily travel would be expensive.
So would the entertaining that a professor is expected to do.
Anderson remained hopeful that Florey would stay in Cambridge, and that suitable promotion for him could be found
or created.

But, though Florey had decided against Liverpool, it was
not because he had decided to stay in Cambridge. A year later
the Bristol Chair of Pathology became vacant, and again
Florey was tempted to apply. Once more he asked Sherrington's advice, going to see him in Oxford. Sherrington, in
a letter dated November 1929, wrote that he thought the
Bristol post would be suitable 'if the conditions were right'.
Perhaps Florey was not sure that they would be, because he
did not apply.

During the next two years Florey devoted himself to his
work on mucus secretion and lysozyme. The study of secretion was widening to include new aspects of the mechanism
normally controlling it. This he studied by the local application of various irritants, together with experiments on blood-
and nerve-supply and on the effect of drugs and hormones.
He also began to study in greater detail the mucus-producing
cells—'goblet cells'—that discharge their contents on to the
surface of the mucous membranes of the respiratory, gastro-
intestinal, and uro-genital tracts. Work on this last site led
to his research with Carleton on the physiology of mammalian
reproduction which, in turn, was to form an important base
for their research on contraception.

The histology of the goblet cell was, even in those pre-elec-
tron-microscope days, becoming complex. The process of
secretion involved the formation of 'mucinogen'—the chemical precursor of the mucin which is the viscous component
of fluid mucus. Mucin is not a single substance, but a group
of polymers whose long-chain molecules and affinity for water
give it a unique capacity to keep the internal membranes of
the body moist, lubricated, and protected from mechanical,

chemical, and probably bacterial damage. Mucinogen is formed within the secreting cells, becomes converted into mucin, which distends the cell to form a 'goblet' and then bursts through the cell wall to reach the surface. Florey wished to understand as much as possible about all these stages and, as usual when faced with a new field, he decided to learn about it from the best contemporary authority. In this case, the authority was Professor P. Bouin, in Strasbourg. With Sherrington's introduction, Florey arranged to spend three months in his laboratory during the summer of 1931.

Howard, Ethel, and Paquita (then nearly two years old) travelled to Strasbourg in August, and remained there until the end of October. They probably enjoyed their stay in the old city, almost on the banks of the Rhine, with its blend of French and German influences. Howard, with his sense of history and his feeling for architecture, must have appreciated the many buildings that had survived from a turbulent past. Bouin's department, in the sixteenth-century University, offered him the most modern approaches to cytology, and his notebooks show that he was soon deep in the cytological study that was his objective. But, as usual, he took the opportunity to visit other centres of research, and Sherrington had also given him letters of introduction to Leriche, Nicloux, and Schaeffer.

During this time the question of his academic future again became an acute one. In Sheffield, the Professor of Pathology had been J. S. C. Douglas, a brother of C. G. Douglas, the respiratory physiologist who worked with J. S. Haldane in Oxford. In June 1931, Professor Douglas died suddenly. The vacant Chair was advertised while the Floreys were in Strasbourg. Florey was interested, and when he returned to England at the end of October he went to Sheffield to see what the professorship there would entail. Dr. H. E. Harding, the acting head of the Department, was his guide. He was becoming experienced in this role, since thirteen pathologists who actually applied for the post and several others inspected the Department.[5]

It is difficult to imagine two more sharply contrasted English University towns than Cambridge and Sheffield. Cambridge, meditating by the lazy waters of the Cam and set in the farmlands of the East Anglian plain, has its material roots deep in the traditions of pastoral England just as its University has its academic roots in medieval scholarship. The streets of Sheffield climb laboriously up the steep valleys of the Don and four of its swift tributaries, valleys that lead up and out on to the open moorlands of the Pennines and the Peak District. Its tradition is industry, a tradition almost as old as that of agriculture, since Sheffield iron, water-power, and abrasive stone had already established its metalwork in Roman times. As coal replaced charcoal, and steel replaced iron, Sheffield kept its lead in the making of the finest cutlery, implements, and instruments, added silver and glass to its specialities, and grew in size and wealth. Steelmasters are, by natural selection, a hard-headed breed, and the Yorkshire strain probably excells in this quality. Yet it was one of them, Mark Firth, who established Firth College in 1879 and thus virtually founded Sheffield University.[6] The original buildings, in exuberant Victorian Gothic style, stand on a pleasantly tree-lined hillside above the centre of the town, and it was these that Florey visited in 1931. They are now almost lost beneath towering slabs of glass and concrete that seem inseparable even from academic growth.

Sheffield University had a purpose as sharp and practical as a piece of Mark Firth's cutlery. It was a tool to fashion the productive life of the town. Metallurgy, geology, engineering, mining, fuel and glass technology, chemistry, physics, medicine, and dentistry—these were the subjects taught and they were directly applicable to the industries and people of Sheffield. The medical school was a small one, and in 1931 the annual intake was fourteen students. The dental school shared the medical science teaching, including pathology. Among the professors of the Medical Faculty, two (both Fellows of the Royal Society) were outstanding as scientists and men of strong character and influence.

The Professor of Physiology was J. B. Leathes,[7] who had worked with Starling. He had been a classical scholar at Winchester and read classics at New College, Oxford. It was only after his graduation that he decided to take up medicine against parental opposition and so was cut off with the proverbial shilling. When he entered Guy's Hospital Medical

School he lived with an aunt at Highgate and, having no money for fares, walked the 12 miles a day to the hospital and back. He was appointed as the first Professor of Pathological Chemistry at Toronto, coming to Sheffield in 1914 at the invitation of H. A. L. Fisher, the Vice-Chancellor. His main scientific work was biochemical, on fats and fat metabolism, but he had a general interest in physiology and pathology, and strong views that they should be integrated with the teaching of clinical medicine. Leathes was, in fact, a man of wide education.

The second outstanding character in the Sheffield Medical School was Edward Mellanby, Professor of Pharmacology.[8] He was a big man, powerful both physically and mentally, and the youngest of three remarkable brothers. Their father, a shipyard manager at West Hartlepool, was an evangelical nonconformist who had been the amateur boxing champion of the North of England. He gave all three sons boxing lessons and strict religious instruction. They did extremely well at Barnard Castle School, each winning the most-sought-after bursary in turn, and Edward in particular was a fine athlete. All three became professors: Alexander, the eldest, in engineering; John, the second son, became Professor of Physiology at St. Thomas's Hospital, London, and later at Oxford; and Edward, like John, read medicine at Cambridge and both were greatly influenced by Gowland Hopkins. Edward was fascinated by the developing work on 'accessory food factors' that led to the discovery of the vitamins, and himself later made important studies on rickets leading to the discovery of vitamin D. He became Professor of Pharmacology at Sheffield in 1920.

Leathes and Mellanby were a powerful pair in Sheffield. They had worked together on several projects, including the preparation of the first insulin to be made in Britain. In scientific circles they stood far above their colleagues in the medical faculty. They were naturally in touch with affairs in Cambridge and Oxford and on friendly terms with Sherrington and Dean.

The Sheffield hospitals served a population of about a million , and they were large and busy. There were four, which had been brought under the same administration in 1924: the

Royal Hospital, the Royal Infirmary, the Children's Hospital, and the Jessop Hospital for Women. Senior staff members were also part-time teachers in the University, and even full-time University medical posts involved clinical work for the hospitals. The Department of Pathology was no exception. The professor was nominally in charge of all the hospital pathology services, but since there were three experienced clinical pathologists with University posts, he could follow his own inclinations towards research, teaching, or clinical work more or less as he wished. The previous professor seems not to have been inclined in any direction. When Florey asked what would be expected of him a University official replied, 'We don't care what you do as long as you make a mess in this laboratory: it has been too clean for a number of years.'[9]

This promise of freedom, combined with the status of a professorship, probably decided Florey. Financially he would not be much better off than in Cambridge—the salary was £1000 per annum. And yet when Drury asked him why he should want to exchange Cambridge, with all its advantages, for a small provincial university in a large, smoky, industrial town Florey replied, 'I'll earn a better pension.' It was a typically laconic remark, but it was also a simple statement of the truth. Ethel's concern for the financial future was having an effect on Howard.

Florey sent in his application for the Sheffield Chair in November 1931. He had strong support from his distinguished friends in Cambridge and Oxford and an impressive academic and research record, his most recent honour being the award of the important Thruston Medal for research by his college in 1931. Against this there was the fact that he was not really a pathologist. He had never 'served his time' in a hospital laboratory, or carried out routine clinical investigations. He had never, for example, done a human post-mortem examination. Nor had he any experience of the teaching of systematic pathology that forms part of the student's clinical studies. There is a story that, when Sir Robert Muir, the Scottish morbid anatomist, was asked for his opinion on the applicants, he eliminated one name immediately. 'There is no pathologist called Florey,' he said.[10] By his own criteria, he was justified. Florey was still, essentially, a physiologist.

But Sheffield, to its credit, was prepared to break with tradition. The short list of candidates drawn up on 3 December contained the names of H. W. Florey, F. E. Reynolds, and B. Duguid. A special meeting of the Faculty Board interviewed them on 9 December and agreed that Florey should be appointed. Undoubtedly Leathes, who was Dean and Chairman of the Board, and Mellanby, its most distinguished scientist, influenced the choice. They would have favoured Florey's vigorous experimental approach. Chapman, in his *History of a Modern University*, writes:

This time the Faculty took a bold step. Instead of appointing one who had served his apprenticeship in some stereotyped laboratory as Assistant to the Professor, it went outside and chose a young man who had shown great promise of research in pathological physiology. The election of H. W. Florey, an Australian by birth and training, brought upon us much adverse criticism from Pathologists throughout the country. It was said openly and freely that Florey was not a pathologist at all and that it was unfair to the younger men, who were devoting their lives to the subject ...[6]

Florey took up his new appointment in March 1932. On the domestic side, Ethel had to cope with moving house for the fourth time in five years. The Cambridge house proved difficult to sell. The depression was taking its toll, and the Floreys had no offers even approaching the price they had paid for it in 1927. Finally, it was rented to an employee at the Post Office. At the Sheffield end they were more fortunate. University professors were, in those days, expected to live in some style and in houses to match their status. The Emeritus Professor of Medicine, Sir Arthur Hall, had such a house, No. 10 Endcliffe Crescent, and he was prepared to rent it to the Floreys. It is a large, stone-built Victorian house with three storeys, stables, and outbuildings and nearly an acre of garden. It stands among other such houses, surrounded by trees and lawns, on a hillside about a mile from the University. Now named 'Florey Lodge' it has been converted into a hall of residence by the University. Then, it was a house that needed a staff, and Ethel, soon after settling in Sheffield, found herself in charge of a cook, housemaid, gardener, and, from September 1934, a nannie, Doris Wheeler. It is not surprising, perhaps, that her financial worries became even more acute.

Florey's new appointment brought inevitable difficulties. The first concerned Kent. After nearly five years together a break in their association would be a near-disaster for both of them. Sheffield University, however, did not provide funds for personal laboratory assistants. The Pathology Department staff included a few technicians mainly doing routine work, and none of these could become a full-time assistant to the professor, even if he were suitable. Florey put the position to Kent with no persuasion or over-optimism. A personal technician could expect no more than 10 shillings per week from the departmental budget, though he could supplement this by doing some private work, such as cutting histological sections. Meanwhile, Florey could apply to the Medical Research Council for a grant. It would mean that Kent, for the first time, would be living away from home, with the extra expense of life in lodgings. Kent did not hesitate for a moment; he too moved to Sheffield. He found lodgings that provided a bedroom, sitting-room, and four meals a day all for £1 a week.[11] With the money he earned from cutting sections he could 'just get by' until Florey's application to the M.R.C. was approved. Florey had a good case for such a grant and Edward Mellanby was a member of the Medical Research Council. But the grant, when made, was hardly princely. It was for 50 shillings a week, and for one year only. Renewals would depend on satisfactory progress.[12]

The Pathology Department divided its energies between teaching, research, hospital work, and private-practice pathology. This arrangement, with its inevitable conflict of interests, did not exist in the Oxford and Cambridge departments. Florey had no experience of dealing with clinicians, who were all-powerful in their own hospitals and a force in the Faculty of Medicine. He had observed, and strongly deprecated at Adelaide and the London Hospital, the prevalent attitude of the clinician to pathology. Too often the laboratory services were used without discrimination. Complicated tests might be 'ordered', not for their specific relevance to a diagnostic problem but to exclude remote possibilities, or because a baffled clinician could think of nothing else to do, or because he was collecting data for his own research. Too often the clinicians did not understand the tests they ordered nor therefore, their real value and limitations. And they seldom took the trouble to consult the person who could have helped them—the pathologist. He felt that, like the hospital engineer or electrician, he was simply employed to keep the wheels turning.

At first, Florey was content to restart his research projects and to leave the routine arrangements of the department more or less as he found them. His total research accommodation was much less than he had had at Cambridge. He had two laboratory rooms on the top floor of the University building, one of which he shared with the bacteriologist, Professor Edington. When he arrived, the pathology staff consisted of a lecturer, Harding, and three demonstrators: Dawbarn, C. G. Paine, and Miss Kirk. Harding, who had read medicine at Oxford and been on the laboratory staff at Guy's Hospital, was in virtual charge of the hospital laboratory services. Dawbarn was based at the Royal Hospital, Paine at the Royal Infirmary, and Miss Kirk at the Jessop Hospital. Research, if any, was a spare-time occupation that competed with work for the clinicians' private practice, for which the laboratory collected and distributed fees. The chief technician, in particular, benefited financially from this system, and he was not greatly pleased by the diversion of even a small fraction of his histological earnings to create a living wage for Kent.

So far as the clinicians were concerned, the routine laboratory services ran fairly smoothly. The hospital pathologists did what they were asked to do, even though they knew that much of their work was unnecessary or misunderstood. Florey was aware of this, and also of the amount of time spent on private work, but he did not immediately try to change things. He adopted much the same attitude to the methods of teaching. Pathology at Sheffield (as in most medical schools except Oxford and Cambridge) was taught as a clinical subject, with systematic lectures and classes that attempted to cover the whole range of human disease. The students were expected to memorize a mass of facts, figures, and 'histological appearances', with no encouragement to question. After his own exposure to similar methods in Adelaide, Florey had little respect for this sort of teaching, but he at first left things as he found them.

Professors are expected to give lectures, and Florey gave them. His inaugural lecture, a ceremonial occasion attended by most of the senior members of the University, was considered to be a success, though much of its subject-matter was probably outside their comprehension. His routine lectures presented a more difficult problem. A series on mucus secretion, the cerebral circulation, or the minute anatomy of the lymphatic vessels, would be hardly appropriate to a course that skimmed over the whole of pathology. Strangely enough, Florey agreed to Harding's suggestion that he should give the pathology lectures to the dental students. Harding felt that these—necessarily basic—lectures would be a good training-ground for the lecturer. Florey accepted the challenge, and with characteristic determination read up the literature on dental pathology, a subject of which he knew nothing. The result was a good series of lectures and the existence in Florey's collection of reprints of an otherwise inexplicable number on dental pathology. Thereafter he took his place among the routine pathology lecturers, covering such selected subjects as inflammation, vascular disease, and the gastro-intestinal disorders. If teaching was a part-time occupation for most of the teachers, so learning seemed to be a part-time occupation for the students. Sheffield at that time did not attract the best, and academic inspiration was not very evident on either side of the Sheffield Medical School, despite the efforts of Leathes, Mellanby, and Naish, the Professor of Medicine.

Two of these three distinguished professors were, in fact, in the process of leaving Sheffield soon after Florey's arrival. Leathes retired to Wantage, near Oxford, in 1933, though later he collaborated with John Mellanby, when he succeeded Sherrington, to work on fats in blood clotting. And, also in 1933, Edward Mellanby left Sheffield to become Secretary to the Medical Research Council on the retirement of Sir Walter Morley Fletcher. Thus Florey lost two senior and sympathetic colleagues and, in the case of Mellanby, a collaborator with whom he had just started an investigation of the effect of vitamin A deficiency on intestinal secretion. But Mellanby's new appointment gave him a position of power unique in British medical research, and he later showed in material ways the respect he had already developed for Florey.

In the same year, Florey nearly decided to leave Sheffield himself. At the end of February 1933, he received a letter from Dr. E. P. Poulton, of Guy's Hospital, London, who was evidently acting as an envoy on behalf of the hospital authorities.[13] 'I am writing to see if you would consider the possibility of coming to Guy's as our Professor of Pathology,' he wrote, and went on to mention that the stipend was £1400 per annum. This tentative inquiry arose from the fact that when the Chair of Pathology became vacant, the selection committee felt that none of the applicants was up to the standard they wanted. Florey was the sort of man they needed, so moves were made to tempt him. It is clear that he *was* tempted, and that he must have given Poulton cause to hope that he would eventually accept. However, several months of negotiation went by without a decision and then, on 15 July, there came an official letter from the Dean of Guy's Medical School. 'At last we have completed all the formalities. The University will send you a formal invitation to accept the Sir William Dunn Chair of Pathology at Guy's.'

Florey had told no one in his department that he was seriously considering leaving Sheffield after having been there for little more than a year. But, with the receipt of this official invitation, he had to inform the Vice-Chancellor, who was clearly dismayed. The loss to the University of three of its best scientists within a few months would be a serious blow to its prestige, and it must have been obvious that, if Guy's had failed to attract a suitable applicant for its Chair of Pathology, Sheffield would be in an even worse position should it have to replace Florey. There was a hurried meeting of the University officials, and on 28 July the Vice-Chancellor wrote to Florey, 'I have seen Colonel Stephenson and the Treasurer. They agree to join me in proposing that, if you will remain here, your stipend should be £1,200 per annum. I hope we shall keep you.'

Meanwhile, through some unexplained misunderstanding, it had been assumed at Guy's that Florey had accepted their Chair of Pathology, and an announcement of his appointment appeared in *The Times* of 26 July. Florey was furious and wrote to the Dean to say so. He received an apologetic letter from him, written on 29 July, regretting the mistake, appreciating his annoyance, and begging that it would not prejudice his decision. But Florey wrote formally refusing the invitation. The Dean replied, on 1 August, 'I was staggered by your letter.' He goes on to say that, during the months of negotiation, Florey had led them to believe that he would accept. Even now they hoped that he would change his mind, and assured him that he could make his own conditions for doing so.

Florey was unmoved by these appeals, and remained in Sheffield. But he was now in a position to make at least some of his own conditions there. He had already improved his own salary by £200 per annum, and he set out to improve his department. He wrote to Turnbull at the London Hospital, McIntosh at the Middlesex Hospital, and Drennan at Edinburgh, asking them for details of the staffing of their departments and of the amount of routine clinical pathology they carried out. Armed with their replies, and with his own figures which showed that routine investigations had more than doubled in the Sheffield laboratories in five years with no increase in staff, Florey asked for additional help with routine work. He got what he asked for, and the result was a greater freedom for research for the staff. He turned one of the classrooms into a research laboratory by dividing it into cubicles (which their inmates called cowstalls) and encouraged his junior colleagues to use it. Various staff changes occurred. C. G. Paine left the Royal Infirmary to work at the Jessop Maternity Hospital, and Dr. l'Hermitte (of French-Mauritian origin) took his place. In January 1934, B. D. Pullinger joined the staff and, being a pathologist of quite outstanding ability and efficiency, she soon became Florey's main ally in his campaign to improve the standard of work in his department.

Beatrice Pullinger, a South African by birth, had qualified from the London School of Medicine for Women and St. Mary's Hospital, Paddington. In 1921 she became Junior Pathologist, and in 1924 Clinical Pathologist, at the South African Institute for Medical Research in Pretoria. In 1929 she returned to London with a Research Fellowship to work with Dr. Mervyn Gordon on the lymph gland disease lymphadenoma, at St. Bartholomew's Hospital. A year later she was appointed Clinical Pathologist at the Mount Vernon Hospital, Northwood, then the leading cancer hospital in the country. Thus, when she came to Sheffield in 1934, she had had a wider experience of clinical pathology and research than anyone else in the department, including the professor. She was, as Harding generously acknowledged, 'the best pathologist we had ever had in Sheffield'.[5]

One of her first moves was against the practice by the clinicians of ordering unnecessary pathological investigations. Harding remembers the day on which she selected twenty request forms from the batch just received and returned them to their senders inscribed with the one word, 'Why?' The outcome of this challenge was a few sensible answers—which earned the investigations ordered—or no answers, either because the clinicians involved could think of no good ones, or because their dignity had been offended. Dr. Pullinger, unintimidated by offended clinical dignity, maintained her refusal to act as a sort of slot-machine from which pathological reports could be extracted at the push of a button. The indignant clinicians then complained to the professor, an unwise move on their part. Florey, of course, had every sympathy with Dr. Pullinger and, in his defence of her attitude, he went much further than the issue of unnecessary requests. He attacked the whole principle of pathological investigating being done in his department for the private practices of the hospital doctors or for their personal research. This raised a real storm and, for once, Florey was not on very secure ground. 'It was a tricky situation,' writes Dr. Pullinger.

The senior physicians of the Royal were a most distinguished lot. I remember particularly Naish and Hall. They naturally wanted clinical pathological research and private work as well. Inevitably conflict arose and reached crisis proportions with Florey and Imrie (a Canadian physician) disputing the allocation of our time. Imrie visited Florey at the University. Florey gave me a vivid description of this encounter, just short of actual assault but including reflections on their respective mothers. The row became publicised. I, myself, on other grounds, was on bad terms with some of the Royal's clinicians and became alarmed at possible consequences. I had to calm Florey (and myself) explaining to him the facts of my appointment and realities of a hospital pathologist's duties which he wished to ignore.[14]

From that time Florey took a closer interest in the clinical and teaching activities of his department. He began to attend Dr. Pullinger's post-mortem examinations and to ask searching questions on her findings. She found his attitude to orthodox explanations disconcerting. He would listen and, at the end of her exposition, merely comment, 'So you say.' Such an attitude is, of course, proper for the research worker, but it can be confusing when applied to elementary teaching. One needs to know at least what is orthodox before looking for its fallacies. But Florey had some success in his efforts to make students think for themselves, and to use what they learned of pathology in their clinical experience in the wards. Professor Hall acknowledged this when he wrote 'He has done a great deal to bring the Hospital pathology laboratories into closer relation to the University and the teaching of pathology. Before his time this relation was minimal. I mention this point particularly, because its successful accomplishment has not been secured without considerable tact and firmness on his part.'[15]

Firmness one can readily accept as one of Florey's characteristics. Tact, as usually defined, was less evident. Roundabout diplomacy and care to tread softly among other people's vanities was not his usual way of getting what he wanted. There was, however, a conscious method behind his direct, frontal assaults, which achieved their object and usually caused little rancour. As Florey himself put it: 'One can get away with any sort of gaucherie [as] just one of these rough colonials. It's a very good line to play'.[9] Sir Rudolph Peters confirms this. 'The outstanding thing about Florey was the Australian pungency of some of his remarks. Very few of us could have got away with them, but from him they sounded all right.'[16] He managed, in fact, to make and keep many friends in Sheffield.

Florey's forays into clinical pathology and teaching caused no slackening of his research. His main interest continued to be mucus secretion. He published in 1932 the results of the work he had done with the techniques learned in Strasbourg, showing that the 'goblet cells' manufacture their mucin in the area of the Golgi apparatus and discharge it from the cell apex.[17] He then carried further the investigations of the antibacterial properties of intestinal mucus. By direct microscopy he showed that inert particles in the trachea or intestine adhere to strands of mucus which then roll up to enclose them, and thus cause their elimination. He also observed that the finger-like processes (villi) of the intestinal mucous membranes extend and contract through this layer of mucus so that they clear their surfaces of adherent particles. When, however, he introduced bacteria known to cause enteritis these organisms invaded the mucous membrane. Thus they somehow avoided removal by mucus, probably by damaging the cells that produced it.

Florey's conclusion that the main antibacterial action of mucus was mechanical did not lessen his interest in lysozyme. Though he published nothing about it from Sheffield, his notebooks show that it remained a major research project. While the chemical nature and mode of action of lysozyme remained the immediate problem, Florey was interested in the possibility that the natural immunity of animals to certain infections might be due to antibacterial substances in their tissues that acted like lysozyme, but on pathogenic organisms. As he had done in Cambridge, he needed biochemical help in any such investigation, but there was no biochemist on his staff, nor did his departmental budget allow him to employ one. In May 1932, he wrote to Sir Walter Fletcher at the Medical Research Council[18] outlining his proposed research and asking for a grant to pay a biochemist. In June he sent a detailed memorandum, in which he described specific examples of natural immunity (such as the resistance of pigeons to infection by pneumococci, and of cats to tuberculosis) and the stages by which he proposed to study the possible involvement of antibacterial substances. 'To start with,' he wrote, 'an attempt would be made to extract from the tissues of the pigeon . . . something of enzymic nature capable of killing pneumococci *in vitro*.'

Fletcher asked for details about the cost of employing a biochemist, and Florey replied, in July, that he had no particular person in mind, but thought that a salary of £400–500 per annum would be needed. This seems to have silenced Fletcher until December, when in reply to further prompting from Florey he wrote: 'It would be difficult or impossible to provide . . . anything like £400–500. Is there no biochemist in Sheffield who might co-operate with you?' He went on to state his view that biochemists and pathologists ought to remain in their proper departments, though free to collaborate with each other. 'Can you not get effective working relations with your sister department in Sheffield?'

After this letter, Florey gave up his attempts to get a full-time biochemist, though he followed, as best he could, the advice it contained. Properly speaking there was no 'sister department' in Sheffield to which he could turn, since biochemistry then formed part of the Pharmacology Department, where Mellanby kept his biochemist, Dr. D. C. Harrison, fully occupied. But Harrison's wife Sylvia was also a biochemist and she was prepared to work with Florey on a part-time basis. There is no record of a grant to cover her salary, but she was accommodated in one of the 'cowstalls' in Florey's research laboratory, and they started on the lysozyme project early in 1933. Unfortunately their work could be advanced only by biochemical techniques that were either not then available or beyond the capacity of Mrs. Harrison. During the next two years no real progress was made and, as with the work with Marjory Stephenson, nothing was published. Florey was well aware that this lack of progress reflected his own inability to energize the biochemical side of the research, but in that field there was nothing that he could usefully do. And when Sylvia Harrison's husband became Professor of Biochemistry in Belfast, her collaboration with Florey ceased.

According to Dr. Pullinger, Florey corresponded with Fleming about lysozyme, but did not meet him to discuss a research topic that was of interest to both of them. Fleming chose lysozyme as the subject of his Presidential Address to the Section of Pathology at the Royal Society of Medicine in 1932. He seemed to have lost all interest in penicillin. A year before he had given a paper, also at the Royal Society of Medicine,[19] on 'The indications for and value of, the intravenous injection of germicides'. He used Almroth Wright's definition: 'a germicide is a substance which will enter into destructive combination with a microbe'. He dismissed most of the chemicals that had been used as more harmful than useful, except the arsenical compounds like Salvarsan, and predicted that mercurial compounds offered the best hope. He did not mention penicillin (or lysozyme) which certainly came within his definition. The omission is highly significant in view of later claims that he had always expected that penicillin would prove to be the 'miracle drug' it later became.

Curiously enough, attempts to use penicillin were being made in 1932, not in Fleming's laboratory, but in Florey's Department at Sheffield. Dr. C. G. Paine, the Pathologist at the Royal Infirmary, who had been a student at St. Mary's, had a culture of Fleming's *Penicillium* and proceeded to grow it and harvest the 'mould juice'. He then used it as a surface application in three cases of chronic staphylococcal infection of the skin. But, though he showed that the bacteria concerned were sensitive to penicillin, the treatment was ineffective, even though it was applied every four hours for a week. Paine then used his crude penicillin to irrigate infected eyes. Here he was much more successful. Two cases of gonococcal ophthalmitis in new-born children were cured in two to three days, and also one case of staphylococcal conjunctivitis. Finally, Paine achieved something of a clinical triumph. A colliery manager had a penetrating injury of his right eye. It became infected by pneumococci, and an operation to remove the foreign body could not be done. The eye was irrigated with Paine's solution, the infection cleared up in two days, so that an operation became possible, and the eye was saved.[20] But circumstances conspired to halt Paine's research. His penicillin culture began to lose its activity. Then, after the death of Miss Kirk, he was transferred from the Royal Infirmary to the Jessop Hospital and began to work on the streptococci responsible for puerperal fever. Like Fleming, he did not consider the possibility of treating this massive type of infection with penicillin, and he was now more interested in other new methods. Leonard Colebrook was beginning to get encouraging results with chemical derivatives of arsenic. The idea of chemotherapy, so long confined to Salvarsan, was beginning to extend, and the first glimmerings of light foretold the dawn of a new era.

In this changing atmosphere, any flicker of interest in penicillin faded. Florey, though certainly aware of Paine's tentative clinical experiments with his home-made penicillin at

Sheffield, since he talked about this work quite freely in the department, was not particularly interested. The results were not dramatic. The crude preparation was effective only when in direct contact with the most superficial infections, such as conjunctivitis. There was little to be had, and soon even this lost its activity. Paine's small project died, and it seems that at the time Florey saw no reason to regard penicillin as a rewarding research project. His interest remained fixed on lysozyme. Any biochemical help that he could get would be for work on that, and an even more difficult and problematical venture was never seriously considered. This attitude was, in truth, probably justified. The little biochemical collaboration then available to Florey was unlikely to succeed where Raistrick had failed.

Lysozyme, moreover, claimed only a part of Florey's research energies. He began to involve some of his staff in projects related to his other interests. Dr. Pullinger, with her experience of cancer research and lymph gland histology, was the obvious collaborator in his work on the structure and function of the lymphatic vessels. They began a joint study that lasted for several years, out of which grew Florey's interest in the lymphocyte, a then mysterious member of the white blood cell population. It had, at that time, no known function and its life-history seemed inexplicable. It posed the sort of problem that Florey enjoyed because it could be tackled by the invention of new techniques. This research, as will be mentioned, was to lead out into a wide field and to discoveries in Florey's department that have changed the whole subject of immunology. The answer to the riddle of the lymphocyte was one of these discoveries: a beautifully simple answer to a problem that had eluded haematologists for decades.

With Harding, Florey extended his work on gastro-intestinal secretion. They studied the mucus-producing glands of the duodenum (Brunner's glands). These secrete an alkaline mucus, which might be important in neutralizing the acid secreted by the stomach. Florey and Harding carried out a long series of experiments in five different animal species using the surgical techniques already perfected by Florey. They created fistulae so that the duodenal secretions could be collected and the mucosal surface studied *in situ*. They also transplanted sections of the duodenum under the skin. These sections were isolated from their original nerve and blood supply, but they continued to function. By these techniques it was shown that the flow of alkaline mucus from the glands is stimulated by the entry of food into the stomach. This effect occurred even in the isolated sections where it could not be caused by the duodenal nerves or blood supply, and the conclusion was that it was due to a hormone released into the general blood-stream. A second stimulus that promoted the flow of alkaline mucus from Brunner's glands was the application of acid to the mucosal surface. This is probably a protective mechanism normally preventing damage to the duodenal surface by the acid gastric contents. It raised the possibility that the occurrence of duodenal ulcers might be due to a failure of this mechanism. Florey and Harding therefore studied the healing of artificial injuries in the duodenum and showed that an absence of mucus delayed it, and might even cause ulceration.

This proved to be another of the many fruitful partnerships in which Florey took the lead and yet his collaborator found himself doing the best research of his career. Harding, describing his work at that time with Florey, writes:

As a researcher he had more ideas (and better ones) than anyone else I have met. Almost all the work we did together (and some that I published alone) stemmed from ideas from Florey. With this went a capacity for hard work—one project involved us spending the nights as well as the days in the University. We did alternate nights for several weeks.[21]

Like the juggler who can keep half a dozen balls in the air, Florey could keep five or six different lines of research in motion without losing track of his objectives. He had not, for example, lost track of his work with Fildes on the experimental treatment of tetanus. He had learned a good deal about the disease and the technical methods involved in its study. The death from tetanus of his brother-in-law, Dr. Gardner (Hilda's husband), in 1929 had given him a personal interest, and in September 1932 he took the problem up again, with a new idea. The usual cause of death in tetanus is exhaustion caused by repeated and very painful muscular spasms. If these could be prevented for a few days, the toxin might be naturally eliminated during the respite. A possible method would be the pharmacological blocking of the nerve impulses that caused the spasms. This might be achieved by deep general anaesthesia, or it might be done with curare—the 'Indian arrow poison'—which causes paralysis by interfering with the chemical contact between motor nerves and their muscle fibres. But in order to prevent the fatal spasms it would be necessary to produce a paralysis so complete that natural breathing would stop.

Florey's new idea was to combine such paralysis with mechanical artificial respiration. The use of curare in tetanus had been previously suggested by other workers, but was then impractical. It was a drug familiar to physiologists, who had used it to study nerve–muscle mechanisms since its introduction by Claude Bernard in 1844. It was the invention of a mechanical respirator for human beings (the so-called 'iron lung') by Drinker and Shaw in 1929 that gave Florey his idea. Its application to tetanus involved a great deal of groundwork and invention beginning in September 1932. Rabbits were chosen as the experimental animals because of their susceptibility to tetanus toxin. Florey had to study the action of curare on rabbits, therefore, and he visited Oxford to do this in Sherrington's laboratory (and with his help) with curare specially supplied by Sir Henry Dale.[22] Then he had to devise a mechanical respirator for rabbits. This was made by S. W. Bush, the supremely skilled and ingenious mechanic in charge of the workshops of the School of Pathology, Oxford.

When these preliminaries were complete, Florey was ready to begin his animal work. Fildes made several visits to Sheffield, and Harding took part in experiments that needed constant supervision for days and nights together. With his usual thoroughness Florey insisted that the techniques for keeping animals alive under prolonged anaesthesia or curare should be perfected before beginning the work with tetanus toxin. These techniques proved difficult. The breathing apparatus worked well and muscular paralysis could be maintained for long periods, but the animals tended to inhale saliva and mucus and develop pneumonia. When the actual experiments with tetanus started, these complications had not been overcome. After nearly two years of experiments Florey, Harding and Fildes published their results in November 1934 in *The Lancet*.[23]

These results were not dramatic. Though several animals had survived what would have been lethal doses of tetanus toxin, they had eventually died of complications. But the authors stressed that the principle employed would give the best chances of recovery in human cases and suggested that special centres should be set up where it could be used. So this paper pointed the way to the modern and often highly effective management of human tetanus. This is based on drugs of the curare type, combined with artificial respiration. What has made this effective is the improvement in anaesthetic technique and artificial ventilation and, of course, the use of antibiotics that prevent the pneumonia that killed Florey's experimental animals. But it was not until twenty years had passed that these practical applications could be made. Florey himself did not follow this research any further. He recognized that it would lead him deep into subjects that were far removed from his main lines of interest.

The other line that he followed at Sheffield was the study of the physiological basis for contraception. He went on steadily with his work with Harry Carleton, tracing the exact mechanisms by which the sperm meets and fertilizes the ovum in the uterus. In conjunction with these studies he investigated a number of spermicidal substances, both for their effectiveness in preventing fertilization and for any possible harmful action on the tissues.

On the domestic side, life for the Floreys in Sheffield was not ideal. Howard was less worried about his family in Australia since, after her husband's death, Hilda Gardner had offered to share her home in Melbourne with his mother, Charlotte, and Valetta and they had been settled there since 1930. But his own home life was unhappy. Ethel's health was still poor. In the summer of 1933 she had developed infectious hepatitis (jaundice) after a family holiday in Spain, and was still unwell in January 1934 when Dr. Beatrice Pullinger arrived in Sheffield. Dr. Pullinger soon became friendly with Ethel and was inclined to take her part in the constant bickerings that went on in the Florey household. Ethel's second pregnancy began early in 1934. Possibly because of the after-effects of her jaundice, she developed sickness to an extent that made her exhausted and miserable. Her baby son, Charles, was born on 11 September 1934. Howard's natural pleasure in this event was marred by anxiety for the child, who from birth suffered from repeated vomiting. This was diagnosed by Professor Naish as pyloric stenosis, a condition usually then treated by immediate operation. Since, in Charles's case, no operation was done, the condition must have been a relatively mild one. Howard had a rather strong aversion to ill-health, both in himself and in other people, and he resented the fact that Charles was not robust during his childhood.

Howard was also out of sympathy with Ethel. They became, in fact, so estranged that they communicated by means of notes left on the hall table. Finally Ethel drew up a memorandum setting out all the complaints and grievances that were poisoning their relationship. This has not survived, but Howard's reply to it, also in the form of a memorandum, was kept by Ethel, as she had kept all his letters. It runs to fifteen closely written pages, answering each of her complaints in detail and then setting out a list of his own. It is indeed a sad revelation of the disappointed or, rather, baseless hopes from which before their marriage he had built his fantasy of a perfect partnership. Most of their complaints seem quite trivial, minor irritations that in a happy marriage would be easily resolved or cheerfully accepted. But, for Ethel and Howard, arguments about money, the housekeeping, the children, mealtimes, the expenses of the car, Howard's liking for the gramophone and radio, and their mutual dislike of each other's minor physical defects assumed intolerable proportions. It seems that almost every grievance throughout their eight years of marriage had been remembered, resented, and now recorded in their memoranda. There was, however, one serious and apparently insurmountable difficulty between them: Ethel's deafness, a source of exasperating failure of communication for both of them. But Howard was genuinely conciliatory towards the end of his memorandum. In hers, Ethel had suggested a legal separation. He replied:

From the children's point of view, this would be lamentable. The children are now a most important link, and I want you to try to live with me, and let us both strive to rectify our mistakes as well as we may. Although neither of us have got all we expected from our marriage, there is still a good deal left which, with a little good will, we can make more. Let us try ... Let us both resolve to do our utmost to heal the breach which has grown up and I am sure we will be successful. Now that we know some at least of the reasons for one another's discontent, it may be easier to be forgiving and charitable.

10

THE RETURN TO OXFORD

Florey's appointment at Sheffield was, from the professional point of view, a fruitful and happy one. He had done a great deal of personal research, publishing twelve papers between 1932 and 1935, and carrying out work on lysozyme, lymphatics, and the physiology of reproduction that was not published. He had revitalized his department, infusing an interest in research and teaching that had been sadly lacking. He had also, not without abrasion, raised the status of pathology in the hospital hierarchy. Clinicians had come to realize that pathologists were not pieces of animated laboratory equipment, but medical scientists to be respected and consulted. It is to Florey's credit that he achieved this minor revolution with little lasting acrimony. His clinical colleagues could hardly fail to respect him, and they also came to like him. Plain speaking is claimed as a Yorkshire characteristic, which soon influences even academic incomers. Florey's Australian directness was better understood in Sheffield than in Cambridge or Oxford. He felt this, and appreciated it. 'They treated me well—very nice people in Yorkshire,' he said, and from him this was high praise.[1]

Despite his success there, Florey had no intention of staying indefinitely in Sheffield. Most of his predecessors in the Chair of Pathology had used it as a step-ladder. The first, Louis Cobbett, appointed in 1906, left for Cambridge a year later. J. M. Beattie stayed for five years, to be succeeded by H. R. Dean who, after three years in Sheffield, became Professor of Pathology at Manchester and then at Cambridge. Florey's immediate predecessor, J. S. C. Douglas, remained for fifteen years and died in office. Florey had no ambition to follow his example. He was determined to make his mark as a scientist, and was well on the way to doing so. It could, of course, be done in Sheffield, but other and larger universities would offer better chances. Harding describes his impressions of Florey's attitude.

He and I got on very well—I liked him and I think he liked me. He had quite a sense of fun, even if it was more often hidden beneath a serious manner. He was self-centred and I think he always knew where he was going. Quite early in our acquaintance I accused him, not very seriously, of working for an F.R.S. and he let it pass with a smile. I don't know when Oxford first came into his sights, but I fancy it was some years before he went there. This, perhaps, sounds disparaging, but is not meant in that way. On the whole I think it praiseworthy: he got where he did because he knew where he was going.[2]

Beatrice Pullinger confirms this ambition to return to Oxford. 'Florey told me in Sheffield that the Oxford chair was what he really wanted but the opportunity came unexpectedly early, in 1934.'[3]

The opportunity was created by the death of Professor Georges Dreyer at the age of sixty-one.[4] Dreyer's standing as a scientist should not be judged from his last few years in office. Throughout most of his career he had shown a vigour and technical versatility not unlike Florey's own. He was born in Shanghai in 1873, the son of an officer in the Danish Navy, educated in Copenhagen, and gained his M.D. in 1900. During his studies he made friends with several Oxford medical graduates who were visiting or working in Copenhagen, including C. G. Douglas, H. M. Turnbull, and H. R. Dean. As a result, Dreyer himself came to Oxford to study with Sir John Burdon-Sanderson. After a period at the State Serum Institute in Copenhagen he visited Oxford again in 1905 to read a paper at the meeting of the British Medical Association. He made such a good impression that he was persuaded by his Oxford friends to apply for the Chair of Pathology when it was created there in 1907.

Pathology in Oxford (as in Cambridge) had started as an academic offshoot from physiology, with no responsibility for clinical work. There was a new building and a nominal professor: James Ritchie, who held the title by courtesy. There was also Dr. Ainley Walker,[5] who had the distinction, in 1903, of being the first medical Fellow to be appointed in Oxford (at University College). When Ritchie left for Edinburgh, Ainley Walker acted as head of the department until the new professor was appointed.

Dreyer was a good linguist, with enough command of

English to give his lectures with the occasional assistance of
Ainley Walker, who used to sit in the front row and interpret
the professor's more obscure constructions. He was indeed
a faithful friend and sincere admirer of Dreyer, despite the
fact that he had himself hoped for the professorship. Dreyer
was by training an immunologist and by nature an experimentalist. He had inherited what was virtually a new subject in
Oxford, and he gave it a life and form that placed it firmly
among the established departments of medical science. This
was a period of fruition—the outcome of fifty years of effort
by Acland and Burdon-Sanderson. Osler, like a magnet, was
drawing pupils and disciples; Sherrington was creating the
best school of physiology in the world; Arthur Thomson was
giving the study of anatomy a breadth that released it from
centuries of rigidity, and James Gunn, though confined to the
attic of the original museum, was helping to create experimental pharmacology. Dreyer, in his own field, was of equal
rank.

Research was Dreyer's main interest. He worked particularly on the formation of the antibodies that clump (aggluti-
nate) infecting bacteria. The presence of such antibodies in
the patient's blood is an important diagnostic sign. Dreyer's
methods for demonstrating them became standard and the
'Dreyer tube' familiar to every bacteriologist and medical
student. He devised the equally well-known 'sigma reaction'
for the diagnosis of syphilis. But he also worked on blood
volume measurement, and on the capacity of the blood to
absorb oxygen—thus overlapping the interest of Haldane and
Douglas. During the First World War he volunteered to go
to the British Military Bases at Boulogne to modernize their
antiquated methods of bacteriology, and Ainley Walker was
left, once more, in charge of the department.

Dreyer was twice mentioned in dispatches and awarded
a C.B.E. for his military services. He was elected a Fellow
of the Royal Society in 1921. A year or so later one of his
research projects failed amid a good deal of unfortunate publicity. This was the 'diaplyte' vaccine that would, it was
hoped, prevent human tuberculosis. It had worked well in
animals, and an unauthorized leakage to the press aroused a
great deal of interest and anticipation. Dreyer was forced to

make a premature publication and to release vaccine in response to urgent clinical demands. The Medical Research Council carried out a controlled trial, which showed the vaccine to be quite ineffective for human use. Dreyer, it is said, bore the consequent discredit without complaint, but he was deeply humiliated, and his research interests declined from that time onwards.

But he retained his ambitions to improve his school of pathology. He and his assistants were well equipped to teach bacteriology and immunology; but for morbid anatomy, histology, and clinical pathology he had to rely on Dr. A. G. Gibson, a physician at the Radcliffe Infirmary who was also a pathologist. Gibson carried out the post-mortem examinations at the hospital, and gave Dreyer's students excellent and well-remembered demonstrations at 8 a.m. every day. One of Dreyer's bacteriologists, Roy Vollum, a Canadian Rhodes Scholar in 1921, had an interest in clinical bacteriology and did some hospital work. Since Dreyer's own research had been largely along clinical lines, there were thus real links between his department and the hospital, besides his nominal title of Honorary Pathologist there. But he had visions of a much larger academic department, one that would compare with Sherrington's.

In 1922, the Sir William Dunn Trustees, prompted by Dreyer, offered Oxford £100,000 to build a new Pathology Department. With their unvarying persistence the antivivisectionists tried to prevent the passing of the Decree in Congregation accepting the money. Dreyer was known for his work on syphilis, and one of the College Fellows spoke earnestly of the dangers of allowing him to keep his animals in a building adjoining the University Parks. Syphilitic goats, he predicted, would escape and infect innocent members of the University during their recreations.[6] His fears were not taken seriously—indeed, less than seriously by the scientists—and the Decree was passed.

The immediate post-war years had favoured the science departments in Oxford. A Chair of Biochemistry was endowed in 1920, though its first holder, Benjamin Moore, died within two years from the effects of his own research on T.N.T. poisoning. He was succeeded by R. A. Peters, and, soon after, the Rockefeller Foundation gave £75,000 towards a new Department. Dreyer also, with what was then a large sum to spend on a new building (an average council house cost £400 to build in 1922; in 1977 it cost £12,000), threw all his energy and talents into its planning. He acquired for it about two acres of land on the edge of the Parks at the eastern end of South Parks Road. It was a site several hundred yards away from his existing department (which would in due course house the Department of Pharmacology). He was thus able to plan a spacious and self-contained department, standing back from the road in its own grounds. In the words of one of his biographers the resulting building was 'in every respect a model of what a pathological department of teaching and research should be. Over its planning he spent infinite care and thought, illuminated by his wide knowledge of Continental and American laboratories, and the furnishings and fittings worked out to scale on paper with minute accuracy and attention to the smallest detail.'[7]

The foundation stone was laid in 1923; Dreyer inspected every stage of the building, climbing ladders and scaffolding, and he constantly sought the advice of Vollum, Gibson, Ainley Walker, and A. D. Gardner, who had worked with him during and after the war. The laboratory was finished by Christmas 1926, and formally opened on 11 March 1927. The ceremony was suitably rounded off by a banquet in All Souls College.

Dreyer had good cause to be proud of his achievement. It was (and still is) not only the handsomest of the Oxford science buildings but the most efficient, since its spacious layout and imaginative planning allowed it to function with perfect adequacy and with no additions or major alterations for the next forty years. The main block is in the Georgian style on three floors. The lowest is a semi-basement, containing most of the service rooms. The main entrance is on the first floor, reached by a curving *escalier d'honneur* in the grand manner, and the hall is paved and faced with marble, with a magnificent oak staircase. There is a wide central corridor on each floor. The western end of the building contains the teaching accommodation: a lecture theatre, museum, a classroom for practical work, and preparation rooms. At the other end are the research laboratories, including the professor's suite on the second floor. This suite consisted of a large laboratory, a study with a Pullman bed behind the panelling, and a bathroom, which allowed the professor to spend his time on extended experiments in comfort. All the furnishings and fittings of the building were made to last, and have done so. Benches, doors, floors, and most other woodwork were of teak. The floor coverings of heavy-duty rubber have stood up to many years of the severe treatment inevitable in a busy laboratory with many generations of students. Behind the main block a covered way leads to a large two-storeyed animal house, complete with post-mortem room and a flat for the resident chief technician. The grounds are laid out as a garden in front of the building. The space behind, which used to grow fresh vegetables for the animals, has now become—inevitably—a car park and the site of a new building and workshops.

Dreyer's second ambition was the academic upgrading of pathology to the status of a Final Honour School subject. But here the difficulties were more subtle. There was already a Final Honour School of Physiology in Oxford and the establishment of another school in a closely related subject (for Dreyer was, like Florey, a physiological pathologist) did not appeal to the University authorities. Dean in Cambridge fought and won a very similar battle; but Dreyer, against stronger opposition, gave up the struggle.

With the completion of the Sir William Dunn School of Pathology, Dreyer's energies failed, as if its creation had exhausted them. The building, a monument to him after his death, became a mausoleum some years before it. The laboratories, rightly described as palatial, did not become filled by the eager young research workers that Dreyer had hoped to attract. Even the student population declined. Dreyer's lectures were so poorly attended that he would send a sort of 'whipper-in' to the physiology and pharmacology laboratories to round up stragglers who were supposed to be taking his course. In his deserted palace, Dreyer merely kept up the outward show of an almost regal authority. No one but himself (or his visitors) was allowed to use the main entrance, or cross the marble hall to climb the monumental staircase. Nor was anyone else allowed to use the passenger lift. Staff and students alike had to enter and leave by the basement and use one of the staircases at either end of the building.

Dreyer did not become a recluse. He attended meetings and social functions, and enjoyed entertaining. The Dunn School became particularly popular for the meetings of the Pathological Society, largely because at his own expense he provided free beer and cider by the barrel. Nor was the Dunn School entirely deserted. One of Dreyer's important and practical contributions to clinical bacteriology was the establishment of a laboratory to select and keep pure cultures of certain pathogenic bacteria for reference purposes. These 'standard' cultures were used to check the antisera employed in other laboratories to identify unknown organisms and to provide known organisms to help to identify antibodies in the patient's blood. The 'Standards Laboratory' thus became a national institution, financed by the Medical Research Council, but it remained in the School of Pathology. Dr. A. D. Gardner was its Director, with a staff of three assistants.

As his research interest and his energy flagged, Dreyer spent less and less time in his Department. Florey, on one of his visits from Sheffield, went to the Dunn School to see the Professor. Dreyer's secretary told him to come back after twelve but before one o'clock, as the Professor usually 'looked in' between those times, but seldom stayed more than a few minutes. The object of Florey's visit is not recorded, but it was possibly to ask for the collaboration of Bush, Dreyer's instrument maker. The splendidly equipped mechanical workshop under Bush's direction was, at that time, perhaps the busiest section of the Department, being largely engaged on work for other laboratories. There seem to have been few other contacts between Florey and Dreyer, despite obvious common interests. One of the obituary tributes to Dreyer praised his operative technique. 'He would quickly and cleanly insert the tiniest cannulae into vessels which appeared almost thread-like.' And it continues:

He had an utter contempt for bad experiments. Having thought long and deeply over the problem which he proposed to attack, he then took the utmost pains and devoted much time to devising a technique which should yield him the most accurate possible quantitative results. Dreyer did not suffer fools gladly. He never concealed his impatience of stupidity and of that kind of obsession with preconceived ideas which refuses to consider dispassionately new evidence and new points of view. In his earlier years he was undoubtedly combative and rather enjoyed trailing his coat.[7]

How easily these words might be taken to refer to Florey! Dreyer certainly knew and admired Florey's work, and perhaps it was his recognition in him of these qualities of his own youth that made him favour Florey as his eventual successor. There is one significant difference. Dreyer's obituary notice concludes: 'But his chief pleasure was found in his ideally happy home life.' It was this pleasure perhaps, and that of convivial occasions, that competed more and more with his research interests and official duties during his last years. Florey had no such distraction.

Georges Dreyer died suddenly during the Long Vacation of 1934, while on a visit to his native Denmark. As on all such occasions, the sense of loss did not prevent a lively interest in the question of a successor. It was a time of change among the heads of the Oxford science departments. Sherrington, then aged seventy-six, was about to retire as Professor of Physiology. Arthur Thomson, who had started his Oxford career as Lecturer in the tin shed adjoining the Museum in 1884, had just retired from the Chair of Anatomy, and Professor Gunn had resigned the Chair of Pharmacology to become the first Director of a new establishment, the Nuffield Institute for Medical Research, the first result of a mounting series of medical benefactions by Lord Nuffield. There was speculation on the direction the new men would take, and much jockeying for position among the possible candidates.

Though Florey had doubts about leaving Sheffield for Guy's, he had none about Oxford. It was, for him, the ideal job in ideal and familiar surroundings, and he set about to make the best of his chances of getting it. As might be expected, his main supporter was Sherrington. The moves that should be made in any Oxford election are well known to the elder statesmen, and Sherrington was a most valuable ally. Oxford Statutory Chairs are assigned to particular colleges, that is, their occupants are automatically Fellows of these colleges. When a new professor has to be chosen the University appoints a special Electoral Board to make the choice. The Head of the College to which the Chair is assigned is *ex officio* a member of this Board and the College can also appoint another representative and thus have two votes in the election. In the case of Pathology, the College is Lincoln.

On 3 October 1934, Sherrington wrote to Florey[8] saying that he would see the Rector of Lincoln and Sidgwick, one of the Fellows and a famous chemist, and he advised Florey to ask Kettle (the eminent London pathologist) to approach Muir. With the Rector of Lincoln, Sherrington could exert a double influence. He could press Florey's claims directly and also suggest as the second College Elector someone he knew to be in favour of Florey. As it turned out, this tactic proved decisive. His advice on obtaining Professor Kettle's support was astute. The greatest opposition to Florey's election was likely to come from Sir Robert Muir, who would be (as Sherrington must have known) one of the Electors. Though Kettle would not be an Elector he would have great influence on Muir. Kettle was both a morbid anatomist and a brilliant experimentalist—a rare combination then. As an experimentalist, Kettle knew and admired Florey's work (he was also a friend of Dr. Pullinger) and, as a distinguished morbid anatomist, his opinion would be respected by Muir.

On 26 October Sherrington wrote again, telling Florey that Lincoln College had appointed Edward Mellanby as their electoral representative. Mellanby, then Secretary of the Medical Research Council, had become a man of great influence; he had a liking and respect for Florey, based on personal knowledge and collaboration. But Sherrington himself, as he told Florey, would not be on the Board. His place would be taken by the Regius Professor of Medicine (Sir Farquahar Buzzard) who, Sherrington thought, would favour the appointment of Matthew Stewart, Professor of Pathology at Leeds and a morbid anatomist of the old school. No doubt as a counter-move, Sherrington proposed to discuss the matter with Dr. A. G. Gibson. Gibson then had the title of Demonstrator of Pathology in the Dunn School of Pathology and Assistant Pathologist to the hospital. New buildings at the Radcliffe Infirmary had included pathology laboratories and a post-mortem room, which had been designed by Dreyer to the same high standard that he had applied to the Dunn

School. Gibson thus virtually had a clinical pathology department of his own, and would not be over-pleased by the appointment of a Professor who might take over its morbid anatomy service. And Gibson's views might be expected to influence the Regius Professor. One may suppose that the other candidates had their own protagonists, who were also engaged in such calculated moves within the University.

Florey sent in his application for the Chair of Pathology in December. It is a printed document setting out his career, experience, and achievements. Dr. Pullinger writes: 'The general opinion was that Matthew Stewart (Leeds) would get it. I remember great excitement in putting Florey's application together, getting it off in time and before it might be delayed by the Christmas mail.'[9] Florey also sent copies of testimonials which spoke of his work and capabilities in the most glowing terms.[10] Sherrington stressed his gift for original research, and his ability to draw the interest of active collaborators: 'In his own special line he has done things of more promise than any which preceded them for years.'

Dean wrote: 'I recognised from his first coming his exceptional originality and ability as a research worker. He was full of ideas and could interest his juniors and colleagues in his research.'

E. H. Kettle (then Professor of Pathology at the newly created British Postgraduate Medical School) stressed Florey's force of character and ability to inspire his associates: 'It is no exaggeration to say that those who know his work regard him as the leading representative of this branch of pathology in this country.'

Hall, the Emeritus Professor of Medicine and Dean at the Sheffield Medical School, referred to Florey's administrative achievements: 'He has done a great deal to bring the Hospital Pathology laboratories into closer relation to the University and the teaching of pathology. Before his time, this relation was minimal.'

Finally, A. E. Barnes, Senior Physician at the Royal Infirmary, Sheffield, took a more specific line: 'I cannot refrain from expressing my admiration for the systematic experimental work which, at long last, is being done upon that same mucus which, for a thousand years, our uncritical profession has been engaged daily in washing away from every accessible surface and cavity of the human body.'

Even allowing for the complimentary bias that is inevitable in an open testimonial, these tributes to Florey are remarkable in their unanimity and enthusiasm.

The Electoral Board met on 22 January 1935. There were five members from the University: the Vice-Chancellor, the Regius Professor of Medicine, Dr. A. G. Gibson, Dr. E. W. Ainley Walker, and the Rector of Lincoln. There were also Sir Robert Muir and Sir Edward Mellanby. When the discussion was over, Douglas Veale, the University Registrar, wrote to Florey as follows: 'I have the pleasure to inform you that at a meeting of the Electors held here today, you were appointed Professor of Pathology, as from 1 May 1935.'

The decision was made, but it was one in which chance played an important role. There is no available official record of what went on at that meeting, but verbal accounts agreed on the essentials. The Board had to begin without Mellanby, who was unaccountably absent. The various merits and demerits of the candidates, some of whom were obviously men of considerable distinction, were discussed. The choice narrowed to two, Matthew Stewart and Florey. One may suppose that Gibson and Ainley Walker would favour Florey because of his pre-eminence as an experimental pathologist. But the Regius Professor and Sir Robert Muir wanted Matthew Stewart—a wholly admirable man, predictable, orthodox, sound, and absolutely devoted to pathology (a devotion demonstrated by his editorship of the *Journal of Pathology and Bacteriology* over a period of thirty-one years). Why gamble? The Regius had the authority of his position, Sir Robert the weight of his great reputation, and his conviction that the best pathologists are necessarily Scottish morbid anatomists. At all events they seemed to persuade a majority, and it was decided that Matthew Stewart should be appointed. Then, just before the meeting was officially concluded, Mellanby arrived. His train from London had broken down and he was two hours late, but not too late to re-open the discussion.[9] Mellanby was a most formidable man. As Secretary of the Medical Research Council he had great power and he knew more about research workers and had stronger views on the future of medical research than anyone else in Britain. He had a positive personality (which later became overbearing) and a coldly determined way of demolishing opposition. He had no doubt that Florey should have the Chair, and gave his reasons. The other Electors, who cannot have been unanimous in their previous choice, revoked it and voted for Florey. Mellanby had won the day, but as the Duke of Wellington said of another momentous victory, 'It was the nearest run thing.' It is perhaps fanciful to compare Mellanby with Blücher, but if either of them had been a little later in arriving, the consequences for mankind might have been equally vast and incalculable.

Florey was naturally elated by his appointment. In reply to congratulations from Allen, the new Warden of Rhodes House, he wrote: 'Thank you very much for your telegram. I am very fortunate to have got this magnificent job with all its opportunities. I am very greatly honoured. We are looking forward to coming to Oxford, which, as you may well understand, I like the best of all the English places I have lived in.'[11] And Ethel in a letter to Howard's sister, Valetta, wrote:

I am sure you will all be delighted with Floss's latest success. The Oxford Chair has been his Mecca for many years and he was buoyed up in his hopes of getting it by the last man's telling him he was the man he wanted to succeed him. Then, unfortunately, he died, three or four years before he was due to retire, so that Floss was very young for the post and also he had one less friend to back his claims. He had almost given up hope of being elected when he heard who were on the Selection Committee, as he felt for some reason or other, that all but one would be against him, or simply neutral. But then he always does think the world has chosen him in particular for its enemy.[12]

Florey had enjoyed his time in Sheffield, and he had learned from it at first hand the conflicts of interest that arise between academic and clinical pathology. They were lessons that he took to heart, and which were to determine his attitude to similar problems in Oxford. His Sheffield colleagues, with a few exceptions, liked him but no lasting friendships resulted, apart from those with Mellanby and Beatrice Pullinger. There seem to have been no close social relations between the Floreys and the other University families. Harding, with whom Florey had collaborated more closely than anyone else in Sheffield, had never been invited into his house, in spite

of the fact that the Hardings lived quite close to it.[13] Dr. Pullinger seems to have been a fairly frequent visitor to Ethel,
and considered that the Floreys' failure to entertain was due
to Ethel's poor health. Her deafness made it impossible for
her to be an easy hostess. There was none of the open-house
atmosphere that Florey had so much appreciated with the
Sherringtons in Oxford. So when the time came to leave Sheffield there were no personal roots to be pulled up for either
of them, particularly since three people who were important
in their lives—Beatrice Pullinger, Jim Kent, and Doris
Wheeler—would be coming with them to Oxford.

The professor's salary in the Chair of Pathology was £1700
per annum—a substantial increase over Sheffield, and Ethel's
financial fears should be even less justified. There had been
difficulty in selling the house in Cambridge, largely because
Ethel could not bring herself to lose money on it (she had
been deeply upset because her parents, when they sold 'The
Red House' in Adelaide had done so at a loss), and Howard had
agreed that half the proceeds were hers. But, with the better
financial prospects at Oxford, she resigned herself, and the
house was sold for £875, a little less than the original mortgage.

The rented house, No. 16 Parks Road, to which they moved
in Oxford was, curiously enough also called 'The Red House'.
It stood near the junction of Keble Road with Parks Road,
then a quiet backwater. It was one of a row of the Victorian
houses that had sprung up in hundreds in North Oxford after
the 'act of emancipation' of 1877 allowing College Fellows to
marry and become non-resident. It was not so large as many
of these, and much smaller than the Sheffield house. But it
was quite big enough, and more convenient. Like 'The Red
House' in Adelaide, it faced an open park (though it lacked
the magnificent view). Howard could reach the Dunn School
from his new home in a ten-minute walk through the Parks
or along the private roads of what were still called the
'Museum Departments'. The Museum itself, the new Radcliffe Science Library, and nine or ten science departments
occupied a roughly square area in the angle between Parks
Road and South Parks Road. The Dunn School, in rather
splendid isolation, was a few hundred yards further to the east.

Ethel moved into 'The Red House' with Paquita, Charles

(then six months old), Doris, the children's nannie, and Phyllis, the golden retriever. Charles, incidentally, was called 'Egg' by Howard (from prenatal days). By a metamorphosis common in the Florey family, 'Egg' became 'Egbert' and finally 'Bertie'. By a similar process the dog, Phyllis, became 'Winnie' (being bear-like), then 'Pooh', and finally 'Poochie'. When she developed uterine trouble, Howard performed a hysterectomy and Kent gave the anaesthetic; never could a dog have had a more skilled surgical team. At 'The Red House', Ethel engaged two maids and a part-time gardener. Howard's involvement in domestic affairs lessened because of his Fellowship at Lincoln, which had prompted a typical first reaction: 'Can one get a decent meal there?'[14] Lincoln, though a little further from the Dunn School than his own home, became a refuge and a place for relaxation. He often lunched and dined and also entertained his guests there. Besides providing a 'decent meal', it was a small friendly college, prepared (like Caius in Cambridge) to accept scientists on equal terms with its arts Fellows, and extending all the amenities of a well-run club.

Florey took up his official post in Oxford on 1 May 1935, and his coming was viewed with some trepidation. His reputation for outspoken toughness had spread from Cambridge and Sheffield, and the remaining staff, uneasily aware of their own slackness, expected drastic changes in the Department. Florey had received a good deal of advice on how he should bring it into working order. Harry Carleton stressed the need to 'get rid of Dreyer's left-overs' and make a clean sweep.[15] But others, including Florey's Cambridge friends, advised against the new broom policy and favoured caution and diplomacy. In the event he chose the gentler and certainly the wiser course. There were no immediate upheavals or dismissals. But, if Dr. Pullinger's account of the state into which the Department had declined is a fair picture, Florey's restraint must have been sorely tried.[9]

The Dunn School could (and, a few years later, did) quite comfortably accommodate a staff of thirty graduate research workers and teachers, twenty-five technicians, and classes of forty or fifty students. In May 1935, its inmates were only a small fraction of these numbers, and most were virtually unemployed. The acting Head of the Department (for the third time) was Ainley Walker. Dr. Gibson was almost wholly occupied with his pathology laboratories at the hospital and his work as a physician. R. L. Vollum, the Demonstrator of Bacteriology, also had clinical interests. B. G. Maegraith, an Australian Rhodes Scholar, had just taken his D.Phil. and been appointed as a Demonstrator. Miss Campbell-Renton and Miss Jean Orr-Ewing, who had been working with Dreyer, continued their researches, but only as a part-time occupation. The most active and efficient unit in the building was not part of the Dunn School organization at all. This was the Standards Laboratory which was financed by the Medical Research Council and paid a rent of about £600 per annum for the space and services it received. It was entirely autonomous under Gardner's direction. The Dunn School itself thus had a staff of six graduates, four of whom were part-time, and there were five technicians besides Bush and his mechanics. The student classes numbered ten or fifteen at most.

The morale among those who were left in the Dunn School was low. Dreyer's head technician had died before Florey's arrival and the man who replaced him was, in Dr. Pullinger's opinion, quite useless. The histology technician, who was largely responsible for preparing the teaching material, seems to have been little better. Slides were wrongly labelled, and the blocks from which they were cut so poorly prepared that the students could have learned little from them. Most of the microscopes were in poor condition, and some were not even complete, since lenses and other removable parts had been pilfered. Technicians could seldom be found when wanted, and they tended to resort to a common room in the basement where a barrel of beer was always available.

This situation was the result of a decline of interest and control on the part of the remaining graduate staff. Until a few years before, Dreyer, Gibson, and Ainley Walker had provided a good course of pathology in their old building. But this once-active interest lapsed at the Dunn School and with it the interest of students and technical staff alike. Gibson's activity had not declined; it was merely transferred to his laboratories at the Radcliffe Infirmary, and Ainley Walker, nearing retirement, was much occupied with college affairs

What remained of the teaching course became fossilized. The same lectures were given year after year, read from typescripts which were distributed to the students, and the practical classes followed the same pre-set ritual, made largely useless by poor material.

Into this limbo came Florey, Kent, and, three months later, Beatrice Pullinger, three people who could be guaranteed to revitalize the Department at every level. Florey, having sized up the situation, began the process with remarkable restraint. The head technician was not dismissed, but he was banished to the animal house, and a more efficient man promoted. Dr. Pullinger took the histology in hand and, once the technician concerned realized that his work was receiving expert interest, he began to take a pride in it and produced good results. Less time was spent in the basement common room, and the barrel of beer disappeared. But Florey made no move to displace the graduate staff he had inherited. Miss Campbell-Renton and Miss Orr-Ewing continued their research. Dr. Ainley Walker retired from the scene. R. L. Vollum for the first time for years found that his teaching abilities were appreciated and he became enthusiastic in the organization of the bacteriology section of a new course of Pathology.

The position of Dr. Gardner[16] was more problematical, and Florey's handling of the situation showed wisdom and restraint. Gardner had fully expected that, when the new professor was appointed, he and his 'Standards Laboratory' would have to move out—probably to London. He himself had been an applicant for the Chair, and it was inevitable that his disappointment should affect his relations with Florey. And the two men were poles apart in their temperaments and background. Gardner's natural courtesy had been strengthened by his upbringing in a distinguished English family, Rugby School, and University College, Oxford. He was cultured, diplomatic, and urbane. Florey must have seemed to him something of a barbarian, an Australian with more of the bush-ranging spirit than was usual even in his countrymen. And for Florey, Gardner had just those upper middle class English qualities of good manners that in his early letters to Ethel he had supposed to be insincere. There was little hope that either would really understand or be in sympathy with the other. But Florey made no attempt to remove Gardner from the Dunn School. In fact, he was active in keeping him, as Reader in Bacteriology.

Florey's reasons for this policy were sound, both politically and scientifically. Though he had every right to ask the Medical Research Council to remove their Unit, there was no immediate point in doing so. The space was not needed—the Dunn School was already half empty Nor was it wise to jeopardize good relations with Mellanby, who would hardly take a notice to quit as a token of gratutude for his support. Gardner, too, would resent being uprooted. He was very much an Oxford man, a Fellow of University College, with influential friends who would certainly take his side in any conflict. Florey, while never afraid of opposition, did not court it needlessly. There was no point in providing ammunition for those who considered him a dangerous radical. But perhaps the deciding issue was Florey's recognition of Gardner's qualities. He needed a good senior bacteriologist, and he needed a counsellor and ally, one experienced in the tortuous ways of the University. In Gardner he had both, an excellent bacteriologist and an astute Oxford diplomat. Florey was essentially practical and he did not allow personal feelings to override his judgement of a man's ability.

The pace of Florey's transformation of the Dunn School was limited by circumstances that he found irksome. After two and a half years, he wrote to Fulton:

My dear John,
    We are struggling along and I hope in the course of time to get a few decent people here. Things are very hectic with a lot of committees and so on. I am not sure we've done very much of permanent use. Perhaps it will come, but I'm not very optimistic on the research side.[17]

This is a letter typical of his talent for understatement, written at a time when he had attracted to his department several first class people, including two future Nobel Prize winners, whose lines of research were well under way, and when he had already created the best teaching course in Pathology in the country. The limiting factors were money and the slow processes of University administration.

In 1935 the total budget for the School of Pathology was

£3432 per annum (excluding the salary of the Professor) for all salaries, wages, and running costs—an absurdly small sum even in those days when money was at least ten times its 1977 value. Stringent economy was enforced by Florey. For example, the use of the lift, which cost about £25 a year to run, was forbidden. The office, which had been in charge of Miss Jorgensen (Dreyer's sister-in-law), was handed over to a new recruit, Miss P. J. Smart (later to become Mrs. Turner), and the accounts, instead of being left entirely in the hands of the University Chest, were checked and scrutinized. Mrs. Turner stayed to become the administrative genius of the Dunn School, retiring in 1976, with an Honorary Degree from the University, after forty years of service there.

Florey's mainstay in the Department during those first few years was Dr. Pullinger, who was appointed Demonstrator of Pathology at a salary of £350 per annum. She had no wish to remain in Sheffield and she was glad to accept Florey's offer of a position. She was an able and vigorous collaborator, not only in research, but in the almost Augean task of cleaning up the class methods and equipment. Gardner writes:

When Florey arrived, one of his first reforms was to institute a very advanced and splendidly planned course of Pathology and Bacteriology, including glimpses into the discoveries on which the teaching was based. This needed much expert preparation, and also more money, and it may well be that he was unable to screw it out of the University Chest continuously once the glamour of the new Professor had faded a bit. For my recollection is that the course had to be greatly reduced. I am, however, in no doubt that it was a superb effort which deserved far more active support than it got.[15]

Florey's research and teaching projects seemed destined to be always hampered by lack of money, and he became almost as concerned about it as did Ethel—with far less reason—in her domestic affairs. He recognized that the University was not obliged to pay for research, but felt that it should be responsible for the cost of teaching. Here he was only one of many claimants, most with greater seniority and more influence than himself. The University funds were, of course, limited. To the traditionalists, who still had the predominant voice in financial matters, the demands of the science departments seemed insatiable. There were senior members of the

University who refused to believe that a Professor of Pathology needed more money to teach his obscure subject than, say, a Professor of Sanscrit. The University Grants Committee did something to redress the balance, but science—and its cost—was moving too fast for the slow pace of the adjustments.

The effect of this situation was that Florey had to spend a good deal of time and effort in a struggle for what now seem to be trivial sums of money. For example, having failed to persuade the University to pay the wages of another technician, he applied to the Rhodes Trustees for a grant of £100. This was refused, and he replied in a letter to the Warden in August 1935: 'The decision of the Trustees did not greatly surprise me, as I agree that it is a function of the University to staff its institutions adequately. I hope I can, in time, convince them of this.'[11] But this hope, despite a constant battle, was never realized to his satisfaction and, as Gardner has stressed, Florey's teaching course could not be maintained to his ideal standards.

It was a strangely tantalizing situation. The School of Pathology was indeed a magnificent laboratory building, and the Chair of Pathology in Oxford did promise almost equally magnificent opportunities. But, in accepting the capital cost from the Sir William Dunn Trustees, the University did not accept the financial responsibility for running a department on a comparable scale. Dreyer had won only half the battle. Florey was described by Drury as 'a great finisher'. It was he who finally made the Dunn School what Dreyer had dreamed it might be, but it was an uphill struggle that lasted for many years. And Florey's financial difficulties were made the harder to bear by comparison with the lavish Nuffield endowment of new clinical departments in Oxford that began soon after his appointment.

The first indication of these developments that Florey received came in a letter from Hugh Cairns written in November 1935.[18] In this Cairns suggested meeting Florey to discuss a scheme for the future of clinical medicine in Oxford and it seems that, apart from the Regius Professor of Medicine (Sir Farquahar Buzzard), Florey was Cairns's first confidant in the medical faculty. Cairns's scheme was, in fact, a most ambitious one, and it needed careful handling of possible supporters and opponents if it were to be launched successfully. Cairns turned first to Florey because they knew each other and had the same sort of background. They both came from Adelaide, and had been Rhodes Scholars at Oxford. Cairns was one of the first people Florey had visited in London when he arrived from Australia, and it was to Cairns that he had gone when he was worried about his abdominal troubles in 1925. Florey's appointment at the London Hospital was certainly backed, if not instigated, by Cairns. But they seem not to have known each other in Adelaide. Cairns was two years older than Florey; he had been educated at Adelaide High School and had entered the University to read medicine in 1914. A few months later he enlisted as a private in the Australian Army Medical Corps, but continued his medical studies and qualified in 1917, when he obtained a commission and served overseas. Thus he barely overlapped at the Medical School with Florey, and immediately after the war he was awarded the Rhodes Scholarship for South Australia and left for Oxford in 1919.

Cairns[19] was a member of Balliol College, and made a considerable impression there. He was a young man of striking good looks and splendid physique, who soon became a fine oarsman, rowing for Oxford in the boat race of 1920. And in the following year he married Barbara, the youngest daughter of A. L. Smith, the Master of Balliol. On the professional side, his career was less spectacular, since he spent his time in gaining practical experience as a surgeon by becoming a demonstrator in the Anatomy Department and House Surgeon at the Radcliffe Infirmary. Then, in 1921, he went to the London Hospital, where he took his F.R.C.S. and was appointed First Assistant in Sir Henry Soutar's surgical unit. Quite soon his meticulous and painstaking operative technique began to attract the attention of his seniors. At that time neurosurgery was being revolutionized by Harvey Cushing in America and by operations requiring the sort of skill that seemed to come naturally to Cairns. It was thus suggested that he should specialize in neurosurgery, beginning with a period of work with Cushing. Cairns readily agreed, particularly since he already had a great regard for Cushing as the

recent biographer of Sir William Osler, who had died i
Oxford in 1920. He obtained a Rockefeller Fellowship, an
went to Cushing in 1926 to work with him for a year.

Cushing was most impressed with his pupil, and when th
time came for Cairns to return to the London Hospital h
used his influence to obtain funds from the Rockefelle
Foundation to establish there a neurosurgical unit unde
Cairns's direction. This, the first such unit in Britain, wa
highly successful—but its accommodation was too crampe
to satisfy Cairns. He had in mind a larger and more integrate
department which was hardly possible at the London Hos
pital—or indeed anywhere in London at that time. H
thoughts therefore turned to Oxford, where there was roor
for expansion, and where he had friends and influence. Cairn
had great ambition and this led him to schemes that went fa
beyond his first objective of a neurosurgical unit. He decidec
in fact, that Oxford would be an ideal location for the develop
ment of clinical science in general—a counterpart to the pre
clinical science begun by Acland seventy years before.

Cairns's crusade had other points of similarity to Acland's
Both were begun by young men who had not yet achieved an
particular professional eminence, but who had boundles
energy, determination, and the ability to convince people i
power. Cairns had a beguiling charm of manner and an imag
nation that created vivid pictures of the future. He was als
politically adroit and could be ruthless when this served h
ends. Having conceived his scheme, he drew up a memorar
dum 'On the Desirability of Establishing a Complete Schoo
of Clinical Medicine in Oxford' and submitted it to Sir Fai
quahar Buzzard in the summer of 1935.[20] As it happenec
rather similar ideas had been lying dormant in Oxford fc
several years. The Rockefeller Foundation had been advise
that Oxford was an ideal place for establishing an Institut
of Clinical Research and had offered finance in 1927, but tl
idea had lapsed through lack of support from either the clin
cal staff of the hospital or from the pre-clinical professor
Though it lapsed, it had lingered in the minds of certai
powerful Oxford elder statesmen (particularly since wha
Oxford had lost in 1927 was about to be gained by Londc
in the shape of the British Postgraduate Medical School

These people were: Sir Farquahar Buzzard, the Regius Professor of Medicine, who had been an eminent London neurologist and a Physician to the King; Sir Douglas Veale, the Registrar to the University, who had been Private Secretary to Lord Addison, the first Minister of Health, and to no less than five of his successors; Sir William Goodenough, Treasurer of the Radcliffe Infirmary, Chairman of Barclays Bank, and financial adviser to Lord Nuffield; and Lord Lindsay, Vice-Chancellor of the University and Master of Balliol College. When Cairns submitted his memorandum to Buzzard, therefore, he was activating the first of four men who between them could most influence the University, the hospital, and, finally, the man whose generosity could make the scheme possible.

Buzzard's reaction to Cairns's memorandum was to arrange a meeting with Veale. Veale became enthusiastic, and asked Cairns to prepare the scheme in detail. Cairns did so, in a thirty-nine-page document that set out plans for departments of medicine, surgery (including neurosurgery), paediatrics, obstetrics, radiology, and pathology. These would supplement the existing clinical departments of the hospital but with an emphasis on teaching and research and with medical staff appointed by the University. He at first planned a small clinical school where the best Oxford students could complete their medical studies instead of going to a London hospital, together with a postgraduate school where advanced students from all parts of the world could spend several years in training and research. (Later the idea for the clinical school was dropped.) Cairns's document was considered by Goodenough and Lord Lindsay, and then shown to Lord Nuffield. Nuffield, like most multi-millionaires, felt a philanthropic urge to share his riches, combined with a well-founded suspicion of those most anxious to help him do it. Buzzard and Goodenough had his confidence, however, and since they were both connected with the Radcliffe Infirmary and Nuffield himself had an interest in medicine, the hospital had already benefited from his generosity. For example, in 1930, the Radcliffe Observatory decided to move to Pretoria because their astronomical work was hampered by the increasing atmospheric pollution caused by the commercial developments around Oxford. Lord Nuffield, whose own factories had

certainly contributed to this, bought the 7-acre site of the
Observatory and presented part of it to the adjacent Radcliffe
Infirmary, thus providing for its considerable future expansion. A year later the hospital did expand—again with Nuffield's help, building new surgical and obstetric wards, a
nurses' home, and a private patients' block. Buzzard then took
the first step to encourage clinical research. On the remainder
of the site the beautiful eighteenth-century building that had
housed the Observatory became the 'Nuffield Institute for
Medical Research' in 1935 with Professor Gunn, the pharmacologist, as its first director.

Cairns's approach to Florey in November 1935 was a political move designed to gain an ally among potential opponents.
The resistance that had defeated the Rockefeller proposition
in 1927 had come from the pre-clinical professors and the hospital clinicians, and the same opposition could be expected
again unless disarmed in advance. Cairns could rely on Buzzard and Goodenough to deal with objections from the hospital because they carried a great deal of weight there. But
they carried little or none with the pre-clinical professors, who
respected scientific status rather than clinical or financial
eminence. And the pre-clinical professors, if (for once) they
acted together, could control the decisions of the Board of the
Faculty of Medicine where they formed a majority. Decisions
by a Faculty Board on an important issue in its own province
would determine the attitude of the Hebdomadal Council, the
University body that could promote (or block) the passage
of Decrees for consideration by Congregation, the University
parliament. Thus, though Lord Lindsay, the Vice-Chancellor, and Veale, the Registrar, would have considerable influence on the later stages of acceptance by the University,
the first step depended on the approval of the six professors
concerned with pre-clinical teaching. Cairns chose the one he
knew best, and who had some cause for gratitude, to be the
first recipient of his plans, and of his powers of persuasion.
If he could persuade Florey, he would have won an ally who
might influence the others.

There is no record of the conversation between Cairns and
Florey, nor evidence of correspondence on the subject during
the following few months. Florey's response can only be

guessed, therefore, from his subsequent actions—or rather lack of opposition. Though he had a low opinion of clinical research in general, he made no official objections to Cairns's scheme, and in the absence of any concerted opposition by the pre-clinical professors it was, in fact, accepted by the University in June 1936. As Cairns would have pointed out to Florey, there was no obvious reason why the pre-clinical professors *should* object, since the major changes would be at the Radcliffe Infirmary. There is also no doubt that Florey believed that he and his pre-clinical colleagues had been promised a share of the expected Nuffield Benefaction to help pay for their own research and to encourage collaboration between the science departments and the new clinical units. Florey referred to this belief in subsequent correspondence.[21] Certainly Sir Farquahar Buzzard in his Presidential Address to the British Medical Association in Oxford, in July 1936, in which he advocated the new clinical research school, plainly suggested a material collaboration with the pre-clinical departments when he said:

It must, I feel sure, have occurred to the minds of many that observations are constantly being made in the wards of a hospital which would bear rich fruit if cultivated in a pharmacological, physiological or biochemical laboratory, and that the results of many laboratory experiments would stimulate new branches of enquiry at the bedside, if the clinician and his scientific colleague were more closely associated in their daily work and had the time and opportunity to discuss their problems. In a medical school where research and training in the methods of research were the chief considerations such an atmosphere could easily develop and many inspired thoughts, now born to perish in the turmoil of teaching and practice, might grow to maturity ... The suggestion has come to me from more than one quarter that here in Oxford are to be found the basic needs for such a school and the ideal atmosphere for its expansion. We have all the ancillary scientific departments, admirably equipped and ably directed; we have one large and several smaller hospitals with ample and varied clinical material and in the Nuffield Institute for Medical Research we have a magnificent structure near which, thanks to the munificence and foresight of its generous godfather, is the space for such further buildings as may be required. An ambitious dream, perhaps, but if it came true, the future growth and development of the Oxford Medical School would be more than worthy of its great traditions.[22]

This combination of eloquence and idealism aroused great enthusiasm among the members of the British Medical

Association and, until they had time to think out the implications, among the members of the University. The pre-clinical staff, with some justice, pointed out that they already provided a school in which 'research and training in the methods of research' were the chief considerations, and they were not pleased that their departments were described as 'ancillary'. And the close co-operation at the bedside between the clinician, whose clear duty was to a sick individual and the scientist, whose aim was to discover general truths, though admirable in principle was difficult to achieve in practice. But such carpings were unworthy of this great occasion and, if they ever reached Lord Nuffield, did not deter him. In October he announced that he was giving £1,250,000 towards the medical school scheme and, at the meeting of Congregation specially convened to receive this benefaction, he made a speech which, since he was not a member, was technically out of order. This lapse was overlooked, however, in view of the fact that the purpose of his intervention was to announce that he was increasing his gift to £2,000,000. It was also revealed that he was giving a further £1,000,000 to the University, to found a Graduate College (now Nuffield College).

Two million pounds was, in those days, a very large sum indeed—enough to build, equip, and endow on a lavish scale five or six departments of the size of the School of Pathology. The comparison is unfair, of course, since the Nuffield Departments must care for patients as well as staff. But the figures indicate the scale of the Nuffield Benefaction as it appeared to contemporary eyes. It was a scale that dwarfed potential opponents and it was only later that the pre-clinical professors realized the consequences of its acceptance. They had been swayed by a supposed promise of substantial funds—which did not materialize. They had lost their predominance in the medical faculty because of the creation of a whole hierarchy of University clinical posts. And these clinicians did not turn to the science departments for collaboration; they built their own laboratories and then competed with the pre-clinical departments for trained staff by offering much higher salaries. But, in 1935, when Cairns first approached Florey, all this was in the future.

A DEPARTMENT COMES TO LIFE

In creative organizations the factors that determine growth or decay tend to be autocatalytic. Reputation, morale, opportunity, financial support, reinforce or undermine each other to quicken their own rise or fall. The saying 'nothing succeeds like success' applies with as much truth to a research department as to an orchestra or a drama company. Good work attracts good workers, and no amount of money or expensive equipment can create talent. At the Sir William Dunn School, the catalyst that set in motion its phenomenal growth into one of the world's best research centres was Florey's own example as a dedicated research worker. In one sense, the state in which he found it was an advantage to him, since he could hardly fail to improve it. But, though the change he brought about seemed all the greater by comparison, it was, even in absolute terms, very great indeed.

Florey seldom allowed a new appointment to interfere with the continuity of his research work. In the previous ten years he had changed jobs six times, done good work in twelve different laboratories, and published important papers from eight of them. One of his characteristics was the ease and speed with which he could settle down to experimental work in a strange laboratory—an ability that often took an unready host by surprise. It was also a characteristic that he could maintain simultaneous lines of research in different laboratories and with several collaborators. As might be expected, therefore, while he lost no time in starting up his experiments in his new and splendid laboratory, he was also visiting Sheffield to finish his work with Harding on the control of secretion in the duodenum, and they published three papers together in 1935.[1]

Florey's appointment to Oxford was really the end of this peripatetic phase. He had found his permanent home and a position for which all his wanderings had been a necessary preparation. He had worked with or visited many great men, now he must become their equal and create a centre of his own. To do this, he must attract first-class research workers. He could offer them spacious laboratories, but no salaried appointments, and only very meagre laboratory expenses. He was limited, therefore, to people who could maintain themselves—Rhodes Scholars, holders of other fellowships or awards, and postgraduate students working for a higher degree. These people would be young—which was all to the good—and they would be attracted by the chance to work with Florey, or at least under his supervision. He would not only gain collaborators but the nucleus of a school of experimental pathology.

Florey had always maintained that, ideally, the experimental pathologist should be able to cope with the techniques that his research line might demand. He himself had mastered most of the arts and crafts of the physiologist, pathologist, and bacteriologist, but in the field of biochemistry he knew that he would soon be lost. And, since the time had come when almost any biological research led sooner or later into biochemistry, Florey had, for the past six years, been trying to enlist a biochemist. He had specified the biochemical study of lysozyme and possibly other natural antibacterial substances in his repeated applications to the Medical Research Council, but he had in mind the more general collaboration of a biochemist as a member of a team of experimental pathologists each with his own special expertise.

Lysozyme, however, remained for the moment the central problem that needed a biochemical approach. Despite the help of Marjory Stephenson at Cambridge and Sylvia Harrison at Sheffield, the essential first stage in the research—the purification of the active agent in crude lysozyme—had not been achieved. So, as Florey arrived at the Dunn School, he looked round for a new collaborator. Sir Robert Robinson the Waynflete Professor at the Dyson Perrins Laboratory (the organic chemistry laboratory) thought that the lysozyme problem would fit in with work being done by one of his young men, Dr. E. A. H. Roberts, and he supported an application by Florey to the Medical Research Council for a grant to enable Roberts to collaborate. This time the application was successful; the Council granted £300 p.a. in July 1935, and Roberts worked on lysozyme with Florey and Brian Maegraith for the next two years.

Meanwhile, Florey continued his efforts to appoint a biochemist to his own staff. He had managed to obtain funds for a salary from the Sir William Dunn Trustees and the University Chest, and the task now was to find the right man. It seems that he considered inviting Macfarlane Burnet, but nothing came of an idea that might have had interesting consequences.[2] Burnet remained in Australia to become Director of the Walter and Eliza Hall Institute in Melbourne and to win a Nobel Prize in 1960. Next, Florey approached Dr. Hugh Sinclair, who was a Demonstrator in the Department of Biochemistry and making a name for himself in the field of nutrition. But Sinclair felt that his own interests were best served by remaining in Peters's Department.

Sinclair's reaction underlines one of Florey's difficulties in acquiring a full-time biochemist. The increasing complexity of science was enforcing specialization at an earlier age. No longer was it possible for a man to become a great surgeon and at the same time to do outstanding work in physiology, bacteriology, and immunology, as Lister had done. The consequence of this specialization was the creation of separate university science departments. In Oxford, for example, the science that had been fairly comfortably housed under one roof in Acland's Museum until 1890 occupied nine large buildings by 1935, with several more to follow in later years. The different branches developed their own languages—or, rather, jargon—published their own journals, and founded their own societies. The necessity to learn, teach, and do research in his own speciality tended to keep a young scientist in what he felt to be his proper department, among those who spoke his language. A biochemist would feel isolated within the four walls of a department of pathology. This mental and physical demarcation was, paradoxically, most acute in the universities—the institutions most committed to the ideal of wide knowledge. It did not occur in the research institutes; in these, problems were attacked on a wide front and by every appropriate special technique. Florey, in fact, believed that experimental pathology needed such an approach, and that the Dunn School should become a self-contained research institute. But he could hardly expect an enthusiastic response from other professors to his recruiting drive in their departments.

There was one biochemist, however, who had asked Florey if there was a job for him at the Dunn School. This was N. W. Pirie, who had been working with A. A. Miles in Cambridge while Florey was Huddersfield Lecturer there. Pirie was in Gowland Hopkins's department, but would be free to move in a year or two and would then like to come to Florey. Florey wanted a biochemist immediately, and thought that Gowland Hopkins might be persuaded to release him at once. The approach would have to be tactful and Florey asked Gardner to go to Cambridge and discuss the matter with Hopkins. Gardner arranged the meeting, and set off for Cambridge, arriving there an hour or two before the time of his appointment. Not being very familiar with Cambridge, he first of all found the Sir William Dunn School of Biochemistry to ensure his arrival there at the proper time, and then set off to explore the centre of the town. It is easy to lose one's way in Cambridge and Gardner did so. When he eventually refound the Department of Biochemistry, he was already fifteen minutes late for his appointment with the great man— a known stickler for precise punctuality. It was not the best of beginnings. 'Of course he was displeased,' writes Gardner. 'But he accepted my explanation and I remember a perfectly friendly interview, in which I think he said they couldn't possibly spare Pirie. I think it immensely improbable that he reacted peevishly to my bloomer by sacrificing Florey and Pirie on my account.'[3] The fact remained, however, that Pirie did not come to Oxford.

A short time later Gowland Hopkins tempered his blank refusal by recommending to Florey one of his pupils as a substitute for Pirie. On 28 May 1935, he wrote: 'I have just had to read his Thesis for the Ph.D. degree here (which he will certainly get) and am struck with the ability it displays. Leathes, who was external examiner, also thought highly of it, and of the knowledge he displayed at our oral examination ... I feel sure that you will import an acceptable and very able colleague in taking him. Incidentally I have found that his remarkable genius as a musician has made him acceptable in certain social circles here—a point which I think is not without some importance.'⁴

Hopkins's pupil was Ernst Boris Chain, a young Jewish refugee from Hitler's regime, the son of a German mother and a Russian father. Gardner had explained Florey's requirements to Gowland Hopkins, and any recommendation from him was to be seriously considered. Florey went to Cambridge to discuss Chain's suitability and was impressed by what Hopkins had to tell him. Hopkins then sent for Chain, who later described the interview.

When Hopkins asked me whether I would like to go to Oxford to join Florey's staff, I was both extremely surprised and delighted. I had never expected such exceptionally good fortune to come my way in my unsettled condition and with a very uncertain future in front of me. He introduced me to Florey in his office immediately after our talk, and I naturally accepted the offer without any hesitation. Thus, I migrated from one Sir William Dunn School to another.⁵

Chain had left Berlin on 30 January 1933, when Hitler acceded to power and Europe, to use Chain's own words, 'was temporarily plunged into a darkness in comparison to which the darkest Middle Ages now appear as a blaze of light'. He worked for a short time at University College London, and then, with the help of J. B. S. Haldane, was accepted as a research student by Gowland Hopkins, for two years. Chain was deeply grateful: 'Hoppy, as he was affectionately called by members of his staff ... was one of the most considerate and kindest human beings I ever had the good fortune to meet.' Chain's work at that time was an attempt to discover exactly how certain snake venoms attack nervous tissue to cause fatal paralysis. He first had to isolate and purify the active principle in the venom, identify it chemically as far as possible, and then discover what chemical structure in nerve tissue it attacked and by what means. It was a line that had obvious similarities to that of lysozyme, and similar techniques would apply. But when Chain first came to Oxford he continued to work on his unfinished snake venom problem, and brought it to a brilliant conclusion. 'For the first time, the mode of action of a natural toxin of protein nature could be explained in biochemical terms as that of an enzyme acting on a component [nicotinamide adenine dinucleotide] of vital importance in the respiratory chain.'[5]

He was then twenty-seven years old: rather short in stature, he had a striking resemblance to Einstein, with a mane of flowing hair (unusual in those days), a high forehead, and large intelligent eyes. He was warm-hearted, excitable, and always voluble. When hot on the trail of some idea he would pace the room, nervously tossing his hair back with one hand, gesticulating with the other, and maintaining a monologue of argument and cries of triumph or frustration. He brought to science an artistic temperament, true inspiration, and originality—an emotional approach that gave joy in achievement and despair in supposed failure. He had, as Hopkins had said, musical genius. He was a splendid pianist, who could have achieved international fame on the concert platform. But he also had a genius for science. He had to choose between music and science when his education had reached the parting of the ways, and decided that science offered a more secure future. So the world lost a possibly great pianist, but gained a new medical era.

Florey was ill and in bed at Lincoln College on the day that Chain arrived at Oxford station. Kent went to meet him, having little idea how he should recognize him. But there was no difficulty. When the other passengers had gone their way, a single, small, and rather forlorn figure remained. 'He looked like a refugee, all right,' said Kent, 'and seemed very depressed in every way. We walked to the School of Pathology and as we passed the Dyson Perrins he caught sight of a Soxhlet apparatus through a window—they were pretty rare and very expensive in those days and I don't suppose there were more than three or four in Oxford. He brightened up at once, "You have Soxhlets in your Department?" he asked. I said that we had one. "One?" he shouted. "I must have six—a dozen!"'[6]

Though the Dunn School of Pathology in Oxford was naturally less well equipped for a biochemist than the Dunn School of Biochemistry in Cambridge, Chain settled into his new laboratory quite happily. he was grateful for his appointment (which was initially for one year only and at a salary of £200), and though he was permanently short of research funds and apparatus, his work progressed rapidly. He found his colleagues congenial and, as Gowland Hopkins had foretold, his musical talents soon made him friends in Oxford. He was not just accepted, he was welcomed. Part of this welcome was due to a natural sympathy for a victim of political tyranny. But mostly it was a response to Chain's own personality. He was warm-hearted and engagingly demonstrative, appreciative of kindness and deeply concerned for other people's troubles. He had, too, something of the aura of the artistic culture of Europe, the imaginative depths of the artist combined, in his case, with the self-critical precision of the scientist.

Those who came to know Florey at that time were surprised at his choice of a biochemist, and even more so when he and Chain became friends. Florey had given the impression that he did not like the Jewish race. He distrusted extravagant ideas and flights of fancy in science, and he abhorred displays of emotion. On all these counts it was predicted that Chain would prove quite antipathetic, and the prediction was proved wrong. This error was due, not to a misjudgement of Chain's character—which was relatively transparent—but of Florey's which was becoming more and more withdrawn behind the façade which he presented to Oxford. The youthful enthusiasm for research—though still as strong as ever—was concealed by a studied understatement. ('We don't seem to be going backwards' meant real progress.) The Australian exuberance of language and action that had earned him such descriptions as 'bandit' and 'bush-ranger' in Sheffield, and at first in Oxford, had largely given way to a laconic scepticism. But his determination to get what he wanted was undiminished, and though he had learned to control his temper, few people cared to put this control to the test.

If one is justified in trying to assess his inner nature from his early letters to Ethel, then it may be seen that Florey and Chain had, in reality, much in common. The main difference was that Chain had no difficulty in expressing his deepest feelings—in fact he was almost compelled to do so—whereas Florey found any such expression nearly impossible. His letters to Ethel are almost completely unemotional. He avoided all endearments and what he had called 'sticky sentimentality'.

Yet Howard's nature was not cold. Music and the arts gave him profound emotional experiences—he confessed to Ethel that he could be moved to tears by them. But all he could say to her, after hearing concerts by Kreisler or Moiseiwich, was: 'He's played before.' He was deeply moved, too, by the romance of Venice or Vienna, and he longed for a companion to share his emotion. But he almost apologized to Ethel for such feelings; he was sure that she was above such mundane reactions. Throughout his research work he professed the detached emotional interest of the scientist, and no particular concern for 'suffering humanity'. But he was horrified by the toll that tuberculosis was taking of his contemporaries in Adelaide, and by his clinical experiences at the Radcliffe Infirmary in Oxford when he worked there during his Rhodes Scholarship. The death rate from infection, particularly of young women at childbirth, appalled him, and he was thrilled by any possibility of an effective antibacterial treatment. Dr. Pullinger remembers the effect on him of the 'horror stories' of post-operative sepsis and cross-infection in the Sheffield hospitals which his colleagues used to tell at the departmental tea table. She is convinced that the search for an antibacterial agent for clinical use was, even then, motivating his work on lysozyme.[7]

Despite his brusque, laconic manner, and his reputation for ruthless behaviour in his personal relations, Florey had a real concern for human troubles. The plight of the starving population of post-war Austria affected him profoundly. He detested militarism, nationalism, and political oppression. And in his personal dealings he was far more humane than he cared to appear. For example, one of the Dunn School technicians was found to have been stealing equipment. Florey dismissed him with no show of mercy, but privately took great trouble to find the man another job, and to persuade him, successfully as it happened, to mend his ways. There are, in fact, many people who, in time of trouble, received a (to them) totally unexpected kindness from Florey. Ruthlessness was his method of dealing with opposition from people of his own size and in particular with bureaucratic or petty tyranny. The reserve that prevented almost all close personal ties was defensive. Only to people in trouble, to the underdogs, could he bring himself to reveal an overt sympathy and kindness that he seemed almost to have regarded as a confession of weakness.

If this is a valid picture of Florey's nature, then his relations with Chain are more easily understood. At the professional level, Chain was clearly a brilliant biochemist with a line of interest that could readily be switched to the problem of lysozyme. Florey had often said that he would work with anyone who was good enough, but this did not mean that he would like them. Up to this time all his collaborators had been his contemporaries or seniors and well able to look after themselves. Florey had channelled their skills into his chosen projects, and confessed that he regarded some of these people as no more than 'pairs of hands'. But his direction usually resulted in better work from his collaborators than they ever did without him. Chain was in a different category. He was Florey's first protégé; a foreign refugee, ten years his junior, to whom he had given a job. And the very qualities in Chain that were expected to annoy or antagonize Florey probably attracted him. Florey had a great admiration and respect for the true artist, and he could not fail to respond to Chain's musical talents. Chain's enthusiasm for his research—so readily expressed—must have been a refreshing reminder of Florey's own enthusiasm, now rather overlain by professorial duties and status. Chain's ability to talk, with deep conviction and no restraint, on almost any subject may have been a vicarious release for Florey, as if this extrovert young man were a sort of *alter ego*.

From Florey, Chain received what he most needed: security; the security of a University appointment, of congenial companionship, and above all, the security of firm direction in his work, the guiding of brilliant ideas into the solid channels of factual achievement. They were much in each other's company, not only in the laboratory but in the walks they took together almost every day through the University Parks. Their conversation, it seems, was mainly about their common research interests, but it was a closer companionship than Florey had previously allowed himself, and when Chain was ill during 1936, he made frequent visits to him in hospital. In their work together their gifts for research were largely complementary. Chain's intuitive brilliance and originality balanced Florey's equally intuitive sense of direction and his genius for picking his way by simple, methodical experiments through a maze of attractive side-issues. And though, in a sense, their characters were also complementary, they had one quality in common. They were both strongly positive—dynamic, forceful people determined to succeed in all they undertook. Their combined energy, directed to a common goal, could (and did) achieve wonders. But it was a potentially explosive mixture and ultimate disruption was inevitable.

Chain was the first of several important new arrivals at the Dunn School. Florey had turned his attention to the depleted bacteriology staff. Gardner was occupied with his Standards Laboratory; Vollum with clinical work; and Miss Campbell Renton and Jean Orr-Ewing with the research they had inherited from Dreyer. Then a friend of Gardner's, Colonel Bridges, retired from the Indian Medical Service and suggested that he might help in the Standards Laboratory. Florey now saw an amicable solution to the problem of this 'foreign-body' organization. He suggested to Gardner that he should give up the Standards Laboratory and become head of the bacteriology section of the Dunn School with, if it could be arranged, the title of Reader. Gardner agreed. He realized that, if he did not, he would soon have to leave Oxford since Florey was determined that the Standards Laboratory should go. So Gardner joined the University staff, becoming Reader in Bacteriology in 1936. Colonel Bridges took over the Standards Laboratory but, because of the war, it did not finally move to the Medical Research Council's laboratories at Colindale until 1946.

The bacteriology section was further strengthened by the arrival, in 1936, of Florey's old Cambridge friend, A. Q. Wells, who had grants to study tuberculosis in animals. As was his custom on changing location, he bought the nearest agreeable manor house, in this case at Shipton-on-Cherwell, where there was stabling for his collection of Rolls-Royce cars, a pleasant stretch of the river, and scope for an extensive alpine garden. These distractions did not diminish his research. He discovered that the field vole is liable to natural infection by a peculiar strain of the tubercle bacillus that has little virulence for man. He pursued the possibility that this 'vole bacillus' might provide an effective vaccine against human tuberculosis, on the lines of Jenner's use of the non-virulent cowpox virus to protect against smallpox. But it was a research that, like so many others, was to be made redundant by antibiotic therapy a few years later. Wells did not solely occupy himself with research. He took an active part in teaching, and personally supervised the replacement of obsolete or pilfered equipment in the classrooms and the proper treatment of the microscopes. His very appearance commanded respect among the more unruly students and technicians. A height of 6 feet 5 combined with an equally lofty manner is an effective combination.

This quality of 'presence' and its effects on lesser mortals is shown by an earlier event in the life of A. Q. Wells. When he was a Demonstrator in Kettle's Department at St. Bartholomew's Hospital, there was an official visit by some important body (perhaps the University Grants Committee). The staff—with one exception—were aware of this impending ordeal and prepared themselves to give an appearance of devotion to duty, overwork, and underpay. The one unexplained exception to this foreknowledge was Wells. On an exceptionally hot day the visitors arrived, escorted by the Dean and Professor Kettle. One after the other, doors were thrown open to reveal hives of industry; searching questions were asked and nervous answers given. The door of A. Q. Wells's laboratory revealed a rather different picture. Wells, who suffered from swelling of the feet in hot weather (probably a penalty of height), was sitting in a chair reading *The Times* and smoking a Turkish cigarette. He had rolled his trousers to the knees, and his bare feet were in a sink full of cold water. The unannounced appearance of important officials under these circumstances would have brought most people to their dripping feet, mumbling explanations and apologies. Wells merely lowered *The Times* by an inch or two and regarded the intruders coldly through a haze of tobacco smoke. It was the visitors who beat an embarrassed retreat.[8]

Concurrently with the build-up of his permanent staff, Florey was attracting research students and Rhodes Scholars to the Dunn School. One of the first was P. B. Medawar whose work on tissue immunity was to lay the foundation of transplant surgery and win him a Nobel Prize in 1960. Sir Peter Medawar describes his experiences there as follows:

I held a scholarship and a senior demyship at Magdalen College, which made it possible for me to take up a career in research ... and sought the hospitality of Professor H. W. Florey's School of Pathology. My intention was to investigate by tissue culture methods a growth-inhibitory factor in malt extract. Florey agreed to this, and let me have a fine room in the path lab. I was a complete beginner in research. I therefore felt very much in need of guidance, and hoped and expected that Professor Florey would give it to me, perhaps in some little piece of formal advice beginning 'It', like this, you see ...' One day he beckoned me into his room and I thought to myself 'Boy, this is it.' Florey asked me what my career intentions were and I said I wanted to go in for scientific research. Florey looked at me in his characteristically piercing way and said: 'Medawar, anyone who goes in for scientific research is a bloody fool.'

He didn't mean this, of course, but I think he had been dismayed by the difficulty of raising grants to carry out his own research and thought intending research workers should be warned of the future that lay ahead of them. In those days, before departmental grants were a common thing research workers had to pay for a lot of their material. I remember vividly going out into the country to buy 'setting eggs' by the dozen for tissue culture work. I also bought myself a little electric centrifuge.

Medawar thought that Florey was basically a kind man who kept an eye on his young and every now and again gave them advice which they would have been foolish not to take and profit by.

Once, for example, I wrote a long-winded and discursive paper about some quite modest observations, which he handed back to me, saying 'This is more like philosophy than science. Try again.' I profited by his advice and if nowadays I am tempted to write in too involved a way about an essentially straightforward subject, I remember Florey's rasping voice saying 'I don't see what you're getting at, Medawar. It doesn't make sense to me.'

I went on working in Florey's lab. until I was offered a post in the Zoology Department but my experience with Florey of 'real life' research left its mark on me and all my later work in the general area of medical biology. In retrospect, I feel a great deal of Florey's extreme toughness and angularity was simulated. He was a good Professor, who helped people and he was, in addition, an extremely clever and amusing man—enough of 'character' to give a lot of fun from parodying himself at times. In spite of his great achievements Florey remained unsure of himself, even at the peak of his fame. I know this because of the enormous pleasure he got from praise of his speeches and lectures when he was President of the Royal Society. I found it indeed very touching when on one or two occasions he came up to me and said 'Was that all right?' It almost always was; Florey in fact filled this office superbly well.[9]

Florey had two distinct ways of dealing with his research workers. He never 'spoon-fed' those who were working on their own problems, even if they were young and inexperienced. They were, to quote one of them, 'thrown in at the deep end' to sink or swim by their own efforts. Some sank, the rest survived, having gained an invaluable experience of real-life research, shorn of the glamour and romance with which the lay public likes to endow it. 'Your first piece of research is to find somewhere to live' would be Florey's greeting to a new recruit. Thereafter he would have to fend for himself. He would have to learn how to acquire apparatus and technical skills, to work with colleagues and technicians, and to use the libraries. But what everyone could learn from Florey was something of his clear-headed and practical approach to a problem, his contempt for woolly, pretentious ideas, and the inescapable necessity for hard work.

With his own collaborators, Florey exercised a closer control. Every move was planned, discussed, and supervised. Dr. Pullinger writes: 'Those on the staff who had their own personal grants were free to pursue their own investigations. Those of us who were dependent on, or sponsored by, Florey were not given the smallest corner of freedom to follow ideas of our own, unrelated to what we were doing with him.'[7] Dr. Pullinger, who was a senior worker in her own right, was to find this restriction irksome, but for the time being she was happy in her research with Florey. They were working on lymph flow, and had mastered a technique for introducing minute cannulae into the lymphatic vessels of rabbits. This allowed the collection of lymph free from the least trace of blood, and the results of measuring the volume and cell content of the fluid had emphasized an extraordinary problem. Since this was to remain a major interest for Florey and a succession of collaborators for many years a glimpse of the contemporary knowledge of the lymphatic system is needed here.

Lymph is the clear fluid that filters through the walls of the blood capillaries into the tissue spaces. From there it passes into the smallest lymphatic vessels, the pumping action of which Florey had already observed in certain animals. These fine lymphatics are tributaries of larger channels each of which passes through one or more lymph nodes where any bacteria, toxins, or foreign cells are filtered off. The nodes produce lymphocytes: small, round cells which enter the lymph stream on leaving the node. This stream is conveyed to the main lymph vessels, the thoracic ducts, which eventually discharge their contents into the blood-stream. These contents consist of lymph from the tissues, an emulsion of fat droplets (chyle) collected by the lymphatics from the intestine, and lymphocytes. Florey had been collecting, measuring, and examining the lymph flowing through the thoracic ducts. It was this sort of work that revealed the mystery of the lymphocyte by the relatively simple process of calculating the normal rate of lymphocyte delivery to the blood-stream and trying to balance the books.

In the circulating blood the lymphocytes form about 20 per cent of the white cell population. The actual number of blood lymphocytes is thus easily calculated, and by measuring the volume of lymph delivered each day and counting the number of lymphocytes it carries per unit volume, one can deduce the life-span of the lymphocyte in the blood. Such calculations led to an astonishing conclusion. In most mammals enough lymphocytes enter the blood-stream via the thoracic ducts to replace completely their blood population several times each day. The inference was that the average lymphocyte survives for only a few hours. But what happens to it? Where does it go and what does it do? All histological and cytological work had suggested that it was rather an inert cell with none of the defensive, antibacterial activity of the other white cells. So its apparently enormous rate of production and destruction was unexplained. As usual with biological problems, when facts are scarce theories abound. In this case there were at least six, each with points in its favour but none that could be either proved or disproved. One suggested that the lymphocytes entered the bone marrow where they turned into red blood cells; another that they were released into the intestine where they provided enzymes to digest the food. A third considered that lymphocytes became transformed into other leucocytes, or tissue macrophages—and so on. Such airy arguments can only be clinched by new factual observations and experiments, not by armchair cogitation. But the armchair does provide a maxim that can help to direct the search for new facts and one which, if it had been applied to this problem, might have hastened its solution. This can be stated as follows: when a number of conflicting theories coexist, any point on which they all agree is the one most likely to be wrong. In the case of this particular problem the one point of agreement was the supposedly high rate of lymphocyte production which, as J. L. Gowans was to show at the Dunn School in 1957, does not exist.[10]

In 1935, however, Florey and Beatrice Pullinger were focusing their attention on the lymph nodes. These become enlarged and hyperactive when the lymph they received had come from an infected or inflamed area; they retain bacteria that have been injected into the peripheral lymph flow, and also abnormal cells, such as cancer cells, which might then grow there to form a secondary cancerous deposit. Their other function is to produce lymphocytes, at a greatly increased rate in response to infection or injury in their drainage area. The absence of any clear idea of the physiological and pathological importance of these reactions was a challenge that Florey could not resist, and he proceeded to tackle it in his customary straightforward manner. One physiological method for discovering the function of an organ is to remove it and see what happens, a method that had been enormously successful in the case of the thyroid, parathyroids, pancreas, adrenal glands, and so on. Florey proposed to apply it to the lymph nodes. Unfortunately this posed the greatest technical difficulties, because lymphoid tissue is scattered throughout the body, not only as discrete nodes, but as diffuse patches in the intestinal wall. Florey was always prepared to follow a simple idea, however complex the techniques that he might have to master to do so. He devised a method for selectively staining the lymphoid tissue with a dye injected into the living animal, and then embarked on the most tedious and protracted operations in which he carefully dissected out every vestige of stained tissue. He, Dr. Pullinger, and others pursued this method with dogged tenacity for nearly two years, but for once, Florey's research judgement had misled him and Dr. Pullinger recognized this before he did.

The extirpation experiments involved prolonged anaesthesia, operations that were shockingly severe, extensive tissue damage, and extreme liability to infection. Dr. Pullinger felt that these factors made the results difficult to interpret with confidence, but Florey maintained his belief that the method would reveal the then unknown function of lymphoid cells and tissues. And in 1936 this project received a new recruit, Dr. A. G. Sanders, a graduate of St. Thomas's Hospital who had been working under Dr. John Mills, the senior pathologist at Reading. Mills had known Florey at Cambridge and, being impressed with Sanders's technical ingenuity and flair for research, suggested that he might work for a D.Phil. at the Dunn School. Sanders was eager to do so, and on Mills's recommendation Florey accepted him as a research student. Florey told him quite frankly that he would have to support himself financially, but that if, after a year, he showed sufficient promise, he would recommend him for a Philip Walker Studentship. Sanders, who had modest private means, was prepared to do this, and also to do exactly what the Professor had planned as a research project for his D.Phil. Florey wrote out every detail of this programme[11] which involved not only Sanders, but also Peter Medawar and later two other research students, J. M. Barnes and Jean Taylor (who was about to become Mrs. Medawar). It was a concerted attack on the lymphocyte problem. Sanders was to acquire, and apply, the technique for the total surgical extirpation of lymphoid tissue. Medawar was to study the lymphocyte in tissue culture, Jean Taylor the possible transformation of lymphocytes into other cells, such as macrophages, Barnes was to deal with lymph flow and composition, while the Professor himself carried out the incredibly delicate operations of inserting glass cannulae into the lymph channels of rabbits, an operation of which he was almost the only exponent.

Meanwhile, Dr. Pullinger told Florey of her doubts about the extirpation method, and her reluctance to continue with it. She asked him if she could take up a line she had begun at the Mount Vernon Hospital before joining him at Sheffield.

This was the use of transparent windows inserted into the skin of the rabbit's ear, to study microscopically the reactions of living tissues to X-rays and other agents used to treat cancer. Florey himself had, of course, been one of the first to use such windows for observing the micro-circulation. He had, therefore, a vested interest in the method, but, though he agreed, he was not pleased. His most reliable pair of hands, and his most dependable ally, was showing signs of independence. From that time, she felt that she was no longer in his confidence.

So, Dr. Pullinger, glad to hand over the lymphatic dissections to Sanders, turned to the project of her choice, the microscopic study of living tissue reactions by the rabbit ear-chamber technique. The original chamber had been designed by J. C. Sandison, in 1924, and subsequently improved in detail by E. R. and E. L. Clark during the next few years. Dr. Pullinger had mastered the technique at Mount Vernon Hospital and introduced her own improvements. Essentially, the Sandison–Clark ear chamber, as it is usually called, consists of two transparent discs about half an inch in diameter which are bolted together through holes punched in a rabbit's ear, one on each side of a central hole. New tissue, complete with minute blood vessels, grows to fill this central hole. The discs then take the place of the skin removed, the edges heal cleanly, and the tissue between the discs continues to live and function normally, with a brisk blood-flow through its vessels. Being almost completely transparent, it can be viewed microscopically under high power, and recorded by still or ciné photography under perfect optical conditions. It is a technique that offers unique opportunities for studying tissue reactions, blood-flow, the behaviour of individual cells, and, of course, the possible transformation of one sort of cell into another. The rabbits, with their ear chambers in place, live, without apparent discomfort, for their normal lifespan.

Dr. Pullinger began the task of establishing this technique as a reliable and reproducible method. The apparatus, though basically simple, was complex in detail and required much fine engineering on the part of Mr. Bush. Many sets of rings, spacers, washers, mica windows, minute nuts and bolts, and equally minute tools for inserting and tightening them had to be made to precise specification. An American Rhodes Scholar, R. H. Ebert (later Dean of Harvard Medical School), joined in the project, and it was not long before Sanders, always fascinated by micro-engineering and microscopy itself, took time off from his endless dissection of lymphatic tissue to help with suggestions and improvements. Florey made no serious objections to Sanders's interest, partly because he, too, became enthusiastic about the potential value of the ear chamber for the further study of inflammation (which had never ceased to be one of his main interests) and also as a way of studying the behaviour and functions of the lymphocyte. So Florey himself joined in the project with Dr. Pullinger and Ebert, and they published a joint paper on their important improvements to a method that Florey and Sanders were to exploit together for many years thereafter.[12]

During Florey's momentous first year in Oxford, his sister, Dr. Hilda Gardner, sent him distressing news of their mother's health. After her husband's death in 1929, Hilda had shared her Melbourne home with her mother and her sisters, Charlotte and Valetta. Hilda was a pathologist at the hospital; she had a daughter, Joan, then aged ten, and a son aged four, so that the new arrivals proved something of a domestic asset. They could look after the house and the children while Dr. Gardner was at her laboratory, and Charlotte took over the household chores and the cooking. It seems to have been a harmonious and advantageous arrangement for all concerned. Then there was a tragedy. In 1932, Hilda's small son developed a mastoid infection and died of meningitis, a not uncommon danger in those days. And, in 1935, it was discovered that Bertha Florey, then aged seventy-two, had cancer. The disease was too advanced for effective surgery, and early in 1936 Hilda wrote to Howard to tell him that his mother had only a few more months to live. He responded by arranging that he and Ethel, with Paquita and Charles, should travel to Australia during the Long Vacation, so that he could see his mother for the first time in fourteen years, and almost certainly for the last.

A complication occurred not long before their sailing date. Paquita developed an acute mastoid infection, and had to have an emergency operation. For a time it seemed that their pass-

age would have to be cancelled, but she was treated with sulphanilamide, then only just coming into use, and she was well
enough to travel. When they embarked, Howard took with
him a supply of the drug in case of a recurrence and this
sample, when he arrived in Melbourne, was the first to be
seen there. It was an incident that much impressed Joan,
Hilda Gardner's daughter, whose memory of her brother's
death from the same sort of infection was still fresh.[13]

The visit to Australia was expensive. Florey was able to
let the Oxford house for three months, which was a small contribution. They travelled tourist class on board the Orient
liner *Orsova* for Melbourne, via the Suez Canal, Colombo,
and Fremantle. Ethel was delighted with the prospect of a
return to Australia and she had made plans to stay with her
family and visit her many friends in Adelaide, most of whom
she had not seen for over ten years. The Floreys enjoyed the
voyage. Howard made a ciné film of its progress, with shots
of the shipboard activities, and their ports of call. They
reached Melbourne in August and went straight to the
Gardners' home in Hawthorne. It was a reunion with mixed
emotions. Bertha, Charlotte, Hilda, and Valetta were delighted to see Howard, and they were proud of his success.
Joan was seeing him for the first time. Paquita and Charles
were, of course, centres of attention. Hilda had arranged for
the visitors to stay at a near-by hotel, and had engaged a nurse-
maid to look after the children. But it was a visit shadowed
by its reason. Bertha was already weakened by her illness and
spent much of each day in bed. She made a gallant effort to
make the visit a cheerful one, and photographs of the family
in the garden show her smiling and in command. But Paquita
remembers sitting by her chair one afternoon while the old lady
was asleep and being terrified that she was already dead.[14]

After a few days, Ethel and her children went to Adelaide
to stay with her parents for a month, while Howard remained
in Melbourne. He spent time with his own family, but he
naturally spent it too in the various laboratories. While visiting Professor MacCallum in his Department of Pathology at
the University he met one of his Lecturers, R. D. Wright,
a young Tasmanian-born experimental pathologist whose
surgical skills and original ideas immediately appealed to him.

Since MacCallum was prepared to allow Wright to have a year abroad on full salary, it was arranged that he should come to Florey in 1937 'at no cost to Oxford except cats'. Wright was remarkable, not only for his research abilities, but for his physique, which made the average rugby forward look relatively puny. His universal nickname, 'Pansy', might have an obvious antithetical origin, but it actually derived from a character part he played in a students' review. It has been suggested that it was a contraction of 'Chimpanzee' but Professor Wright states firmly: 'That the other spelling has some zoomorphic excuse may perhaps account for the sporadic revival of this variant.'[15]

After leaving Melbourne, Florey joined Ethel and the children for a week in Adelaide, reviving old memories, visiting 'Coreega' and renewing his friendship with Mollie—now Mrs. Bowen. Then they returned, again by Orient Line, to Oxford in October. Bertha Florey died in Melbourne on 27 November. She was buried in her husband's grave in the churchyard at Mitcham, in Adelaide. She had been a domineering, determined, clever, and, since Joseph's death, a rather unloved woman. Charlotte and Valetta continued to live with Hilda and Joan Gardner and, after Anne's retirement from her post in England, she too joined them in Melbourne.

Meanwhile, by repeated applications to the University and to the Sir William Dunn Trustees, Florey had managed to increase his budget for teaching expenses, and was thus able to offer one or two temporary demonstratorships with opportunities for research, but with very small salaries and no permanence of appointment. One of the first recruits was Dr Margaret Jennings, the wife of Dr. D. A. Jennings, a gastroenterologist. She had been strongly recommended to Florey as a good teacher and an able histologist by his old friend Robert Webb, who had become Professor of Pathology at the Royal Free Hospital in London, and in whose department she had been a demonstrator. She was the daughter of Lord Cottesloe, third baron and an indirect descendant of the Captain Fremantle after whom the West Australian port is named.

Dr. Margaret Jennings came to the Dunn School in October 1936, while the Floreys were still on their way back from Australia. She was soon involved in the new teaching

ourse and, when the Professor returned, she and R. D.
Wright (from Melbourne) worked with him on mucus secre-
ion. They studied the action of drugs and the effect of nerve
timulation on the colonic secretion of mucus in the cat, and
O. A. Jennings, who then had an appointment on the staff
of the new Nuffield Medical Unit, collaborated with them in
a study of the action of gastric juice on intestinal function in
he pig.[16] This work had a direct bearing on human gastro-
nterology and thus provided one of the first (and one of the
ew) examples of the contact between clinical research and
pasic science that the Nuffield scheme had been designed to
promote.

Of the cost of this research, only £50 was contributed by
he Nuffield 'Fluid Research Fund'. The rest came from
Florey's grants from the Medical Research Council, and Mel-
anby had a habit of 'dropping in' unexpectedly to see how
his funds were being used. On one of these inspections Florey
took him to meet Pansy Wright in the room he shared with
O. A. Jennings. Wright was tinkering with a mass of home-
made electrical equipment, and Mellanby asked him what he
was doing. 'Testing apparatus,' said Wright. Mellanby replied that anyone who fiddled with apparatus instead of doing
xperiments got nowhere. Wright retorted, 'Anyone who
tarts an experiment with crook apparatus is a bloody fool.'
Describing the occasion he writes: 'I was surprised that this
erminated the exchange instead of starting a conversation.
Florey and Mellanby practically shot out of the door and Jen-
ings collapsed on to the table. He indicated that Mellanby
eld his future in his hands and that I had wrecked it.' Actu-
lly, outside the door, Mellanby had said, quite mildly, to
Florey: 'Another of your bloody Australians, I suppose.'[15]

12

LYSOZYME LEADS TO PENICILLIN

After his return from Australia in 1936, the difficulty of financing his teaching and research plans became one of Florey's main preoccupations. The energetic young workers he had attracted to the Dunn School demanded increased laboratory supplies and services that could not be met from a small, fixed budget. One of the more serious consequences of this financial stringency was the scaling-down of the teaching course; another was the diversion of his own time and energy. He had to engage in the unpleasant task of grant-begging from all likely sources, with endless applications, descriptions of projects, reports, estimates, letters, interviews, and the final frustration of waiting—perhaps for months—for the decisions of those who dispense this form of charity. Nothing could have been more irksome to someone of Florey's independent and energetic character.

If he had expected financial help from the new Nuffield scheme, his hopes must have soon faded. By the end of 1937 most of the Nuffield posts had been filled and the funds earmarked. Cairns had, predictably, become Professor of Surgery and was building the neurological unit that was to be the best in Britain. The Nuffield Professor of Medicine was L. J. Witts, with whom Florey had once worked in Cambridge. There was a professor of obstetrics and gynaecology and—at Lord Nuffield's personal request—a professor of anaesthetics, a subject not before regarded as of academic status. (Objections to the establishment of this latter chair and to his nomination of its holder were dealt with quite simply by his Lordship; he declared that if he did not get his way the University would not get their £2 million.[1]) Two other chairs were also established: one in therapeutics for J. A. Gunn, Director of the Nuffield Institute; the other for orthopaedic surgery at the Wingfield Morris Hospital. Below professorial level, there were Nuffield additions to the staff an accommodation of the hospital departments of paediatrics, pathology, and biochemistry. A. G. Gibson became Nuffield Reader in Morbid Anatomy and Director of Pathology, with A. H. T. Robb-Smith (who had worked in Kettle's department at St. Bartholomew's Hospital) as Assistant Director. R. L. Vollum was appointed Clinical Bacteriologist, though still retaining his post at the Dunn School, and clinical biochemistry was put in the hands of J. R. O'Brien, a demonstrator in the University Department of Biochemistry.

The University, perhaps daunted by the very size of the Nuffield Benefaction and the complexity of its medical consequences, had vested the funds in a board of trustees, and their administration in a special Nuffield Committee (on which the Nuffield staff had a majority) that was responsible only to the General Board of the University and not to the Board of the Faculty of Medicine. This gave the Nuffield Committee a degree of autonomy that was disliked both by the hospital staff and by the medical faculty, and it was thought to be unsympathetic to the basic medical sciences on the one hand and to ordinary clinical medicine on the other. Instead of becoming a bridge between these two sides, the Nuffield organization tended to become a self-sufficient island.

While the Nuffield professors, with their readers, lecturers, and assistants, were being provided with new buildings at the Radcliffe Infirmary—wards, operating theatres, laboratories, and offices—and with salaries and research funds far above those applying to equivalent posts in the pre-clinical departments, Florey was having to impose the most petty restrictions on his own colleagues. Lights must be turned off, heating reduced, all laboratory expenses limited to bare essentials. From time to time there was no money to pay for the barest, and Florey would send round an order that nothing—not even glass tubing—was to be issued from the stores until further notice. It is, perhaps, a reflection of these financial difficulties and of the time and energy spent in trying to overcome them, that Florey published only two papers during 1936 and 1937, the smallest output for a two-year period in the whole of his long career.[2]

Chain was as much affected by these economies as anyone.

When the year of his £200 appointment was ended, in 1936 Florey had to find another source of income for him. Cance research was better endowed than most other lines, and it wa already being done in the department by Beatrice Pullinge and the Beit Fellow who had joined her, I. Berenblum. Chai was an enzyme chemist and Florey thought that he could par ticipate in this cancer work by studying the differences—i any—between normal tissue enzymes and those of malignan growths. The project suited Chain. It also suited Florey because any study of tissue enzymes might lead into the fiel of natural antibacterial substances and immunity that ha long been an intriguing prospect. So he submitted an applica tion to the British Empire Cancer Campaign in 1936, an Chain was awarded a personal grant of £300, with expense of £100, per annum for the next three years.

The biochemistry of tissue enzymes involved the develop ment of micro-methods, and Chain felt that he needed a colla borator who was expert and inventive in the field of laborator engineering on a minute scale. He already knew the man h wanted, N. G. Heatley, who was in Gowland Hopkins's de partment completing his scientific education, which ha begun at Tonbridge School and continued at St. John's Col lege, Cambridge, where he had taken the Natural Scienc Tripos. Heatley had no immediate prospect of a job in Cam bridge, and he was glad to come to Oxford. Florey had suc ceeded in getting a grant from the Medical Research Counc to pay his salary, and in his application he had stated tha Heatley would work with Chain on cancer tissue metabolisr and that he would also study the biochemistry of lymphocyte

At that time, the research that was most occupying collec tive attention at the Dunn School was the lymphocyte projec The programme that Florey had written out in detail fc Sanders, Medawar, and Jean Taylor (by then Mrs. Medawa finally involved eight graduate research workers, each wit a particular aspect of the problem to investigate. Whil Sanders continued his work on lymphoid tissue extirpatio Jean Taylor, using the tissue culture techniques develope by Peter Medawar, and R. H. Ebert, with the rabbit ea chamber, studied the possible transformation of lymphocyt into other cells. The problem of the fate of lymphocytes the intestine was assigned to two new recruits: Alan Cruik-shank, a B.Sc. student, and J. M. Barnes, a Nuffield Fellow, who worked directly with Florey. Heatley applied his micro-methods to the study of lymphocyte metabolism. Finally, by the most delicate and difficult techniques, Florey, Sanders, and Barnes examined the effect of depriving an animal of its normal supply of lymphocytes by making the thoracic ducts discharge their lymph outside the body.[3]

The whole project was an example of the sort of team work that is now common but was then almost unknown in Britain. Each member of the team solved his own problems in his own way, but the results contributed to the larger pattern. It was a realization of Florey's ambition to direct research on a wide front by simultaneous attacks along different lines, each yield-ing information that he could then build up into a complete picture. This grand scheme, though interrupted by sub-sequent events, yielded much information and several published papers. And the parts of it that were resumed when the Dunn School returned to its normal academic life after the war led directly to discoveries of fundamental importance in immunology.

There were incidental consequences of this programme. Sanders, Ebert, and Florey took advantage of the improved rabbit ear-chamber technique and the perfection of ciné photomicrography to produce a magnificent colour film of the micro-circulation. This showed the normal flow of the red and white blood cells in minute detail, and the dramatic changes that occur with local damage or inflammation, and it is rightly regarded as a classic. But, on the debit side, Dr. Pullinger, who had done so much to make this work possible, was no longer happy at the School of Pathology. She found the anti-podean contingent there rather overpowering. One of her friends (who was later one of the first two women Fellows of the Royal Society) exclaimed after a visit to the Dunn School: 'I can't think how you can work with those Australian bandits!' More important, perhaps, she was losing touch with the main lines of research. They were becoming too biochemi-cal for her to follow. W. E. Gye had told her that if ever she decided to leave Oxford, he would be glad to offer her a research post. So, in July 1937, she joined the staff of his

Imperial Cancer Research Fund laboratory at Mill Hill, where she remained for the next fourteen years.[4]

One of Florey's research lines came to an end about thi time, because it had reached a practical conclusion. This wa the work on contraception. With Carleton he had established many physiological facts about the process of fertilization and the action of spermicides. This work, with a clinical applica tion in mind, was continued at the Dunn School by Carleton and a zoologist, Dr. (later Professor) J. R. Baker, who wa given laboratory space by Florey because the Professor c Zoology considered the work inappropriate to his own depart ment. Baker and Carleton tested many spermicides and thei possibly harmful effects, and the result (in 1938) was the pre paration commercially called 'Volpar' (from Voluntar Parenthood)[5] which remained in wide use for three decade:

Meanwhile, the biochemical work on lysozyme had been making rapid progress despite the ever-present financial diffi culties. The Medical Research Council had made a gran in 1935 to pay Roberts's salary, but when Florey applie for money in April 1936 to buy him an essential piece c chemical apparatus, the request was refused. Florey the turned to the Rockefeller Foundation for help. He was o friendly terms with Dr. Warren Weaver of the Foundation' Natural Sciences Division, and it was to this division that h applied for an expenses grant. The work was, indeed, chemi cal, and therefore not quite appropriate to the Medical Re search Division. Another reason, perhaps, for Florey's choic of approach was that the Medical Research Division of th Rockefeller Foundation had an arrangement to work throug the British Medical Research Council, so that any applicatio would automatically be referred to Mellanby, whose Counc had just turned down a similar request. Florey's tactics wer in the event, successful. He was promptly granted $125 (about £320)—a very useful sum. In consequence, Robert using material provided by Florey and Maegraith from di ferent animal sources, was soon able to purify lysozyme 1 a degree that enabled E. P. Abraham, working with S Robert Robinson, to crystallize it in 1937.[6] Thus the poir had been reached from which an attack could be made on th final problem, that of its precise mode of action on bacteri

The answer would, of course, involve, or perhaps provide, knowledge of the structure of the bacteria themselves, and it was such knowledge that Florey hoped might lead to the discovery of other ways by which they might be destroyed.

The reasons for Florey's sustained interest in lysozyme are not immediately obvious. He was almost alone in pursuing its study, since even Fleming had abandoned it. It presented, of course, a scientifically interesting problem. Its ability to dissolve bacteria was a challenge to the chemist and the bacteriologist alike, and its wide distribution in nature posed the obvious possibility of a physiologically important function. Florey often declared that his researches were motivated only by scientific interest: in fact, it seems that he was more aware of the clinical possibilities of his work than he cared to admit. For example, his experiments on the treatment of tetanus had a direct clinical bearing; he had worked on the causes of duodenal ulcers with Harding; on colitis, and on epileptiform convulsions. But the control of bacterial infection was a constant interest which gave a positive objective to his lysozyme studies.[4] In Oxford, R. L. Vollum tried, at Florey's instigation, to discover ways by which lysozyme could be made to attack pathogenic bacteria.[7] The advent of chemotherapy did not reduce Florey's interest in biological antibacterial substances, but in fact increased it. When he had visited Professor MacCallum's laboratory in Melbourne in 1936, bringing with him the first sulphonamide to be seen there, he gave an informal lecture on its therapeutic triumphs, but also on its serious limitations.[7] Only certain organisms were susceptible. The drug would not work in chronic or localized tissue infections, because the products released by the leucocytes in pus inhibited its chemical action on bacteria. The sulphonamides, too, had unpleasant or dangerous side-effects, such as persistent vomiting, skin-rashes, liver and kidney damage, and, most serious of all, a tendency to destroy the ability of the bone marrow to produce leucocytes—then a usually fatal condition. Florey stressed the point that the sulphonamides, life-saving though they were proving to be in certain acute infections, were only the beginning of the attack on bacterial disease. An ideal antibacterial agent would be non-toxic and effective under all conditions. He expressed very much the same views in a lecture to the pathology class at the Dunn School in 1937, in which he discussed the possible development of biological (as opposed to chemical) antibacterial agents.[8]

With the purification of lysozyme in 1937, Chain was able to take up the problem that Florey had had in mind for him when he first came to Oxford. He had completed his work on snake venoms, and the cancer research that provided his salary involved tissue enzyme studies easily integrated with experiments on lysozyme. He had acquired an able collaborator, L. A. Epstein, an American D.Phil. student and Rhodes Scholar who joined him in October 1937, and their first task was to determine from a study of the kinetics of its action whether or not lysozyme really was an enzyme. Research of this sort needs special equipment, and though Florey's frequent applications to Mellanby were beginning to illustrate the law of diminishing returns, the Medical Research Council provided in April 1938 a grant of £350 for the salary of a biochemical assistant (R. V. Ewens) and for the expenses of 'work on the biochemical action of bacterial toxins', which would also cover the sort of equipment needed for research on lysozyme. One of the bacterial toxins investigated was the so-called 'spreading factor' produced by virulent streptococci and gas-gangrene organisms and which allows them to invade the tissues with alarming rapidity. Chain and Edward Duthie (who had been appointed a demonstrator in 1937) studied an enzyme produced by such organisms that was able to liquefy a viscous component of mucus. It was therefore called 'mucinase'[9] and it seemed that its lowering of the viscosity of tissue fluid might allow the rapid spread of the infection. Later it was found that the bacterial enzyme attacked hyaluronic acid, a long-chain polysaccharide that acts as a tissue cement, and it was re-named hyaluronidase. This research like other important lines, had stemmed from Florey's original interest in mucus. He had shown that one of the factors in its antibacterial function was simply the viscosity that prevented microbial penetration. Chain's interest in the enzymic nature of toxins, derived from his snake venom studies, was therefore directed to a possible action on mucin in the case of 'spreading factor'.

During the next academic year (1938–9) Chain and Epstein

had satisfied themselves that lysozyme was, indeed, an enzyme—a polysaccharidase. The next step was to identify the precise substrate—the chemical structure—in the bacterial cell wall that was broken down by this enzyme. This involved culturing about 150 grams of *M. lysodeikticus* per week. Neither Chain nor Epstein, being biochemists, had much experience of bacteriology, and they were faced with the necessity of using large-scale culture techniques, which involve special problems. They had, of course, the advice of A. D. Gardner, who was now head of the bacteriology section with the title of professor, and with his help they grew the bacteria in large Winchester bottles that had been coated inside with a layer of agar by rotating them on horizontal rollers. The harvest of bacterial cells was dried, and then fractionated into its various chemical components. One of these was a polysaccharide specifically attacked by lysozyme, and it was a fairly straightforward biochemical problem to identify it, and then the chemical bonds in its molecule that were broken by the enzyme. Chain and Epstein showed that the substrate for lysozyme in the bacterial cell wall was N-acetyl glucosamine.[10] They published their results in 1940. Writing of them in 1972, Chain considered that they 'marked the beginning of, and laid the foundation for, a chapter of biochemical research ... the chemical nature of the bacterial cell wall and its enzymic hydrolysis and synthesis'.[11] For Chain, therefore, the lysozyme work was complete.

Chain's brilliant solution to the problem of lysis by lysozyme did not of itself lead to any significant advance in the search for natural antibacterial substances. It was, therefore, something of an anticlimax for Florey since, however interesting it might prove to be for the chemist or the bacteriologist, it then seemed to indicate a physiological dead-end. However, the work did provide a quite unexpected diversion of the greatest possible importance, and one in which chance, yet again, seems to have played a decisive role. Different accounts have been given by different writers (and even by the same writers) of the events that led Florey and Chain to investigate penicillin. There is little to be gained by discussing academic details, or accounts for which there is no factual basis. For the present, the best course is to rely on accounts by the people actually involved. It should be made clear that, at the time, the decision to work on penicillin seemed neither very surprising nor significant; it was merely one of the several research options taken up in parallel with the closing stages of the lysozyme project. There was nothing so remarkable about it that its beginnings were firmly and accurately fixed in people's memories or notebooks.

In 1945, Florey wrote as follows: 'As this work [on lysozyme] proceeded, Dr. Chain and I had many discussions on the possible interest of a systematic and thorough study of the known natural antibacterial substances, and it seemed to us that such a survey might prove well worth while from many academic as well as practical points of view.'[12]

From a detailed account by Chain, published in 1972,[11] it seems clear that his own interest in the antibacterial substances produced by various micro-organisms was stimulated by his reading of the historical literature relevant to his work on lysozyme. In general, research workers can be divided into those who like to follow a single line, to dig down deeper and deeper into the foundations that underlie all biological processes, and those who like to take a wider view, to find parallels and general principles from some particular discovery. Chain was the latter sort of research worker. Instead of following lysozyme into the deepest recesses of microbiology, he preferred to see it as a special case of the general phenomenon of bacterial lysis by natural substances. Since these might be enzymes—hitherto unrecognized—of biochemical interest, he searched the literature for examples of such activity, and also of bacterial inhibition, a lesser effect of lysozyme. He began this literary research 'in the years 1937–38', that is, when his work on lysozyme with Epstein was starting. He certainly discussed his findings with Florey who had himself published a good example of the inhibition of one sort of bacterium by another in his paper with Goldsworthy in 1930 and had also referred to two reviews covering the whole field of microbial antagonism. Florey drew Chain's attention to one of these reviews, the monograph by Papacostas and Gaté, published in 1928, which has a sixty-page chapter on 'antibiosis' (a word coined by Vuillemin in 1889 and gives several hundred historical references.[13]

Chain collected 'about 200 references on growth inhibitions caused by the action of bacteria, streptomycetes, fungi and yeast on one another. It was evident that in many cases the growth inhibition was caused by specific metabolites produced by various micro-organisms. However, next to nothing was known about the chemical or biological nature of the inhibitory substances, and it seemed an interesting and rewarding field of exploration.' Then Chain found Fleming's paper on penicillin:

I had come across this paper early in 1938 and on reading it I immediately became interested. The reason was that, according to Fleming's description, the mould had strong bacteriolytic properties against the staphylococcus ... When I saw Fleming's paper for the first time I thought Fleming had discovered a sort of mould lysozyme, which, in contrast to egg white lysozyme, acted on a wide range of gram-positive pathogenic bacteria. I further thought that in all probability the cell wall of all these pathogenic bacteria whose growth was inhibited by penicillin contained a common substrate on which the supposed enzyme acted, and that it would be worth trying to isolate and characterise the hypothetical common substrate. For this purpose it would, of course, be necessary to purify the supposed enzyme, but I did not foresee any undue difficulties with this task for which I was well prepared from my previous research experience.

Chain's discovery of Fleming's paper was a piece of great good fortune, since it fired an enthusiasm for a definite research project that he was able to convey to Florey. But the chance that led him to find it was not a very remote one. Fleming's paper was in the *British Journal of Experimental Pathology* (all the volumes of which were on the shelves of the Dunn School library), and in this same journal were four papers on lysozyme that Chain would have had to look for: two by Fleming and two by Florey. Fleming's lysozyme papers were in Volumes 3 and 8, and both of Florey's in Volume 11, so it was quite by accident that Chain, glancing through the contents of Volume 10, came across Fleming's name and the title of a paper that caught his eye. It is interesting that it should be the (mistaken) impression given by Fleming that penicillin was a bacteriolytic enzyme that attracted Chain's attention and suggested to him that its investigation would be a logical parallel to that of lysozyme.

Florey, of course, was already familiar with much of the field about which Chain was becoming so enthusiastic. He had read the historical reviews on antibiosis that he had quoted in his paper with Goldsworthy. It is almost impossible to believe that he had not read Fleming's penicillin paper, though he had not quoted it. He habitually read every paper of interest that appeared in the *British Journal of Experimental Pathology*, and he was already one of its editors when it published Fleming's paper in 1929. It is certain that he heard a good deal about penicillin from Paine in Sheffield though, in Paine's words, 'He did not seem very interested at the time.'[14] But he knew enough about Paine's work to remember that he had devised a method for testing the toxicity of anti-bacterial agents on leucocytes, and to write to him for details when a similar test was needed for the Oxford penicillin work in 1940. But he had never shown any inclination to follow up his own single observation of antibiosis, nor Paine's work on penicillin. His attention was fixed on lysozyme and on the possible existence of similar tissue enzymes. It had needed Chain's promptings to redirect his attention, as he himself acknowledged: 'Eventually Chain proposed, and I agreed to go along with it, that we should make a thorough investigation of antibacterial substances. This was why we looked at peni-cillin.'[15] And Chain writes: 'My part of this project was the isolation and study of the chemical and biochemical properties of these substances, Florey's the study of their biological properties.'[11]

Of the many historical observations that had prompted this joint venture by Chain and Florey, only a few examples need be mentioned here. (A full account can be found in *Antibiotics*, the two-volume book produced in 1949 by Florey and his colleagues.[16]) The first published observations on antibiosis seem to be those of Pasteur and Joubert in 1877. They noted that an unidentified but common airborne bacillus inhibited the growth of anthrax bacilli in culture and, when mixed with the virulent organisms, prevented them from producing anthrax in animals. They recognized the possible use of such bacterial antagonism, writing: 'Tous ces fait autorisent peut-être les plus grandes esperances au point de vue thérapeu-tique.'[17] Ten years later, in 1887, the Swiss bacteriologist Garré published other examples of bacterial inhibition, and the description of a method by which such phenomena could be demonstrated on culture plates—a method subsequently used in principle by most workers, including Fleming and Florey in their experiments on lysozyme, and Fleming in his celebrated work on penicillin. In 1889 Emmerich and Löw, in Germany, found that *Bacillus pyocyaneus* inhibited a variety of pathogenic organisms, and extracted from its cultures an antibacterial substance they called 'pyocyanase'. This proved too toxic for human injection, but it was produced commercially as a local antiseptic for over forty years. In 1907 the French worker Maurice Nicolle clearly demonstrated the existence of a bactericidal and bacteriolytic agent produced by the soil organism *B. subtilis*, and both he and others in France employed it to dissolve bacteria for use as vaccines. Greig-Smith, working in Sydney in 1917, made the first clear-cut observation of the antibacterial power of certain actinomycetes, soil organisms with affinities to the class of fungi. This work was taken up in 1924–5 by the Belgians Gratia and Dath, who performed many experiments on the bacteriolytic action of an actinomycete on pathogenic organisms and an agent called 'actinomycetin' or 'actinomycin'. They also observed the antibacterial effect of certain strains of *B. coli*, as Goldsworthy and Florey were to do in 1930.

Besides these and many other observations on bacterial antagonism, various moulds and fungi were found to inhibit bacterial growth. Such things had been used in 'folk-medicine' for centuries throughout the world to treat infected wounds. Lister's experiments in 1871 on the antibacterial action of a *Penicillium* mould seem to have been the first scientific observations in this field, but as described in Chapter 1, they were not published. Lister's notes on the subject in his 'common-place books' were discovered by Guthrie,[18] and Florey referred to them in his Lister Oration in Vancouver in 1954.[19] The first published experiments were by Gosio, an Italian, who described in 1896 the production from a *Penicillium* of a crystalline substance (now called mycophenolic acid) that inhibited the growth of anthrax bacilli. But he was unable to prepare enough material to allow animal experiments to be made.* In 1897, Duchesne, a twenty-three-year-old French army doctor, submitted a thesis in which he

*The author has since learned that in 1895 V. Tiberio of Naples published work on the inhibition of pathogenic bacteria by extracts of *Penicillium* and *Aspergillus* moulds and made some (inconclusive) animal tests. *Annali Ig. sper. Roma*, **5**, 91 (1895).)

described the protective action of *Penicillium glaucum* when inoculated into animals at the same time as highly virulent pathogenic organisms. He pointed out the obvious therapeutic possibilities, but his early death from tuberculosis cut short his work.[20] Vaudremer, also in France, used a product from the mould *Aspergillus fumigatas* in 1913 to treat tuberculosis in animals and also in man; though the results were inconclusive, the material was at least non-toxic.

Finally there was the contribution by Fleming on penicillin. Seen against a background of numerous and rather similar observations, one can understand why penicillin did not stand out with any startling clarity. Its essential superiority to the other antibacterial agents was not apparent. Running through earlier work was the awareness of possible therapeutic value, but this had been frustrated by factors that were then decisive. There was the difficulty of producing enough material for adequate testing. Then there was toxicity; the substances with the greatest bactericidal power had been too toxic to be injected, while those that were non-toxic had little effect on bacteria in the body. Fleming had shown that penicillin was both non-toxic for animals and lethal for pathogenic organisms in culture. But he had never tested its bactericidal power in the animal body. Even a crude extract of penicillin was difficult to prepare in any quantity; it was unstable, and the half-hearted attempts at purification by several biochemists had been unsuccessful. The therapeutic potential of penicillin thus remained unrecognized through lack of data, and it remained just one of many antibacterial substances that had been observed, briefly investigated, and then forgotten. Though penicillin had been effective as a local antiseptic in a few cases of superficial infection, it remained in being only because it was useful in the laboratory as a selective bactericide in the culture of certain delicate organisms such as *B. influenzae*.

If it is instructive to see penicillin in its contemporary context in 1938, it is necessary to see Chain and Florey's interest in it in the same perspective. Chain was an extremely active research worker, full of ideas and enthusiasms. He was already working on several different projects—lysozyme with Epstein and Roberts, cancer tissue metabolism with Heatley and

Berenblum, 'mucinase' with Duthie, the biochemistry of venoms and toxins with Goldsworthy and Ewens, some chemical aspects of gastric function with Goodfriend and Florey—and he had published seven scientific papers during the past two years. He was, in fact, building up a biochemical subdepartment in the School of Pathology, a development that was exactly what Florey had planned. Chain, by then involved in several research projects and with several collaborators and assistants, was no longer the penniless refugee who had come to Oxford three years before. He was anxious to expand his research and his laboratory organization. And with this expansionist urge came the same frustration that had beset Florey for so many years—the lack of adequate research funds. Thus, as will be seen later, one of the reasons why Chain favoured the antibacterial project was that he thought it might attract a long-term grant. Both he and Florey, of course, were basically interested in the scientific aspect of the subject. They selected substances for study that seemed likely to be enzymes and which would lead on to the study of bacterial structure. In his 1945 statement Florey wrote: 'Eventually three were chosen for the first investigation—the products of B. subtilis, B. pyocyaneus and penicillin.'[12] (In fact, in the first mention of this project to the Medical Research Council in January 1939, the substances specified were penicillin and actinomycetin.[21]) The least consideration, at that time, was the possibility of therapeutic value, as Florey recorded:

There are a lot of misconceptions about medical research. People sometimes think that I and the others worked on penicillin because we were interested in suffering humanity—I don't think it ever crossed our minds about suffering humanity; this was an interesting scientific exercise. Because it was some use in medicine was very gratifying, but this was not the reason that we started working on it. It might have been in the background of our minds—it's always in the background in people working in medical subjects ... but that's not the mainspring.'[15]

And Chain wrote: 'The only reason that motivated me was scientific interest. That penicillin could have a practical use in medicine did not enter our minds when we started work on it.'[11] However true this may have been of their attitude at the outset of their work, it was an attitude that was to change profoundly. And one wonders how deeply in the background the idea of 'suffering humanity' really was in Florey's case, when his letters to Ethel are remembered.

Chain made a start, in the summer of 1938, with penicillin and pyocyanase. Work on the latter was largely carried out by Miss R. Schoental, a Polish refugee biochemist. She extracted from it three separate antibacterial substances, which all proved to be highly toxic. At the same time Chain and Epstein began work with penicillin. Their first requirement, of course, was a culture of Fleming's mould, but this turned out to be no problem. There was already such a culture in the Dunn School, where it had been maintained for several years by Miss Campbell-Renton. Its origin is in some doubt. In most published accounts it is stated that it had been obtained by Dreyer from the National Type Collection for use in his research on bacteriophage, but there is a suggestion that R. L. Vollum had asked Fleming to supply the culture for his own bacteriological work.[22] In any case the mould was available, alive and active, and it proved to be an excellent producer of penicillin. But it is clear that not many biochemical discoveries were made with it that year. Persuading the mould to produce enough 'juice' for useful experiments was a messy and uncertain process. As Epstein (who had by then changed his name to Falk) described later: 'With Florey's permission, Chain and I recultured this strain and tested the antibacterial properties of the medium on several cocci. The results were not impressive. Preliminary experiments rarely are. It was a particularly busy moment in the lysozyme research ...'[23]

Some slight progress was made, however. Chain's supposition that penicillin was 'a sort of mould lysozyme' capable of dissolving bacteria was soon disproved when he failed to reproduce Fleming's original observations. He did not know then that the lysis observed by Fleming was not a direct effect of penicillin, but was due to autolysis (self-digestion) that occurs only when susceptible organisms are exposed to penicillin at a particular stage of their growth. It was only later that the mystery of Fleming's observation of lysis (which neither he nor anyone else had been able to repeat) was solved. Meanwhile, its non-occurrence in Chain's laboratory surprised, but did not deter him.

The main source of his biochemical information came from the paper by Clutterbuck, Lovell, and Raistrick, and Chain followed their methods in the hope of taking them further. He grew his mould on their synthetic medium, tested the anti-bacterial power of the filtrate, and then began to concentrate what was a watery solution of numberless components—including penicillin. Alcohol, as all but its most hardened consumers know from experience, mixes with water. Most proteins are soluble in water but not in alcohol; so if alcohol (or acetone) is added to a protein solution in water, the protein is precipitated. Fleming (or, rather, Craddock and Ridley) had used this obvious first step towards separating penicillin and had found, surprisingly, that it was not precipitated with the proteins, but remained dissolved in the alcohol. This was a useful result, but it suggested that penicillin was not a protein. Fleming could not dissolve it in other solvents, such as ether and chloroform, that do not mix with water. However, Raistrick and his colleagues showed that penicillin *is* soluble in these solvents in the presence of acid. Penicillin can therefore be extracted from its watery mixture of impurities by acidifying and then shaking with ether. But, having got the penicillin into the ether, the problem was to get it out, since attempts to evaporate the ether or re-extract the penicillin with water resulted in a loss of activity. Though they abandoned their work at this point, Raistrick and his colleagues had shown the importance of pH (the degree of acidity or alkalinity) on the solubility of penicillin and had shown that low temperatures were essential to preserve its activity.

Chain used these observations as starting-points. What neither he, nor those who joined him later, knew was that Lewis Holt had solved the problem of re-extracting penicillin from solvents in Fleming's laboratory, but he had not published his results, and Fleming himself did not mention them in his papers or lectures.[24] Chain, with only the published biochemical findings to go on, and with some new observations of his own, felt that penicillin presented just the sort of challenge he found stimulating. Its curious solubilities, peculiar instability, and other biochemical indicators all suggested a molecular structure of a most unusual sort. He had found that the active substance would pass through micropore filters that retained proteins. Its molecule, therefore, must be relatively small and it was almost certainly not an enzyme. This, though biochemically disappointing, meant that if the substance were to prove therapeutically valuable, it could be given by injection without causing the allergic reactions produced by foreign proteins. And the smaller the molecule, the better the chance of identifying its structure. Florey and Chain discussed all these points and decided on their course of action.

In the strong light of subsequent events one naturally seeks the details of any deliberate decision to make penicillin the subject of a major research effort in Oxford. Certainly such a decision was taken, because the resources of a university department became heavily committed to the project; and it must have been taken by Florey, since only the head of the department could have authorized such a commitment. But was the decision a sharp one—as some accounts suggest—taken on the spur of a particular moment? And, if so, when did this moment occur? Or was it more a gradual change of direction induced by the build-up of promising evidence? Chain would support this latter view. Although he discussed his preliminary experiments on penicillin with Florey over a period of several months, he felt that he did not fully engage Florey's enthusiasm for the project until the autumn of 1939. But there is other evidence suggesting that Florey had already made up his mind on penicillin as a worthwhile subject about a year earlier. Two published accounts describe this decision as being reached at a particular moment. One of these is attributed to Florey himself. Towards the end of 1938 he was dining at home with his family—an event rare enough by then to be itself memorable—and during the meal he told them of the proposed work on antibacterial substances. He then said, so Bickel quotes Paquita, 'Tonight I came to a final decision, and the moment I reached that decision I was standing beneath the big old chestnut in the park at the back of the building.'[25] Masters, in his book *The Miracle Drug*, has this account: 'They [Florey and Chain] were walking home one evening discussing the problem on what to start on first when, just as they were passing under a great elm at the entrance to the park, they decided to start on penicillin.'[26]

This anecdote, which is now part of the penicillin mythology, raises some interesting points. Whether it was an elm or a chestnut, why should Florey have remembered the tree at all? It is a psychological commonplace that a sudden significant or surprising mental experience tends to fix an equally vivid picture of the subject's physical environment at that time firmly in his memory. But what would have been so remarkable at that time in a decision to make penicillin a major research project? Scientists are not given to premonitions, and Florey had made scores of similar decisions on research projects in the past. That he did, indeed, make such a decision at that time is confirmed by R. D. Wright, and in a letter on this subject he may have given a clue as to why Florey might have considered it significant. After describing his first meeting with Mellanby (see p. 271), Professor Wright continues:

Ever afterwards Mellanby and I met on equal footing—North Country style. In Zurich, at the Physiological Congress, mid-1938, we saw quite a lot of each other and it came out that he was having trouble keeping funds available for Florey's antibacterial biological materials programme, and was telling Florey that he must make one full-blooded throw. When I was going to the train [at Oxford] to come back here [Melbourne] in late November 1938, Florey told me of his plan to go all out for penicillin.[7]

Florey, in fact, had then decided to gamble, realizing the penalties of failure rather than the undreamed-of consequences of success.

If Florey had, indeed, made up his own mind on penicillin as a promising research project in the autumn of 1938, he did not begin to apply his energies to it until several months later. Penicillin was briefly mentioned in January 1939 in his application for funds from the Medical Research Council, but finance for a definite project was not sought until September 1939. There were reasons for this interval. The lysozyme work was still being completed by Chain and Epstein, and the lymphocyte project, among other research interests, was still occupying Florey himself. But the main disturbing influence was on a global scale, one that was beginning to cast an ominous shadow over all plans for the future. Europe was by then plunging towards a war that was becoming accepted by most realists as inevitable.

With Hitler's annexation of Austria in March 1938, the

unholy alliance of the Rome–Berlin Axis presented a grimly united front. The next blow was threatened against Czechoslovakia, and there seemed to be every chance that this would fire the general conflagration. Britain, after years of economic decline and inactivity, had barely awoken to the danger. There were belated efforts to re-arm and to face the fact the the civil population had no protection against the bombs and gas attacks from the air that could be expected within a few hours of the opening of hostilities. Then, on 30 September, when these hostilities seemed imminent, Chamberlain went to Munich and returned with Hitler's signed promise of 'peace in our time'—a pact that proved to be about as dependable a protection against the Luftwaffe as Chamberlain's much-publicized umbrella. Meanwhile there was relief, though it was marred for many by a sense of shame. They realized that the time bought for Britiain had been at the expense of Czechoslovakia and was merely a respite that must be used to prepare effectively for an unavoidable conflict.

In every city, town, village, and household in the country, plans were being made for Civil Defence. They ranged from the central organization of the fire services, air-raid wardens and the provision of shelters, gas-masks, and first-aid posts down to the domestic details of blackout curtains, sand buckets, stirrup pumps, and protection from flying glass. There were plans for the evacuation to country areas of a large part of the population of the more vulnerable cities. On the medical side, the major hospitals would be placed under an almost military command—the Emergency Medical Service. Those near city centres would be largely evacuated to provincial hospitals, sanatoria, and asylums. One of the medical innovations was the setting-up of a blood transfusion service, made possible by the technical advances of the previous few years and the practical experience gained in the Spanish Civil War. The development of the gravity drip-feed method of transfusion, coupled with the introduction of new fluids that preserved the blood and prevented clotting, meant that blood could be bottled and stored at 4 °C. for two or three weeks, so that a 'bank' was always available. Largely instigated by Dr. (later Dame) Janet Vaughan in London, the transfusion service was later co-ordinated by the Medical Research

Council through a committee under the chairmanship of Florey's old Cambridge colleague, Dr. A. N. (later Sir Alan) Drury. The hospital pathology services of the whole country were to come under central direction, and the man designated to command was Sir Philip Panton, of the London Hospital. An Emergency Public Health Laboratory Service was also planned, and when established had its headquarters at the School of Pathology in Oxford.

The Oxford medical plans conformed to this general strategy. The Radcliffe Infirmary would become the Area Emergency Medical Service Hospital, with Major-General Sir Robert McCarrison, who had recently retired from the Indian Medical Service, as Group Officer. Dr. Gordon Sanders would be Area Transfusion Officer, and the original idea that he would be based at Reading was dropped in favour of Oxford, largely through Florey's intervention. Florey then managed to acquire for the transfusion service the use of the large reading room and two adjoining rooms in the New Bodleian Library, which had not yet been commissioned for its academic purposes.

Under Sanders's direction the arrangements for the Oxford area transfusion service were worked out in detail. The preparation and sterilization of the equipment and solutions would be carried out at the Sir William Dunn School of Pathology. A van for transport between the laboratory and the New Bodleian and outlying centres would be shared with the Emergency Public Health Laboratory, when it became established. The enrolment of blood donors was organized by Dr. Sybil Creed, the wife of a physiologist, and the V.A.D. nursing service by Mrs. Hobson, the wife of a senior hospital physician. Teams of doctors from the hospital staff to bleed the donors were organized by Ethel Florey. When the scheme finally came into operation she was very much in charge of the blood-collecting sessions. Good-looking and invariably well dressed, she impressed her teams of young Radcliffe doctors with her undoubted charm tempered with a steely determination that made for efficiency through trepidation. She wore a double tortoise-shell ear-trumpet that curved beneath her hair, and there was a general agreement that her hearing was selective, in the sense that, in the event of an argument, it was her opponent's most telling points that seemed inaudible.[27]

Ethel was meticulous in her work, and she obviously enjoyed it. Probably for the first time since her marriage she was in a position to take an active part in affairs beyond her purely domestic world. This latter world was, if anything, contracting, since Howard was withdrawing more and more from it. Ethel had her own rooms at The Red House, which he hardly ever entered. When their friend Mrs. Thomson (Nan Mitchell) came to stay in 1938, Howard joined her one evening for a glass of sherry with Ethel. But he soon left them, and Ethel remarked, 'That's the first time he has been in my room for two years.'[28] The Florey children had happier memories of their parents' relationship, and took their bickering less seriously than did the guests and visitors, who were distressed by it. There had been family holidays in Cornwall and a bicycle tour of Brittany, during which Howard and Ethel evidently enjoyed the company of each other and their children. There also remained the bond of being Australian and a shared amusement at the strange pretences of the English. But Paquita was aware of strained relations. A few years later she remarked to her mother's sister, Emmeline, who was staying with the Floreys in Oxford: 'How extraordinary it is that the two most difficult people in the world should have happened to marry each other!' If by 'difficult' she meant 'determined', she had identified the source of conflict.

During the first months of 1939 Florey, like the other heads of University departments, was himself occupied to an increasing extent with the planning of emergency arrangements. And being concerned with medical teaching he would also have to budget for the possibility of many more students. It was planned to evacuate the London hospitals with their medical schools in the event of war, and Oxford could expect a much larger intake of pre-clinical students who would then proceed to complete their clinical training at the Radcliffe Infirmary, where special arrangements would be made to accommodate them. Such distractions diverted time and energy from Florey's research plans, but probably not to any great extent. Within a few months Europe would be committed to a desperate struggle against the bid by the Fascist dictators for world domination, while in Oxford a few people would be beginning their own struggle to produce penicillin, a struggle destined to save humanity from the domination of many bacterial diseases. But, during that short period of 'peace in our time', life in Oxford and at the Sir William Dunn School was not much disturbed by the prodromal manifestations of great events soon to come.

13

THE DEPARTMENT BECOMES
A FACTORY

Despite the distractions of university committees, the continued struggle to raise research grants, and the quickening tempo of preparations to meet the threat of war, Florey completed a good deal of personal research during the years 1938–9. He published eight papers: four on mucus secretion with Wright, Dr. Margaret Jennings, and others; one with Roberts and Maegraith on lysozyme; one on the relation of gastric secretion to red blood cell production with J. Goodfriend and Chain; and two on the rabbit ear-chamber work with Beatrice Pullinger and Ebert.[1] There was also the film on the micro-circulation. It was a productive period in which Florey had picked up the lines that had slackened during his first two years in Oxford. Now, in 1939, his own part in the new penicillin project—the investigation of its biological properties—could not begin until there was enough active material to work with.

Chain, in fact, was making very slow progress, but he remained optimistic that his success with lysozyme would soon be repeated with penicillin. Florey, perhaps, was more aware of impending difficulties. These were evident from Raistrick's failure; and Raistrick, as Florey is reported to have acknowledged, was 'no slouch'—high praise in Florey's language. And before Chain could tackle the biochemistry of penicillin, he too had to have enough material. Production needed experience in fields that were foreign to him. The mould had to be grown free from contamination in a liquid medium under the conditions that would favour its production of penicillin, and on a larger scale that created its own new problems. It was a fermentation process like alcohol production. But whereas alcohol can be quickly measured chemically, penicillin could be measured only bacteriologically. So Chain had to become familiar with the techniques of the mycologist and the brewer, and he also had to become something of a bacteriologist—none of which came easily to a trained biochemist. And all the processes were slow. The mould took about ten days to produce penicillin and the bacteriological tests, which needed another fifteen hours at least, were not quantitatively precise. Chain was using the 'well-plate' method developed for lysozyme, but it gave only a very rough measure of activity and he was unsure of how much he had gained or lost by experimental changes in mould culture technique. It was something of an impasse. Until he had enough penicillin, he could not learn about its biochemical properties; and until he knew about those he could only hope to increase production by trial and error experiments, each of which took two or three weeks to complete. Florey kept in touch with this work during the first half of 1939, but it could claim only part of his attention.

There were, at this time, other things to worry about. On 15 March Hitler invaded Czechoslovakia, and the Munich pact was revealed as a worthless piece of paper. Mussolini invaded Albania three weeks later. It was clear that the next move would be by Hitler against Poland, and the integrity of Poland had been guaranteed by Britain and France supported, it was hoped, by Russia. Then, to the consternation of the West, Russia signed a non-aggression pact with Germany, and war in the very near future became a practical certainty. Britain introduced conscription, and all the plans for Civil Defence began to move into operation.

Oxford, for the first time since the Civil War—a disturbance still fresh in the corporate memory of its older colleges—began to take steps to protect its treasures from the high-explosive and incendiary bombs that were expected. At the Dunn School, Maegraith took charge of the digging of an air-raid shelter in the gravel soil of the vegetable garden behind the building. Every able-bodied member of the staff took turns with picks and shovels, supervised for much of the time by Charles Florey, then aged four.

On 24 August Germany invaded Poland and triggered the chain-reaction of the Second World War. France, and then Britain on 3 September, declared war on Germany, and a few days later Russia staggered its many Western sympathizers by invading Poland from the east. Most people in France and Britain expected immediate and devastating air raids on their major cities, and a large proportion of their civilian populations was quickly evacuated. Oxford adapted itself with a surprising lack of academic dislocation. The city and its surrounding countryside absorbed its share of evacuees. The Florey children went, with Doris, to St. Austell in Cornwall, and their place in The Red House was taken by six schoolgirls and two teachers from London. The staff of the Radcliffe Infirmary, now charged with the clinical teaching of sixty or more students transferred from London, and with a readiness to deal with possibly massive air-raid casualties, was augmented by senior doctors who emerged from retirement. Army medical units were set up in the area, and one of these, the Head Injuries Unit, was commanded by Hugh Cairns, with the rank of Brigadier. The headquarters of the Emergency Public Health Laboratory Service, under Professor G. S. Wilson, was established at the Dunn School, in rather cramped accommodation for which they paid a rent of £150 per annum. But apart from the blackout (perhaps the greatest nuisance) and the rationing of food, clothes, and petrol, life in the University went on very much as it had done during the previous year of uneasy peace. Florey was not alone, therefore, in pursuing his research plans without obvious interruption.

As usual the factor limiting his activities was money. He had approached the Nuffield organization and described the outcome in a letter to Mellanby on 11 June:

I applied to the Nuffield Fluid Research Committee for a grant of £200 (p.a.) for Mrs. Jennings to enable her to continue her work with me on gut function. It was turned down in contemptuous terms (Committee consisted of Buzzard, Cairns, Witts, Ellis and Le Gros Clark). I had a row on the full Nuffield Committee yesterday and got an apology from Buzzard and had their statements removed from the record. However, Lindsay took the line that it was inadvisable to upset decisions of the advisory committee which, as a general proposition, is right. However, I am now faced with the prospect of having to close down the intestinal work altogether for lack of assistance.

The financial difficulties of trying to keep work going here are more than I am prepared to go on shouldering as it seems to me that I have acquired a reputation of being some sort of academic highway robber because I have to make such frequent applications for grants from all sorts of places. To crown the lot, I have just received from the University an intimation that they are contemplating cutting my grant on the grounds that a new heating plant will effect economy on heating the building. I have struggled to keep the place warm on money I ought to have devoted to research. It would seem that the University almost wish to make it as difficult as possible to carry on. You may gather that I am fed up.

Florey then went on to point out that Heatley would relinquish his grant in September, having been awarded a Rockefeller Fellowship, and he suggested that Heatley's grant might be transferred to Dr. Margaret Jennings.

Mellanby replied a few days later:

You know that I have always had a great deal of sympathy with you in your attempts to get the Pathology Department at Oxford into an active state, and this sympathy has been shown in a number of cases in a practical way. I don't think you ought to be disturbed, your efforts have had great success, but I am quite sure that if you want to keep the place going you will always have some kind of a struggle, so that you had better resign yourself to it. It is no good getting agitated because people are difficult, for this is the natural condition of the average man and it is only the rare person who is not obstructive.

Mellanby then indicated that the Council might agree to transferring Heatley's grant to Dr. Jennings in September, and with that Florey had to be content.[2]

But on 21 July Mellanby sent to Florey an obviously official rebuke.

I daresay you know that Gardner recently applied to the Council for an expenses grant in order to cover the cost of ordinary laboratory material ... This matter came up before them on Friday last and was turned down ... I thought you ought to know that the opinion was expressed that the state of affairs from which Gardner is apparently suffering is 'frankly deplorable' and it seems to me your immediate task is to remove this cause of criticism, even if it means a reduction in the total volume of research being carried on in the laboratory.

Florey replied:

Many thanks for your letter about Gardner's application for a grant. Unfortunately he did not show me the form in which he applied, nor did he tell me what he wrote to the Council. I gather from your second paragraph that the Council think I am seriously at fault in the matter. Perhaps this is so, but I hardly think they know all the relevant facts. However, I am needing no urging to reduce here. It is a great deal easier to stop than to start research; five people have gone this year and by the end of next I expect to have reduced the numbers by twelve, including Maegraith. It will then be possible to run the rump with what the University provides.[3]

It is clear from these exchanges that Florey's constant research demands were straining Mellanby's financial resources and his patience, and it is in the context of this atmosphere that later developments in the financing of the penicillin work must be viewed. Mellanby had a great respect for Florey as a research worker, which he had shown when he proposed him as a candidate for election to the Royal Society in October 1937.[4] But his criticism that Florey's research drive was being made at the expense of his department's proper functions and should therefore be reduced must have caused Florey dismay and some resentment, and his previously friendly relations with Mellanby began to deteriorate.

Within two months, however, Florey was again asking Mellanby for help, being urged on by two considerations and also probably, by Chain. The first was the penicillin project which, as it slowly took shape, began to look the more attractive. The second consideration was Chain's financial future, since his three-year personal grant would cease at the end of 1939. Florey took the obvious course of making the work on antibacterial substances the basis for a grant application for Chain. On 6 September (three days after Britain had declared war) he sent details of a proposed research, and a covering letter to Mellanby:

I enclose some proposals which I think might be profitably carried out by Chain and myself. There is little doubt that Chain has a great flair for dealing with enzyme proteins and the proposals now made have a very practical bearing at the moment. I can get clinical co-operation from Cairns for any products we produce and I have tested on animals. Chain has a grant, which will cease, from the British Empire Cancer Campaign. We are closing the cancer work here. Would it be possible to transfer Ewens' grant of £300 p.a. to Chain for the work outlined? [Ewens had taken a job in London.] Would it be possible to furnish £100 for expenses, such as media for growing organisms and special chemicals and apparatus? I have long had the feeling something might be done along the lines proposed and I hope you can view the proposal favourably.

The proposal Florey enclosed described the work on lysozyme and then continued:

Although chemically most suitable for use as an antiseptic *in vivo*, its practical application is limited by the fact that its action is almost entirely confined to non-pathogenic organisms. There are, however, accounts in the literature of substances with chemical properties very similar to those of lysozyme which act powerfully on pathogenic bacteria, especially on staphylococci, pneumococci, and streptococci. These substances have been obtained from filtrates of certain strains of penicillium, actinomyces and of certain soil bacteria. Of these substances only the bactericidal principle from soil bacteria which is especially effective against pneumococci has been studied in some detail. Recently prominence has been given to this substance in the American medical literature since it has not only a strong bactericidal effect on most types of pneumococci *in vitro*, but can cure and protect animals from infections with virulent pneumococci. The chemical properties of the bactericidal substance are very similar to those of lysozyme; there can be no doubt that it is an enzyme belonging probably to the same group of enzymes as lysozyme.

Filtrates of certain strains of penicillium contain a bactericidal substance, called penicillin by its discoverer Fleming, which is especially effective against staphylococci, and acts also on pneumococci and streptococci. There exists no really effective substance acting against staphylococci *in vivo*, and the properties of penicillin which are similar to those of lysozyme hold out promise of its finding a practical application in the treatment of staphylococcal infections. Penicillin can easily be prepared in large amounts and is non-toxic to animals, even in large doses. Hitherto the work on penicillin has been carried out with very crude preparations and no attempt has been made to purify it. In our opinion the purification of penicillin can be carried out easily and rapidly.

In view of the possible great practical importance of the above mentioned bactericidal agents it is proposed to prepare these substances in a purified form suitable for intravenous injections and to study their antiseptic action *in vivo*.[2]

This document is of some historical importance, since it contains the first specific proposal for work by Chain and Florey on penicillin—the 'one full-blooded throw' mentioned by R. D. Wright. And it also raises points of interest. The obvious one is that the possible therapeutic value of such work was by this time certainly in the minds of both Florey and Chain to the extent that they had set it down on paper, and proposed actual clinical tests. It is not clear why or when this departure from pure scientific interest had occurred, since the laboratory work in Oxford had justified no more hopes of clinical value than could have been gained at the outset from the literature. It is true, of course, that the possible medical importance might have been stressed to improve the chances of a grant from the Medical Research Council, but nevertheless the issue had come to mind. The second point is the airy optimism with which the production and purification of penicillin were dismissed as problems that would be easily and rapidly solved, an attitude so uncharacteristic of Florey that it must have been inspired by Chain. The statement that 'penicillin is non-toxic to animals even in large doses', which could only have been based on Fleming's experiment with one mouse and one rabbit, was hardly justified as a generalization, and again one supposes that Chain's enthusiasm overcame Florey's scientific caution. Finally the reference to recent American work on the bactericidal action of a soil organism concerns the observations by Dubos, published in 1939, on a substance later called 'gramicidin' that proved to be too toxic for human systemic use.[5]

Mellanby responded promptly, if not very generously, to these proposals. On 8 September he wrote:

Your suggestion that Ewens' grant should be transferred to Chain seems to be a reasonable one, but naturally I cannot give any definite answer until the matter has been considered by the Council ... Provisionally, and under the same conditions, you can assume that we will give you £25 for expenses for this work and will remember your application for £100 for expenses when the time comes for a proper decision ... It seems to me that the line of work you are suggesting will be interesting and may prove to be of practical importance.[2]

So the Secretary of the Medical Research Council gave his official approval to the penicillin project, and £25 with which to launch it. There cannot be many Government-sponsored schemes that have had a smaller initial investment and a larger return.

In October Florey was told by the Medical Research Council that they had approved Chain's grant of £300 per annum, with £100 expenses, for three years. His immediate future was thus assured. But the general financial situation had scarcely been improved, merely stabilized. There was no additional money for expanding the penicillin project and Florey felt that expansion was essential if it were to succeed at all. Yet he had reached the limit of what he could expect from the Medical Research Council and, indeed, from any of the other sources in Britain he had been trying to tap for the past five years. It was the visit of Dr. H. M. Miller, of the Natural Sciences Division of the Rockefeller Foundation, to Florey on 1 November that crystallized a tentative idea for obtaining a substantial backing that had been strongly advocated by Chain.

Florey's relations with the Rockefeller Foundation had been good ever since he had been one of their Travelling Fellows in 1925, and the Natural Sciences Division (through Dr. Warren Weaver) had already supported his work on lysozyme. Florey had kept in touch during 1937 and 1938, but nothing definite had emerged except the fact that the Foundation was more likely to support a biochemical than a medical project. This was now confirmed by Dr. Miller, who listened to Florey's proposals and urged him to make a formal application without delay. So Chain and Florey set out their mainly biochemical programme, drafting it in wide terms that would cover the likely twists and turns of the research. The application described all Chain's researches since 1936 and then the plans for future work. One line was the biochemical study of a number of natural antibacterial substances, including those from *B. subtilis*, other soil organisms, *B. pyocyaneus*, and *Penicillium notatum*. Therapeutic possibilities were mentioned, also the intention to prepare substances suitable for intravenous injection. The second line was the investigation of the mode of action and chemical nature of mucinases and their physiological significance.

Chain explains the diversity of these proposals on the rather mundane basis that it would be more likely to attract a long-term grant and thus remove for several years their constant financial worries.[6] But Florey, at least, had probably recognized the possibilities of penicillin and realized that the crucial experiment—a protection test—had never been done. One of his attributes was his gift for seeing and exploiting the gaps and fallacies in previous work, and here was the most important gap of all. And if Florey was prepared to put most of his energies into penicillin there is another reason why it had not much emphasis in the application. He was, by nature and experience, anxious to keep his major ideas to himself until he had worked them out. He had made this clear in his letters to Ethel and, as will be seen later, he kept the early penicillin work on the secret list.

Whatever may have been the reasons behind the order of

priorities listed in their application to the Rockefeller Foundation, Chain and Florey produced a very effective document.
Florey was tired of the unrewarding efforts he had put into
a multitude of applications for trivial sums and on this
occasion he took the plunge. He asked for £1670 ($5000) per
annum for salaries and recurrent expenses, and £1000 for the
initial cost of equipment, and he stressed that the grant should
be for not less than three years. He sent one copy of this application to Dr. Miller in Paris and another to Dr. Weaver in
New York, mentioning in his covering letter that 'postal communication is now so uncertain'.[7] The applications arrived
safely and they were approved. The Foundation actually
granted more than had been asked, since they extended the
period of the grant of $5000 per annum to five years. Florey
and Chain were hardly able to believe their good fortune.
They had received a 'princely sum' that would underwrite
an all-out effort unhampered by niggling economies. But, in
making his appeal direct to Dr. Weaver, Florey had once more
by-passed the Medical Research Council which, by established custom, dealt with applications by British medical
workers to the Rockefeller Foundation. Florey had cut a
corner, and unknown to Mellanby—for the time being.

The war brought one positive—and as it turned out
very important—advantage to Florey. Heatley had been
awarded a Rockefeller Fellowship to work in Copenhagen
with Lindestrøm Lang the great Danish biochemist. But
though Denmark was neutral, it seemed unwise to become
even a temporary resident in a foreign country on Germany's
border. Heatley consulted Florey, who advised him to stay
in Oxford, and suggested that he should work on penicillin.
Heatley was exactly the person most needed at this stage
when none of the biochemical and biological problems could
be tackled because of the practical difficulties of production.
He was a most versatile, ingenious, and skilled laboratory
engineer on any scale, large or minute. To his training in biology and biochemistry he could add the technical skills of
optics, glass- and metal-working, plumbing, carpentry, and
as much electrical work as was needed in those pre-electronic
days. Above all, he could improvise—making use of the most
unlikely bits of laboratory or household equipment to do

job with the least possible waste of time. Heatley agreed to Florey's proposal, but he did not wish to work under Chain's direction again. Instead, he became Florey's 'personal research assistant' while retaining his Rockefeller Fellowship.

Now that, almost for the first time in his career, there was no financial handicap to his research, Florey set about organizing the work itself. It was exactly the sort of venture he enjoyed, one that presented several different aspects each of which could be tackled by its own specially qualified expert. The project did indeed develop along such lines, and by people who were best able to carry them forward, but the organization was not planned overnight; it evolved as unforeseen problems were met and surmounted. Certain aspects were clearly defined from the outset. One was the research needed to find the ways of growing the mould that would give the best yield of penicillin, and on the largest scale feasible in what was, after all, a university department. A second aspect was the bacteriology. Penicillin could be tested only through its antibacterial activity. The accuracy of such tests had to be improved, since they provided the only measure of success or failure in the production and purification experiments. And since the effect of penicillin on bacteria was the reason for the whole research, it was necessary to find out which organisms were vulnerable and under what conditions. A third line was the wider biological study of penicillin—its effects on other living cells and tissues, its actions in the body and any reactions it might cause there. Finally, and most fundamentally, there was the biochemistry involved in the purification of penicillin, the identification of its structure and the chemical nature of its action—the sort of problems that had been posed by lysozyme.

All these lines began to yield results at an increasing rate during the years 1940-1, a multitude of observations relevant or irrelevant, helpful or misleading, from which the difficult and intermittent growth of useful knowledge and technique finally led to success. Writers of scientific history tend to select from such a jumble of data the points that—with hindsight—can be seen to have been significant and, ignoring their true chronology, rearrange them in time to form a coherent, readable narrative of scientific detection, exploration, or treasure-hunting. They can scarcely be blamed for such historical distortion since scientists themselves are often equally guilty when they publish their own results. 'We all know', said Florey in his Dunham Lectures in 1965, 'that when we compose a paper setting out ... discoveries we write it in such a way that the planning and unfolding of the experiments appear to be a beautiful and logical sequence, but we all know that the facts are that we usually blunder from one lot of dubious observations to another and only at the end do we see how we should have set about our problems'.[8] These remarks apply very aptly to many of the published accounts of the Oxford penicillin work. A complete and factual record would occupy many volumes and be, for the layman, incomprehensible and unreadable. Here, one must again pick out the highlights and ignore the confusing mass of detail, but at least some attempt will be made to keep these highlights in their proper time sequence.

Possibly the first significant advance was Heatley's invention of the 'cylinder-plate' for penicillin assay, a great improvement on the well-plate method, the inaccuracy of which had hampered progress on all fronts. In the new method culture plates were made in the ordinary way, and sown evenly with the test organism, usually the 'standard' staphylococcus. Heatley then placed short lengths of glass tubing on the surface of the agar, having bevelled the lower edges to sharpen them. These little cylinders, usually six or eight to each plate, were then filled with the test solution, the cover put on, and the plate incubated while the solutions in the cylinders diffused into the agar in a widening circle. If it contained penicillin the organisms failed to grow within a certain distance of the cylinder, and the diameter of this circle of inhibition was an index of the strength of the penicillin solution. At first potency was expressed simply in terms of these diameters. Later in the research, when it became possible to preserve penicillin activity, a 'standard solution' was kept and used periodically to check the method, and an arbitrary 'unit' was defined.

The next part of the project was the biochemical research—the ostensible scientific goal to which the other aspects were directed. Chain, of course, was in charge of this. Heatley took on the task of growing the mould and testing the results of cul ture experiments, and Chain could revert to his proper sphere of pure biochemistry and make rapid progress. Indeed, the limit to his work was the amount of material that Heatley could provide, a limit that was rapidly rising as methods of mould culture were improved. During the first six months of 1940, Chain discovered a number of facts about the penicillin molecule, but no easy way to extract it in a relatively pure and stable form. One way would be to take advantage of its solubility in solvents such as ether or chloroform, since most of the impurities in the crude preparations would be insoluble. But then, as Raistrick had found, it seemed impossible to get rid of the solvent without destroying the penicillin. Yet this solvent had to be removed if any biological work was to be done on penicillin. It was a difficulty to which Florey's method of handling his team brought the answer.

Florey, who could be very firm with his own collaborators, adopted a liberal attitude to people who were involved in their own parts of a general project. Chain, Heatley, and others working on penicillin all knew what was required of them, and Florey left them to solve their individual problems with no direct interference, and certainly no nagging. He made it his business to know exactly what was happening, dropping in for an informal talk in each laboratory almost every day and often making a suggestion, just as he was leaving, that the person concerned later found to be very much to the point. Though the penicillin project was already involving several experienced research workers, Florey did not call formal conferences to discuss progress and plan future work. There was nothing resembling a 'committee'—that blunt instrument that often produces nothing more original than the date of the next meeting. But he did invite the people concerned to discuss their problems with each other, and him, from time to time. At one of these conversations, in March 1940, Chain was discussing with Florey the effect of pH on the solubility and stability of penicillin. Heatley, as usual, said very little. A thought had struck him which, once conceived, seemed so obvious that he was almost reluctant to speak of it. If penicillin passes from water to ether only when the water is made acid, perhaps it would pass from ether back to water if it were made alkaline. Rather diffidently he suggested this idea. Chain, however, was reluctant to change his plans and, when Florey insisted that Heatley's idea should be tested, agreed that Heatley himself should do the experiment.[9]

Heatley was only too pleased to make the attempt and soon found that he was right. He shook a mixture of acidified mould filtrate and ether and then allowed time for the two liquids to separate. The ether floated on top of the water, which could be drawn off. The next step was to shake the ether, containing its dissolved penicillin, with a fresh lot of water which was slightly alkaline. As he had hoped, the penicillin then passed back into solution in the water. This discovery marked a technical advance of the greatest importance since it had removed the obstacle that had blocked Raistrick's progress.

It is strange that the idea of 'back-extraction' had not occurred to Raistrick or, indeed, to the Oxford workers earlier. Perhaps their minds were directed along other lines of thought, and it is often forgotten that no idea can appear obvious before it appears at all. Heatley was not, it seems, the first person to have thought of it. According to Hare, Holt working in Fleming's laboratory in 1934, had recovered penicillin from an acid solution in amyl acetate by shaking it with alkaline water.[10] Holt thus achieved two technical advances, back-extraction and the use of amyl acetate as an organic solvent. Holt did not publish these observations nor did Fleming ever refer to them. In consequence, the Oxford worker had to rediscover these things for themselves, including the value of amyl acetate, which Chain suggested using in 1941.

The main practical problem continued to be the rate of production of crude penicillin. The fungus had to be grown under aseptic conditions on the surface of a layer of Czapek-Dox liquid. In a few days it formed a yellowish gelatinous skin later covered by green spores. The liquid medium beneath then became yellow and began to contain penicillin, which reached a maximum in ten to twenty days. Although many attempts to increase the yield of penicillin failed, two useful discoveries were made. The first was the result of a visit by Paul Fildes who suggested adding brewers' yeast extract to the medium. This did not increase the yield, but reduced the incubation time by about a third. The second discovery was that if the fluid beneath the mould were drawn off when the maximum penicillin concentration had been reached, and replaced by fresh culture fluid, a second harvest of penicillin could be obtained. Up to twelve such changes (all carried out aseptically) could be made with little reduction in the penicillin yield obtained at each. Not only did this manœuvre save time but also the labour of preparation.

Experiments had shown that the best yield of penicillin was obtained when the liquid medium was not more than 1·5 cm. deep. To meet this condition posed difficulties. Most sterilizable laboratory containers did not provide a large flat area without being very uneconomical of space in the incubator and the autoclave, when volumes of 50 to 100 litres of medium per week were to be dealt with. Later, special containers would have to be designed and made. In the meantime the best that could be done was done with bottles laid on their sides.

The next stage was the extraction of crude penicillin. The first step was filtration—through discarded parachute silk—to remove pieces of mycelium, spores, and other solids. The filtrate then had to be cooled, acidified, and shaken with ether (or later, amyl acetate). Cooling was necessary because of the instability of penicillin in acid solutions, and the process was carried out in the cold room. The discomfort for the operator made a mechanical method highly desirable, and Heatley devised one that, with later modifications and developments, was used on an ever-increasing scale. The acidified filtrate was made to flow down a long glass tube while a stream of ether or amyl acetate flowed upwards past it in the same tube, so that the two liquids had a large, moving surface of contact. This method evolved into the final one in which the filtrate was sprayed by fine jets into an upward-moving column of amyl acetate through which its droplets sank. The solvent, now containing the penicillin, was collected from the top of the column and the exhausted filtrate was drawn off from the bottom. It was an effective and continuous process, but one involving a maze of glass tubing, junctions, pumps, cooling coils, warning bells or lights, and much associated plumbing. It was a creation that might have appealed to Mr. W. Heath-Robinson—and it worked.

The last step in Heatley's process was the recovery of the

extracted penicillin from its solution in acidified ether. For this he could use the same process of counter-flow exchange, the rising column of ether giving up its penicillin to the falling droplets of water containing buffer salts to render it slightly alkaline. This time it was the watery solution collected from the bottom of the column that contained the penicillin. Having got this solution, the next problem was to purify and preserve it. Purification itself improved the stability of penicillin in solution, which could then be stored for months without significant loss of potency. But a surer method of preservation was 'freeze-drying'—a process developed in Sweden in 1935 that was, by 1940, revolutionizing the preservation of unstable organic materials ranging from milk and blood plasma to living bacterial cultures. It was the logical extension of the principle of vacuum distillation, in which the boiling-point of a liquid is lowered by reducing the atmospheric pressure. If a near-perfect vacuum can be maintained, ice will evaporate rapidly—and stay frozen during the process—provided the water vapour is removed. In practice this is done by condensing it in a vessel kept at a much lower temperature by immersion in liquid nitrogen or solid carbon dioxide in acetone. So Heatley's solutions of penicillin could be converted into a dry and stable brown powder—the first highly potent (though still very impure) preparations to be made safely captive.

By the middle of March 1940 Chain had accumulated 100 milligrams or so of this brown powder. It was more potent than anything they had produced before. A solution of one part in a million inhibited the growth of streptococci, indicating that, weight for weight, it was at least twenty times as effective as the latest sulphonamide. Clearly, therefore, Chain was making good progress. But on Tuesday 19 March he did something that was quite out of step with pure biochemistry. He took a large proportion of his stock of powder and dissolved in in 2 ml. of water. (The exact amount seems to be in doubt, and published figures range from 40 to 80 milligrams.) What is strange is that Chain's intended experiment was biological. He proposed to inject the solution into mice to see what happened.

Not having an animal licence, Chain had to find someone

who would give the injections for him, so he went to J. M. Barnes, who was working in another part of the building, and was well used to testing possibly harmful substances in animals. Chain asked him to test his solutions, so Barnes injected two mice intraperitoneally with 1 ml. each, and then took little further interest. Chain, as may be imagined, was far from being uninterested. But the mice showed no obvious ill-effects and were still quite well next day. Barnes remained in ignorance of the fact that he had done something historic, but Chain was elated. In one account he is quoted as saying, 'To my mind this was the crucial day in the whole development of penicillin.'[11]

Any such assessment would have been extravagant. Fleming himself had already shown that his (admittedly less powerful) crude penicillin was harmless to a mouse and a rabbit. And Chain's reasons for doing the experiment at all are far from clear. He claimed that his only interest in penicillin was biochemical—the sort of research he was doing on lysozyme. But no biochemical object would be served by injecting a large part of his hard-won stock into a living animal. The object must have been progress towards clinical use. Chain's action was also strange because there was an understanding (according to his own writings) that he would be responsible for the biochemical work and that Florey would carry out the biological experiments. However, Chain has stated that he took matters into his own hands because, when he had asked Florey to test his extracts, Florey had been too busy to do so.[12]

Naturally Chain told Florey of the results of the toxicity test, and it may be imagined that Florey's feelings were mixed—an obvious relief that the material seemed harmless, combined with annoyance that an experiment he would have done himself should have been done by Chain, or rather by Barnes. At all events, Florey insisted on repeating the experiment and Chain had to hand over what remained of his stock for Florey to use. Curiously enough, with all his skill in minute animal operating, Florey had not acquired the technique of giving intravenous injections in mice. So, since he wanted to test the toxicity of penicillin directly in the blood-stream, and since there was only enough penicillin for an experiment on mice, he had to ask J. H. Burn, Professor of Pharmacology, to make the injections and to show him the method. Burn writes: 'I remember the day when he and Mrs. Jennings brought a sample which they wanted me to inject into a mouse's tail vein. But when I had done it and the mouse was unaffected, they did not explain what it was, or why they wanted it done.' Burn thought that he was owed some explanation, but he did not then realize what lay behind Florey's silence.[13]

During the next two months Florey, assisted by Dr. Jennings and, of course, the indispensable Kent, settled down to do just the sort of steady, painstaking animal experiments on the absorption, excretion, and possible toxic effects of penicillin that had marked all his previous research work in other fields. He tried in rats, mice, rabbits, and cats, various ways of giving penicillin: by intravenous, intraperitoneal, and subcutaneous injections; giving it by mouth, and via a stomach tube. Correlated with these routes were assays of the levels of penicillin reached in the blood, and their duration, and a search for the ways in which penicillin was eliminated. The results showed clearly that it was destroyed in the stomach, so that administration by mouth was useless. It was absorbed from the small intestine, if given via a duodenal tube to by pass the stomach. All forms of injection were effective. But the blood levels soon declined, and within one or two hour had sunk to only a fraction of their peak value. Penicillin was thus either rapidly destroyed in the body, or excreted. It was, in fact, mostly excreted unchanged in the urine. All this was spade-work, but the sort of spade-work that digs essential foundations.

The toxicity testing was equally thorough. One of the most delicate cells in the body is the blood leucocyte, and a study of its reactions had the advantage that it can be done *in vitro*. Florey remembered that G. C. Paine had been working on the effects of sulphonamides on leucocytes in Sheffield, and obtained from him details of his methods. These were suitably modified by the Oxford workers, who then found that penicillin was astonishingly harmless to these cells. Though it was lethal to bacteria when diluted one in a million, a solution two thousand times stronger (1 in 500) had no detectable effect on the white cells. These experiments were extended by Peter

Medawar, who tested the toxicity of penicillin on his tissue culture preparations, and Dr. Jennings then took on this work. Again a concentration of 1 in 500 had little effect on the health of the growing tissue cells.

In the meantime Gardner had become personally involved in the bacteriology. He and Miss Orr-Ewing were making a complete survey of what penicillin could and could not do to the more important pathogenic organisms. The gonococcus was the most sensitive, followed by the meningococcus, staphylococcus, streptococcus, the anthrax bacillus, actinomyces, and the organisms that cause tetanus and gas gangrene. What was interesting about this group was their diverse nature. Such a mixed bag suggested that penicillin interfered with a process fundamental to widely differing organisms. Gardner watched bacteria while they were attempting to grow in the presence of penicillin, and saw that they became swollen or elongated, but did not divide and eventually burst or died.[14] Penicillin was, therefore, neither an antiseptic that killed the organisms immediately, nor an enzyme that dissolved them. It was something that blocked the normal bacterial process of cell-division.

Gardner's list of susceptible organisms included most of those that caused serious and often fatal infections. Among them, those that caused lobar pneumonia and diphtheria were inhibited only by concentrations of penicillin of about 1 in 200,000. The typhoid group were far less susceptible and inhibition required concentrations near the toxic level. Naturally, one of the first concerns was the tubercle bacillus. Tuberculosis remained one of the greatest scourges of mankind, with 30,000 deaths per year in Britain alone, and hundreds of thousands of sufferers living in all grades of ill health. But penicillin proved to have no effect on the tubercle bacillus in culture. It was not an easy experiment to do, since the bacillus takes two to three weeks to grow, and the penicillin concentration had to be maintained by continuous additions. But, even without the enteric fevers and tuberculosis in its range, the potential clinical value of penicillin was beginning to dawn on the Oxford workers.

One final point needed to be cleared up before the definitive protection tests began. Penicillin had been proved to be non-

toxic to healthy animals and tissues, and it had been prove
to be immensely lethal to most of the pathogenic bacteria. B
would it work where there were oedema fluid, serum, pu
and breaking-down tissue? It was these conditions that ha
neutralized the sulphonamides in localized areas of infectio
It was a question easily answered. Penicillin was tested in tl
presence of these things and none of them inhibited or d
stroyed it. It was observed, however, that certain contamina
ing bacteria could destroy penicillin. Chain and Abraha
later investigated this phenomenon, and showed that it w
due to the production of a specific enzyme by the bacteria-
a penicillinase.[15] But this problem was neither immediate n
important at that stage of the research.

What had become immediate—and absolutely crucial-
was the testing of penicillin as a protection against fatal infe
tion in the living animal. The achievement so far was on
an extension, though a very considerable one, of what ha
been done before by Fleming and Raistrick. The potency
penicillin and its harmlessness to normal animals and tissu
had been confirmed by thorough and impeccable exper
ments. The major practical advance had been its productio
in larger quantities and in a far purer state than had been do
before. But inhibiting microbes in artificial culture is a ve
different process from inhibiting them in the living body th
are attacking. There are countless and unknown factors the
that might render germs immune, or penicillin ineffecti
Only the actual experiment could provide the answer. So,
the morning of Saturday 25 May 1940, Florey did the exper
ment.[16] This is the day, surely, that does mark one of the h
toric turning-points in medical history.

It was a time of turning-points in another field—the fie
of war. A few days earlier the German Army had launch
the offensive so long awaited that many had come to belie
that it would never happen. To the inexplicable surprise
the Allied Command the Germans, instead of dashing ther
selves to pieces on the Maginot line, drove their Panz
columns round the northern end of it, and swept between t
British and French armies against almost no resistance un
they reached the coast near Abbeville. Into this corrid
poured an irresistible wave of tanks and artillery, while

second wave further to the north engulfed Holland and Belgium—defenceless neutrals who had, it seems, learned nothing from the events of 1914. A brief resistance in Holland was repaid by the destruction of Rotterdam as a first demonstration of the power of the Luftwaffe. Thereafter, organized Dutch and Belgian opposition ceased and the British Army, having lost Boulogne and Calais, was trapped in an ever-decreasing area around Dunkirk.

Against this gigantic backcloth, the fate of eight white mice may seem of ridiculously small consequence. And yet, it might be argued that, in terms of human lives and suffering, what followed was of greater importance than all the wars that had been won or lost. It is not a point that need be argued here, nor is it the sort of hypothetical exercise that appealed to Florey—his main concern at that time was the planning of an experiment to the last detail so that it would give a clearcut answer with the greatest economy in material. All the spade-work done before was designed to provide a solid base for this assault. Kent had the eight mice ready in Florey's laboratory when he arrived. Gardner had, by careful experiments, determined the least dose of virulent streptococci that would kill 100 per cent of mice of standard weight. Eight lethal doses were prepared and waiting. At 11 a.m. Florey started the experiment by injecting each of the eight mice intraperitoneally with the dose of about 100 million streptococci. Four of the mice he put back into their cages without further treatment. Of the four remaining mice, two were labelled 'A' and two 'B'. One hour after the infecting dose, the group A mice each received 10 milligrams of penicillin by subcutaneous injection, and the group B mice 5 milligrams. Group A had no further treatment, but Florey planned to give four more injections each of 5 milligrams of penicillin to the group B mice during the next ten hours. Those interested in the design of biological experiments will appreciate the amount of information provided by this admirably economical plan (see Fig. 1).

Several hours had to elapse before any results could begin to appear and, apart from the injections for the mice in Group B, there was nothing more to be done. Florey watched the mice during the afternoon, but they all appeared normal. At

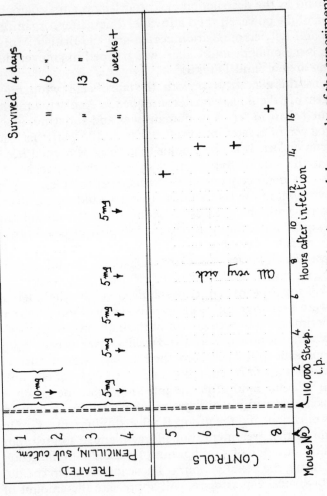

Fig. 1. Chart showing the timing of injections and the results of the experiments drawn by Dr. N. G. Heatley from his notes of the first mouse protection test

6.30 he told Kent to go home, but remained in the laboratory himself, where Heatley joined him. They looked at the mice and decided that the controls, the four mice which had received no penicillin, seemed to be definitely sick. All the others looked normal. At seven o'clock they gave the penultimate injection to Group B, and then Florey went home for a meal, and Heatley out to a restaurant. Florey came back at 10 p.m. to give the final injection. He left a note for Heatley, who intended to come back later: 'Both mice in Group A are very well. Of the Group B pair one is in good condition, cleaning itself and eating biscuit. The other is not quite as lively. The four controls are the worse for wear and their breathing is laboured and they look very sick.' He did not stay longer, but went home to bed. Heatley came back at about 11 p.m., shortly before the first of the control mice died. The other three had all expired by 3.30 a.m. He recorded the details and then cycled home in the blackout.

Next morning, Sunday 26 May, Florey came into the department to discover that the results of his experiment were clear-cut indeed. All four control mice were dead. Three of the treated mice were perfectly well; the fourth was not so well—though it survived for another two days. Chain arrived, and then Heatley, who had had very little sleep. They all recognized that this was a momentous occasion. What they said is not recorded, but memory has supplied subsequent writers with various versions. One might suppose that Heatley said very little, that Chain was excited, and that Florey's reported comment 'It looks quite promising' would be entirely in character. But the word 'miracle' seems to have been used by one of them—a strange word for a scientist confronted by the expected (or at least hoped-for) result of a careful experiment. And Florey telephoned to Dr. Jennings to tell her of the results and said, 'It looks like a miracle.'[17] (When she repeated this to a good Roman Catholic some years later he said, 'How surprised Florey would have been if it really had been one.') But the word 'miracle' conveys the profound impact of this experiment, and a realization of the magnitude of what lay ahead. Typically, Florey wasted no time on speculations. Dr. Jennings joined the group in the laboratory and there and then on that Sunday morning they planned the experiments needed to prove that this success was not a freak result. Florey also pointed out, rather dauntingly, that a man was 3000 times the weight of a mouse.

On Monday, Florey and Dr. Jennings repeated the mouse experiment, using ten animals. All six treated mice survived, three of the four controls died. On Tuesday they used sixteen mice, with twice the dose of streptococci, and varying amounts of penicillin. All six controls died within twenty-four hours, but though the treated mice survived for several days, they were all dead by the nineteenth day. Clearly there was a correlation between the concentration of penicillin maintained in the animal body and the final outcome, that could be determined only by systematic experiments which would consume far larger amounts of material than the present production methods could provide.

The factor limiting production was by now the shortage of suitable containers in which to grow the mould. Heatley experimented with various shapes and sizes of containers from trays, pie-dishes, and petrol cans to lacquered biscuit tins and hospital bed-pans. The standard sort of bed-pan, with its protruding spout, proved to be the most practical. It formed the model for the final, specially made containers, but these did not become available until the end of 1940, and all the experiments done during that year used the material produced by Heatley's astonishing assortment of makeshift equipment.

The material itself was divided between the biochemical and the biological laboratories. Chain now had a first-class colleague, E. P. Abraham, who, with Sir Robert Robinson, had crystallized lysozyme. Abraham joined the research staff of the Dunn School in 1940. He had been in Sweden when war broke out but returned to Oxford in November 1939. Robinson, who was impressed by Florey's chemical approach to pathology, then advised him to accept the post that became available in Chain's laboratory with the advent of the Rockefeller money. But, even with Abraham's collaboration, the biochemistry of penicillin was proving to be a succession of almost insoluble problems. There were two practical difficulties that slowed progress. One was the extreme lability of the active material, which might suddenly disappear toward the end of a long and difficult experiment. The second was the lack of an immediate test for penicillin: it could be recognized and measured only by the biological test on staphylococci which needed twelve hours to produce results. The biochemists were working, so to speak, with their hands tied and their eyes blindfolded. It is an astonishing feat that, in the end, they isolated penicillin as a pure substance and discovered the structure of its molecule. But that was not until 1943. In 1940 all they could do was to remove, little by little, the various substances in the crude extract that were not penicillin. This they achieved by trying all the known (and several novel) methods for separating organic substances from each other: differential precipitation and extraction; absorption and elution; and endless attempts to find salts or other compounds of penicillin that might be more stable or easier to handle. Only when they managed to apply column chromatography did they begin to make more rapid progress, and this was not until 1941.

During 1940 therefore, the penicillin extracts produced by Heatley and partly purified by Chain were the only source of material for the biological work and it was far less pure than anyone at that time could have believed possible. The filtrate of the mould culture—the starting material—usually contained 1 or 2 units of activity per millilitre. By extraction with ether (or amyl acetate), back-extraction into water, and freeze drying, a brown powder was obtained which might have as little activity as 5 units per milligram. This was the sort of material with which Florey did his biological experiments in 1940. It was thought astonishingly potent, since a solution of one part in a million had bacteriostatic activity, compared with the original filtrate which could only be diluted 1 in 800 with the same effect. The process had, in fact, yielded a product more than a thousand times as powerful as the starting material. But the impression that something like purity had been achieved in 1940 was rather far from the truth. By 1941, Chain and Abraham had produced material with 50 units per milligram, and felt that the goal was near. But when penicillin was, at last and in reality, purified, it had an activity of 1800 units per milligram.[18] In other words, more than 99.7 per cent of the 1940 material, and 97 per cent of the 1941 product, was rubbish. It was fortunate that none of this 'rubbish' was highly toxic; if it had been, further work might have been discouraged.

During June and July 1940, Florey pursued his systematic animal experiments designed to show how much penicillin had to be given, and how often in order to protect animals from various doses of virulent organisms. Because he was working to fine limits he had to use large numbers of mice to get significant results. After Gardner and Miss Orr-Ewing had done the initial experiments Dr. Jennings took over the preparation of suspensions of living bacteria and their injection into mice for therapeutic tests. On 1 June she began, early in the morning, to inject a batch of 75 mice each with a lethal dose of streptococci. One hour after each injection Florey gave a dose of penicillin to 50 mice, the remaining 25 serving as controls. The doses were graded, in order to determine the minimum protecting dose. He repeated the penicillin injections at two-hourly intervals for twelve hours, finishing at 2.30 a.m. Within two days, 17 of the 25 controls were dead, but only 1 of the 50 treated animals. But, by ten days, 25 of the 50 treated animals had died, and 21 of the 25 controls. It was clear that penicillin treatment could protect, but that it must be continued for a longer period.

On 5 June, Mellanby appeared in the department, saying that he had 'just dropped in' while in Oxford. Unexpected visits from him were not unusual, he felt—quite rightly—that he was then more likely to see what the people on his payroll were really doing. He was an alarming person, and the knowledge that he might appear at any moment kept many a young research worker busier than he might otherwise have been Florey took him round the department, avoiding the rooms in which the large-scale production of penicillin was going on. Mellanby might have noticed the peculiar smells that pervaded the building at that time, but he saw nothing to suggest an activity highly unusual in a university department. He knew, of course, that Florey and Chain were working on natural bacteriostatic substances, including penicillin, but he left Oxford with no impression that an all-out drive in any particular direction was in progress. When Florey wrote to him on 11 June, he said: 'As I told you recently, we have been working with a substance that gives the greatest promise of being an important chemotherapeutic substance. We don't want anything said about it at this time, but I need further help from you.' He went on to ask permission to transfer Dr. Jennings, Abraham, and Miss Orr-Ewing, who were on M.R.C. grants for other work, to this chemotherapeutic research, and the Council agreed.[2]

At about this time Florey decided to ask for help outside Oxford in the large-scale production of penicillin. He approached Dr. J. W. Trevan, Director of the Wellcome Physiological Research Laboratories at Beckenham, a distinguished physiologist and a friend of Sir Henry Dale. Trevan was interested, and he and his chief biochemist, Dr. Pope, came to Oxford to see the penicillin process on 15 July. The Wellcome Laboratories were well equipped for large-scale work and in peacetime might have made an important contribution. But they were already heavily committed to the war effort in the preparation of vaccines and antitoxins, and in attempts to preserve blood plasma for transfusion—a problem solved eventually by freeze-drying. Though some preliminary experiments were done by Pope on penicillin they did not progress far. It was not seen then as important to the war effort, and it was not given high priority.

On 1 July another experiment had been done using 50 mice. This time the dose of streptococci had been increased threefold, the severest test so far. Twenty-five animals received no treatment, the rest were given injections of penicillin every three hours for the next two days and nights. Florey and Kent lived in the laboratory throughout this time, grateful for Dreyer's forethought in planning for such an occasion, though Kent had to sleep on a camp bed. Every three hours each night an alarm clock woke them, and they gave 25 injections. This time the results were beyond question. All 25 of the control mice died within sixteen hours. All but one of the treated mice were alive and well ten days later. During the next few weeks they did a series of such extended experiments on batches of 50 to 75 mice, but using different lethal organisms that had been shown to be sensitive to penicillin *in vitro*. They showed that animals given otherwise lethal doses of staphylococci and the organisms of gas gangrene could be protected by penicillin as effectively as they had been protected against streptococcal infection. Florey had at last satisfied himself that penicillin was, within the limits of his experiments, indeed the near-perfect chemotherapeutic agent. He felt that the time had come to publish the results of all these labours and began to set them out for a paper to the *Lancet*. The final test—the most difficult—was still to come. How would penicillin work in man? What may be harmless in one species might be toxic in another. Above all was the problem of supply. A man 3000 times the weight of a mouse would need amounts of penicillin in the same proportion. But it was the sort of practical difficulty that Florey took pleasure in overcoming, and to which Heatley could apply his initiative and genius for improvisation.

Meanwhile it was initiative and improvisation on a larger scale that had saved the British Expeditionary force—but not their arms and equipment. At the beginning of June the astonishing makeshift fleet of hundreds of small craft, from yachts and motor boats to paddle steamers and fire floats, had somehow materialized off the beaches of Dunkirk. Manned by amateur sailors, they snatched over 300,000 men from under the guns of the Wehrmacht and the Stukas of the Luftwaffe, while the gallant support of the Royal Air Force covered the evacuation. The miracle of Dunkirk masked the bald facts of defeat—the fact that Britain was now alone, with no effective army, an air force minute compared to Germany's, and only the Channel and doubtful naval superiority preventing immediate invasion. Throughout June and July the Battle of Britain raged in the air and the outcome was in the balance. Even the most optimistic realized that German tanks might soon be rumbling through England against no more opposition than could be mounted by the Home Guard.

Academic Oxford had made various preparations for such a catastrophe, and one of them affected the Florey family profoundly. This was a scheme worked out by University parents for an evacuation of their children to America. John Fulton, then Professor of Physiology at Yale, had been active in the arrangements and in July the outlook for Britain seemed so bleak that the plan was put into effect with much understandable anxiety and heart-searching. The party of 125 children and 27 mothers boarded ship at Liverpool, bound for Canada. Charles and Paquita Florey were among them in the care of Mrs. Duthie, the wife of Edward Duthie who had worked with Chain on mucinase. Ethel was determined to stay in England and Paquita remembers how their parents saw them off at the station. 'Dad didn't like fuss or emotional bother. He just kissed us goodbye and told us to be good, but mother was terribly cut up and cried because she felt it was a dreadful thing for us to go alone.'[19]

The arrangement had been that the children should disembark at Montreal and John Fulton had travelled to Quebec to board the ship for the last stage of its voyage, during which he could sort out the affairs of the party. But he arrived to find that all had already disembarked there amid mountains of baggage and great confusion, with no one to meet them except outraged immigration officials and a flock of reporters. One of the latter discovered Paquita sitting placidly on a trunk with five-year-old Charles.[20] She was immersed in a comic paper and disinclined to answer questions on her first impressions of Canada. 'You seem to be enjoying that paper, anyway', said the reporter. 'Oh I am', said Paquita. 'You see my parents never allow me to look at comics in Oxford.'

Fulton wrote to the Floreys on 24 July describing the safe, but confused, arrival of the whole party. In this letter he offered to give the Florey children a home 'for the duration'.[21] He and his wife Lucia were childless, and would be pleased to look after them. The Floreys accepted with relief and gratitude. But it was an emotional dislocation on both sides. Paquita, in particular, resented 'being sent away' and wrote 'terrible letters' accusing her mother of putting her work before her children. Ethel, in fact, was deeply distressed by the separation—the result of her own difficult choice—and the energy with which she threw herself into the later work on penicillin may have been as much a reaction to this as it was a result of her greater freedom.

One other precaution that Florey took brings home the mood of apprehension of those days; and also his certainty that, in penicillin, he had something of immense value. If the

Germans were to reach Oxford, they would learn nothing o
the penicillin work because the essential records an
apparatus would have been deliberately destroyed. But th
mould itself must be preserved, undetected. Florey, Heatley
and one or two others smeared the spores of their strain o
Penicillium notatum into the linings of their ordinary clothe
where it would remain dormant but alive for years. One o
more members of the team might escape to start their wor
again.[22]

On 24 August *The Lancet* published their paper 'Penicilli
as a chemotherapeutic agent' by Chain, Florey, Gardne
Heatley, Jennings, Orr-Ewing, and Sanders—the name
being arranged in alphabetical order. It was a very brie
account, amounting to only two pages, but it gave th
facts.[23]

'The results are clear cut', the paper concluded, 'and sho
that penicillin is active *in vivo* against at least three of th
organisms inhibited *in vitro*. It would seem a reasonable hop
that all organisms inhibited in high dilution *in vitro* will als
be found to be dealt with *in vivo*.' The authors thanked th
Rockefeller Foundation and the Medical Research Counc
for financial support. There was also a short and cautious edi
torial comment on penicillin which ended: 'What its chemica
nature is, and how it acts, and whether it can be prepare
on a commercial scale, are problems to which the Oxfor
pathologists are doubtless addressing themselves.'

The secrecy which Florey had exercised before this publi
cation suggests that he expected active interest when i
appeared. If so, he was disappointed. Indeed, it is ironica
that, for the next two years, he was to spend most of his tim
and energy in trying to persuade people to take notice of wha
he felt to be the medical discovery of the century. The firs
official reaction to the *Lancet* paper came in a letter fror
Mellanby:

I read your penicillin paper with great interest and it seems clear that yo
are on to a very good subject. I hope it develops as well clinically as
promises. I noted at the end of the paper that you gave a great boost i
the Rockefeller Foundation for having supported this work and that th
Medical Research Council had to play a very minor role. I doubt wheth
this is in accordance with the facts. [Mellanby then tabulated grants fror

the Council to Florey and his staff totalling £1200.] I should be surprised if the Rockefeller Foundation are supporting the work to anything like this extent. It seems to me that your method of dealing with this matter is wrong tactics, partly because most of the grants that the Rockefeller Foundation give for medical research in this country follow discussion between O'Brien [Rockefeller Medical Division] and me, and partly because if you have got a good thing in your own country, you might as well give it proper credit and not follow the people who, in cases of research, find it more convenient to give foreigners boosts than their own colleagues.[2]

This was a pretty sharp rebuke, occasioned as much by Mellanby's annoyance that Florey had made his approach to the Rockefeller Foundation without consulting him as by the implied minor role of the Medical Research Council. Mellanby did not know that the Rockefeller contribution (including Heatley's Fellowship) was over £3000 that year, and he was by no means mollified when Florey pointed this out in his reply and stressed that the money had come from the Natural Sciences Division, with which he had always had good relations. Mellanby then retorted:

It seems strange that you are claiming to use ... the Rockefeller money for the penicillin work, which is as pure a piece of medical research as can be imagined. In fact, I should imagine the Medical Research Division of the Rockefeller Foundation feeling aggrieved at this intrusion of the Natural Sciences Division into this particular field.

The second reaction to the *Lancet* paper was the appearance of Alexander Fleming at the Dunn School on Monday 2 September. He had, it is true, telephoned Florey on the previous evening, but his arrival was a surprise to several of the penicillin workers. Chain, for example, had not realized that he was still alive. But there he was, wearing his usual bow tie, and asking what they had been doing 'with my old penicillin'. Florey and Chain took him on a tour of the whole production, extraction, and testing laboratories, and gave him a small sample of their most purified material. Fleming said almost nothing during this inspection, and returned to London without comment or congratulation on what had been achieved.

Florey was now determined to press on with the truly formidable task of testing penicillin in human subjects in a way that would leave the least possible room for either doubt or disaster. As always, supply was the first difficulty. He had

hoped that the *Lancet* paper might stimulate commerci
firms to undertake large-scale production—that their repr
sentatives would be calling at the Dunn School anxious to tak
advantage of the experience gained there. But again he wa
disappointed. There was no increase of interest at the Wel
come Laboratories, and certainly no large-scale production
which was not even seriously considered until six month
later. And there were no approaches from any other con
mercial firm until the middle of 1941. It was, after all, th
time of the blitz with all its disruption and damage, and th
firms were struggling to maintain their essential production

Undaunted by this lack of interest, Florey became the mo
determined to produce the penicillin in the Dunn School. I
order to supply the minimum for a satisfactory trial in ma
the output of filtrate must be more than doubled, to at lea
500 litres per week. Since this was quite impossible with exis
ing culture vessels, new ones would have to be designed an
made. Heatley produced various models, and the design fir
ally chosen was an obvious derivative of the hospital bed-pa
It was a shallow rectangular box about 11 inches long, 9 inch
wide, and 2½ inches deep. It had a sloping spout at one en
and a litre of fluid filled it to the required depth of just ov
1·5 cm. It was decided that these containers could be mo
readily and cheaply made of pottery, glazed inside so that the
were impervious, and easily cleaned and sterilized. Florey ha
no personal contacts in the pottery trade, but he knew a con
sultant physician, Dr. J. P. Stock, who practised in the Stok
area, and he sent him sketches of what he wanted, and aske
him to suggest possible manufacturers. Stock took immediat
action; three days later he telegraphed that James MacInty
and Co. of Burslem would do the job. Heatley visited then
next day, and within a few hours the potters had made a s
of prototypes ready for firing. These were tried out in Oxfor
and found satisfactory. Florey placed an order for 600, at
cost of £300, and two days before Christmas 1940, Heatle
collected the whole batch in the van borrowed from the Eme
gency Laboratory Service. Florey, Heatley, and Georg
Glister (who had joined the penicillin team in May 194
washed, sterilized, charged, and seeded all 600 vessels durin
Christmas Eve and Christmas Day. Then they stacked the

for their ten-day period of incubation in the animal operating room. It was the beginning of penicillin production at the Dunn School on the scale needed for clinical trials in man, and a University department had become a factory.

The Medical Research Council had, on this occasion, agreed to Florey's appeal for the cost of this work and granted him the £300 for the pots, and rather less than £2 per week each for two girls to act as factory hands. The matter had been considered at a meeting of the Council, and not all of the distinguished members (some elderly) were clear what Florey was doing. One, in particular, who misheard the word 'mould', was puzzled by all this interest in moles—but Florey got his grant.[24] He also, in the end, recruited four extra girls, making six in all. These 'penicillin girls', as they came to be called, Ruth Callow, Claire Inayat, Betty Cooke, Peggy Gardner, Megan Lankaster, and Patricia McKegney, deserve their place in history. In 1943 he asked the Medical Research Council to increase their wages to £2. 10s. per week. 'They have now had nearly three years of experience, are good technicians and extremely willing workers, fully entitled to a good wage. I think that, but for their enthusiasm for the work, they would have gone long ago as it is quite easy for them now to obtain £3. 10s. 0d. a week or more serving in canteens and such-like.' In fact they had worked with a steady devotion under conditions that no factory inspector would now sanction for a moment. The practical classroom became the inoculation department, where the washed and sterilized pots were charged with medium and then inoculated with mould spores by paint spray-guns. Everything had to be done under strict aseptic conditions to prevent contamination by bacteria, and the girls were dressed in sterile caps, gowns, and masks. The inoculated pots were then wheeled on trolleys to what had been the students' preparation room, now converted into a huge incubator kept at 24 °C. After several days of growth, the penicillin-containing fluid was drawn off from beneath its mould by suction, using the special equipment designed and made by Heatley, and the pots recharged with medium. After filtration, the crude penicillin solutions went through Heatley's extraction contraption. The air was full of a mixture of fumes: amyl acetate, chloroform, ether. These dangerous liquids were pumped through temporary piping along corridors and up and down stair-wells. There was a real danger to the health of everyone involved and a risk of fire or explosion that no one cared to contemplate. But, by February 1941, despite several serious setbacks, including a sudden break in the ability of the mould to produce penicillin, Florey had available enough material to begin his first trials in human beings.

14

THE PROVING OF PENICILLIN

By the beginning of January 1941, the 'factory' established in an Oxford University department had produced enough penicillin to begin trials in human beings. The production of the crude mould filtrate had been increased to 500 litres per week during the previous few months, and the purification process had been greatly improved by Chain and Abraham. The solutions derived from Heatley's counter-flow extraction process were good enough for the animal experiments. But the introduction of chromatography yielded a material ten times stronger, weight for weight, and thus ten times purer. The principle of column chromatography is simple, though its practice can be highly complicated. If a solution containing various organic compounds (some of which may be coloured) is made to flow through a glass tube packed with an insoluble absorbent powder, such as alumina, the components that are most actively absorbed will saturate the first layer of the powder and those less actively absorbed will be removed from the solution by the further layers. Thus quite a sharp separation of the components can be achieved; and if these are coloured, it is necessary only to extrude the column, separate with a knife the different coloured layers, and then to elute the components from the powder by a chemical technique. Hence the name 'chromatography'—not entirely appropriate since many substances separated in this way are colourless. In the case of penicillin at that time, the most active material was found in a light-yellow band bounded on either side by dark-orange bands which contained some penicillin but many unwanted impurities.[1] In practice, this process of extraction with amyl acetate, back-extraction, column chromatography, and elution was repeated several times on each batch. Purification was increased at each repetition, but material was also lost. These gains and losses had to be carefully balanced since the ideal of chemical purity might be incompatible with the reality of a clinical trial.

In his letter to Mellanby about the penicillin project written in September 1939, Florey had said that he could 'get clinical co-operation from Cairns'. But when he went to see Cairns in January 1941, the best that Cairns could do was to refer him to L. J. Witts, the Nuffield Professor of Medicine. Witts was sympathetic, though unsure of the precise way in which he could help. While he and Florey were talking, Dr Charles Fletcher (the son of Sir Walter Fletcher), one of the Nuffield Research Students, happened to knock at the Professor's door and open it. Witts welcomed the interruption, because he had seen a way of helping Florey that would also provide an interesting project for Fletcher. He introduced the two, and asked Fletcher if he would like to find suitable cases on which to try penicillin. Fletcher agreed, and Witts turned to Florey and said, 'Here's your man.' So chance gave Fletcher a splendid beginning to a long and distinguished career, and to Florey the most able and co-operative of young clinical colleagues.

The first step was clearly a thorough investigation of the absorption and excretion of penicillin in human beings and of any possible harmful effects—a repetition of the work already done in animals. But now there was an ethical problem. As far as was then known in Oxford, no human being had ever received an injection of penicillin, and the side-effects of doses large enough to kill infecting bacteria might be quite unexpected. It is a problem that (like vivisection) involves unanswerable and conflicting arguments. It would have been easy to have found volunteer 'guinea-pigs' among the Dunn School staff. But would Florey and Witts have been justified in subjecting a healthy young man or woman to an experiment that might prove disastrous? On the other hand, a clinician has a responsibility to his patients. Can he, with a clear conscience, ask one of them, perhaps already old and ill, to submit to an experiment that might do harm? As the lesser of two evils it was the latter course that they took, and Fletcher was asked to find a patient whose life could not in any case be long maintained, and who would agree (knowing all the relevant facts) to receive an injection of penicillin.

Fletcher soon found a suitable patient. There seems to be some doubt about her identity, because the injection of penicillin she received was not recorded in her notes. Bickel,[2] from evidence supplied by Dr. Fletcher, suggests that she was a Mrs. Elva Akers, of Oxford, who was dying of inoperable cancer and had only a month or two to live. Fletcher explained the position to her, and she agreed to receive the injection. So, on 17 January 1941, Fletcher injected 100 milligrams of penicillin intravenously, with Witts and Florey standing at the bedside to see what would happen. The result was not entirely satisfactory, because within two hours the patient had a rigor, that is an attack of shivering and a rise of temperature. (It was this reaction, recorded on the patient's temperature chart, that is the evidence for the date of the injection.) Rigors of this sort were then not uncommon reactions to the intravenous infusions and transfusions that were being given in hospitals with increasing frequency. They were due to minute traces of 'pyrogens'—bacterial breakdown products that could contaminate even the sterile distilled water used for making up infusion fluids. Florey realized that the patient's reaction was probably of this sort. Fletcher remembers 'a slightly annoyed grunt by Florey, who started tests for pyrogens on rabbits'.[3] But it was a setback and the possibility that penicillin might itself be toxic for human beings caused some dismay. This proved to be unfounded, since pyrogens were soon detected in the penicillin solutions, and they were removed by more selective chromatography.

Since this first recipient, apart from her brief feverish attack, had suffered no other ill effects from her penicillin, the way was clear to give pyrogen-free material to other human volunteers by various routes, and then to determine the blood levels achieved, their duration, and the way in which penicillin was eliminated. It was soon confirmed that the slow infusion of penicillin by the intravenous drip method gave the best-maintained levels. Penicillin was rapidly removed from the blood by the kidneys and excreted in the urine. It was destroyed in the stomach, so that oral administration was almost useless. In one case, a large dose was given by intraduodenal tube (which by-passed the stomach), and satisfactory blood levels were maintained for three hours. In all volunteers

there were no ill effects, apart from a febrile reaction in o[?]
of the early subjects before the process for removing pyroge[?]
had been perfected.

With these data at his disposal, together with the precedi[?]
animal work that had determined the blood levels of penicill[?]
needed to kill invading organisms, Florey felt that they we[?]
ready to treat a case of bacterial infection so severe that oth[?]
remedies were useless. Fletcher found such a case on [?]
February: a forty-three-year-old policeman, Albert Ale[?]
ander, who had been in hospital for the past two months figh[?]
ing a losing battle against a spreading infection. This h[?]
started with a small sore at the corner of his mouth, whi[?]
had become infected by staphylococci and streptococci. The[?]
progressively invaded the subcutaneous tissues of his face [?]
reach his eyes and scalp. He had been treated with massi[?]
doses of sulphapyridine, without effect, and a series [?]
abscesses had been drained surgically. His left eye had to [?]
removed on 3 February, and it was found that the infecti[?]
had involved his right shoulder, and then the lungs. It w[?]
indeed, a desperate case.

On 12 February, Fletcher gave the patient an intraveno[?]
injection of 200 milligrams of penicillin—the largest dose y[?]
given—followed at three-hourly intervals by doses of 100 m[?]
Within twenty-four hours the man was obviously bett[?]
The discharge from his suppurating wounds was less; h[?]
temperature dropped to normal during the next four day[?]
and his appetite returned. By 17 February his right eye ha[?]
become almost normal. But the supply of penicillin w[?]
exhausted. For the last three days of his treatment it had be[?]
maintained only by collecting all his urine, taking it to t[?]
Dunn School, and extracting every milligram of the penicill[?]
contained in it. Then that, too, was gone, and there was [?]
hope that the best rate of new production could mainta[?]
treatment any longer. For the next ten days his improv[?]
health continued, and there was hope that the five days [?]
treatment had turned the scale. But, at the beginning of Marc[?]
the lung infection returned and he died on 15 March 19[?]
of a widespread staphylococcal infection.

This first case, therefore, was not the triumph that wou[?]
have attracted immediate clinical interest and support. The[?]

was a suggestive, but not conclusive, improvement in the patient's condition during his treatment with penicillin and for ten days afterwards. But Florey, always scientifically objective, recorded the only valid conclusion: 'The attempt to treat this forlorn case was chiefly valuable in that it showed that penicillin could be given over a period of 5 days without significant toxic effect.' The obstacles to the maintenance of bacteriostatic concentrations of penicillin (i.e. enough to prevent the growth of bacteria) in human subjects were the small output of the Dunn School 'factory' and the rapid excretion of penicillin by the patient. It was, as Florey put it, 'like trying to fill a bath with the plug out'. He might also have added that he had only a small tap with which to do it. For future cases he decided to accumulate a larger supply before starting treatment, and to choose children when possible, since they would need smaller doses.

The second patient was Arthur Jones, a boy of fifteen who, following an operation to insert a pin in his hip on 24 January, developed a streptococcal infection of the wound. He had become gravely ill with septicaemia, despite treatment with sulphonamides. On 22 February, when his temperature had been reaching 103 °F. every evening for the past two weeks, he was given an infusion of 100 mg. of penicillin, which was repeated three-hourly for five days. Within two days of starting treatment his temperature dropped almost to normal, and remained there for the next four weeks, when a further operation was done to remove the pin that had caused the infection. The published conclusion states: 'Penicillin therapy was followed by great improvement in the patient's general condition, in spite of the dose being insufficient to maintain a detectable concentration of penicillin continuously in the blood.'

The third patient was an adult named Percy Hawkins, a labourer aged forty-two, with a 4-inch carbuncle on his back, enlarged axillary glands, and a raised temperature. Carbuncles are deep, localized staphylococcal infections of the skin which usually take weeks to heal, and were often treated by surgical excision. Beginning on 3 May 1941, this man was given 200 mg. of penicillin every hour for five hours, then 100 mg hourly. These larger doses maintained a bacteriostatic blood level continuously. On 7 May the carbuncle was clearly better, and the dose of penicillin was reduced to half. By 10 May the carbuncle had almost completely disappeared, the area was painless, the enlarged glands were shrinking, and the temperature was normal. This was indeed a striking success, and the first case in which an adequate dose of penicillin had been given. Carbuncles simply do not heal of their own accord within a week and without a scar.

The fourth case was John Cox aged four and a half. He had developed septic spots on his left eyelid following measles five weeks earlier. These had caused a spreading infection of the orbit, and thrombosis of the cavernous sinus (one of the great veins in the skull bones behind the eyes)—then an almost invariably fatal condition. The boy had been treated with sulphapyridine for two weeks, and when penicillin was started on 13 May he was desperately ill, semi-comatose with gross swelling of his face and eyelids and signs of meningitis. He was given 100 mg. of penicillin hourly for two doses, 50 mg. hourly for four doses, and then 25 mg. hourly thereafter. By 16 May he was obviously better, and the swelling of the face was subsiding. By 19 May he was 'vastly improved' and by the 22nd he was talking and playing with toys. The penicillin was stopped, and improvement continued. Then, on 27 May, he suddenly had convulsions and became unconscious. He died on 31 May. At post-mortem examination, Dr. Robb-Smith found that the cause of death was a ruptured artery in the brain, one that had been weakened by the infection. But the infection itself had completely cleared. It was a sad ending to what would have otherwise been a clinical triumph, for here there was no room for doubt. Penicillin had defeated an infection that no other treatment could have touched. But the patient was dead.

The fifth case was that of a boy aged fourteen and a half, who had been admitted on 6 May 1941 with staphylococcal septicaemia and osteomyelitis of the left femur. He had been treated with sulphathiazole, and the left hip joint (which had also become infected) had been opened surgically. By 6 June he was extremely ill, with signs of kidney involvement. On that day he was given 500 mg. of penicillin by slow infusion, which was then continued for ninety hours during which 3·5 grams were given. By these means a continuous bacteriostatic level of penicillin was maintained. By 10 June he was much improved. On 16 June continuous administration was stopped, and intermittent doses were given until 20 June, by which time he had received 17·2 grams of penicillin. His fever had subsided some days before and did not return. His hip joint, though damaged by the infection, improved, movement became painless, and he completely recovered. It was a case that began to prompt the use of the word 'miracle' by people who were familiar with the usual outcome of such infections, about which Ethel, for example, had written in despair to Howard when she was House Officer in the Children's Hospital in Adelaide.

The last case in this group of patients was a baby boy aged six months who was seriously ill with a staphylococcal infection of the urinary tract and impending kidney failure. Sulphonamide treatment had been tried, but was stopped when there was a disastrous fall in the white cell numbers in the blood. He was given penicillin by mouth on 5 June together with sodium bicarbonate to prevent its destruction by the acid in the stomach. Penicillin appeared in the urine, and the treatment was continued for a week, during which time the urine became sterile, the general condition improved, and continued to do so after treatment was stopped. There was a complete recovery.

The final task in this clinical experiment was to publish the results. Florey and his colleagues combined to write a paper in which the production methods, the biochemical work, the animal experiments, and all the clinical cases were described in detail.[1] The series of patients, which had begun rather disappointingly, showed that increasing success had followed increasing dosage to become really dramatic. But the number of these successes was too small to convince a statistician—or Florey. A full-scale trial was needed, and this would demand amounts of penicillin far beyond the production capacity of the Dunn School. Florey had therefore turned his energies from the practical problems of his department to the political one of persuading people who could make penicillin on a commercial scale to do so.

This task proved to be more difficult—and certainly more frustrating—than the pioneer work on penicillin itself.

Despite Florey's own conviction and the influential people who respected his judgement, commercial interest had remained minimal. On 6 February 1941, Fleming had again visited the Dunn School, bringing with him E. J. King, the biochemist at the Postgraduate Medical School, and Douglas Maclean from the Lister Institute. They saw the extraction process, but there was no subsequent move from the Lister Institute. In March, Trevan from the Wellcome Laboratories came with Sir Henry Dale, who was then President of the Royal Society. But progress at the Wellcome Laboratories had been almost negligible; they had hardly succeeded in getting the assay method to work, let alone large-scale production. Perhaps at Dale's prompting, the Wellcome organization approached a small chemical firm, Kemball, Bishop, who specialized in fermentation processes in their East London works, with the suggestion that they might undertake the large-scale growing of the mould and provide 10,000 gallons of filtrate as soon as possible. But Kemball, Bishop were fully committed to other work, the blitz had devastated large areas around their factory, and supply difficulties were formidable. So the proposal was refused—but not forgotten, since Kemball, Bishop became, in September 1942, producers of crude penicillin filtrate for processing at the Dunn School.

Meanwhile there were other sparks of commercial interest, which at that time seemed to Florey unlikely to fire productive enthusiasm. Trevan had investigated the possibility of large-scale spray-drying with the Kestner Company at New Cross; and Sir Robert Robinson, of the Dyson Perrins Laboratory in Oxford, had hopes of persuading commercial firms who sought his advice on chemical matters to take an interest in penicillin. In June, Dr. Thurlow and Dr. Martin of I.C.I visited the Dunn School to spend a day being shown details of the production process, and Dr. Hobday of Boots was another visitor for the same purpose a few days later. But these visits were purely exploratory. There was no sign of any commitment, and this is quite understandable. What the visitor saw was an extremely complex and temperamental process depending essentially on the growth of a mould that might change its habits at any time. The yield of active material was minute in proportion to expenditure of equipment, chemicals

and human energy, and the clinical results so far available
were promising, but not conclusive. To embark on full-scale
production would be enormously expensive and commercially
hazardous. Penicillin might have unexpected ill effects, or lose
its beneficial ones. Above all, its chemical structure might
soon be discovered and cheap synthetic methods could then
make the expensive biological process obsolete almost overnight. Finally, Britain was fighting for survival against the
heaviest odds since Napoleon, and against a far more destructive enemy. The diversion of any large resources from essential
war work was hardly to be contemplated. No one in authority
seriously considered that penicillin might prove to be the contribution to that war effort which it became three years later.

Florey did recognize this potential. He realized that the
rapid and effective treatment of battle casualties would be a
military advantage and that, even if Britain did not exploit
the possibilities of penicillin for this purpose, Germany might
do so. The German chemical industry was the best in the
world, and had already introduced chemotherapy. Florey was
convinced that in penicillin he had something far superior to
the best existing chemotherapeutic products, and that the
Germans, if they realized this, would act quickly and effectively. When he heard, from a colleague in Switzerland, that
the Germans were already interested, he wrote on 16 April
to Mellanby: 'I have had the enclosed letter sent to me. It
seems to me very undesirable that the Swiss and hence the
Germans should get penicillin, and I think it would be well
worth while to issue instructions to the National Type Collection Laboratories not to issue cultures of *P. notatum* to anyone
with possible enemy connections, and to send a letter to Fleming to the same effect.' But Mellanby seems to have missed
the point. In his reply of 23 April he told Florey not to worry:
'It does not seem to me that this is a serious matter because
I expect you are miles ahead of any possible competition ...
I hope you have been able to arrange for the increased production of penicillin by bringing in industrial interests or otherwise.'[4]

(The German interest that worried Florey involved Hitler's
personal physician, the notorious charlatan Theodor Morell,
who claimed to be the original discoverer of penicillin. But

his standing in his own profession was so abysmally low that his claims probably retarded any serious work on his 'discovery' in Germany.)

Another issue that arose out of Florey's hopes for the commercial development of penicillin was the matter of patents. The production process used various inventions and innovations that could be protected by patents, and Chain, as the son of an industrial chemist, was very much in favour of applying for them. His motive was not personal financial gain, though he felt that possible royalties could be used for research: it was the recognition that if patents were not taken out in Britain they would be taken out abroad by anyone who had a mind to do so. Florey was undecided. There had been some talk of patenting 'Volpar', which had not concerned him directly, and it was felt then that medical research workers should not be involved in the commercial exploitation of their discoveries. But Chain insisted, and Florey put the matter to Mellanby and Sir Henry Dale. Their reaction was immediate and emphatic, though unfortunately not on record. They both agreed that the suggestion of patent applications by the Oxford workers was unethical. In their view medical research workers were paid to follow their chosen vocation, and this vocation was to make discoveries and to give them freely to the world. As in most ethical problems the conflict is between idealism and realism. Florey did not press the realistic argument against the ruling of Mellanby and Dale, because he probably agreed with them and he accepted it. But he then had to face Chain, who was furious. So Florey suggested that Chain himself should see Mellanby. He did so, but Mellanby so violently rejected his arguments that the idea of patents was dropped.[5]

At that time worries about patents seemed premature since there was no immediate prospect of commercial exploitation anywhere. Florey had realized that the chances of procuring full-scale production in Britain at that time were slim, and his thoughts naturally turned to America; a country not at war, with every resource available, and where he had many friends. Chain too, it seems, was greatly in favour of a visit to the United States to stimulate interest there. On 14 April 1941, Florey saw Dr. Warren Weaver of the Rockefeller

Foundation, who had come to London. The result of their meeting was an offer by the Foundation to pay the expenses of a visit to America by Florey and a colleague. Florey put the proposal to Mellanby, who replied:

After discussing the matter with you last Friday, I have come to the conclusion that the only way in which your most important work on Penicillin can be made to go forward is for you and Heatley to go to the United States of America for three months. It is quite clear to me also that you cannot get the substance made by manufacturing firms in this country under present conditions.[4]

(Chain states that he was not consulted about this plan and knew nothing of it until the day of Florey's departure with Heatley.[5])

There were, of course, many official clearances and permits to be obtained before such a journey could be arranged. Britain was in a state of siege; sea travel was slow and hazardous; there were in those days no direct transatlantic flights; and marauding German fighters might intercept any aircraft within range of the Continent. But, on 30 May, Florey telephoned Heatley from London to tell him 'to begin to polish up the bags'. He was anticipating, however, because it needed another three weeks of permit-hunting and the censorship of all the written material they proposed to take before the journey could begin.

During this interval Florey sent in the *Lancet* paper for publication, all the various authors having read and approved the final draft. For many years it was generally supposed that the patients described in this paper were the first ever to have received penicillin by injection, but this was not the case. On 17 May, Dr. Weaver sent Florey a cutting from *The New York Times* reporting the presentation, by Dr. M. H. Dawson, of a paper to the Society for Clinical Investigation, in which he described his efforts to treat cases of a fatal heart infection with penicillin. No technical details were given in the newspaper report, and Dawson's paper was not published. It was only many years later, and long after his death, that further details were retrieved from his laboratory notes by one of his associates, Dr. Gladys Hobby, for Lennard Bickel, who describes them in his book *Rise up to Life*.[2]

Dawson's was a gallant but unsuccessful effort. Despite the

fact that he himself was suffering from a progressive disease, he devoted his energies to the treatment of subacute bacterial endocarditis, a streptococcal infection of the valves of the heart. From this site the organisms could not be eradicated, so that they not only slowly destroyed the valves, but chronically infected the blood-stream. Dawson had hoped that the sulphonamides would cure this otherwise fatal condition, but they did not. Then he read, in August 1940, the first *Lancet* paper on penicillin published from Oxford, and determined to produce the material and use it. He obtained a culture from Dr. Reid, who had worked with Fleming's mould in Pennsylvania in 1935, and proceeded to follow the Dunn School process, even to the extent of turning classrooms at Columbia University into the same sort of factory. After overcoming similar difficulties, Dawson, with Karl Meyer, a biochemist, and Gladys Hobby, a microbiologist, produced enough crude penicillin to confirm its activity and lack of toxicity in animals. On 15 October 1940, the first test injection of penicillin was given 'intracutaneously' to a patient in the Presbyterian Hospital, New York. The dose must have been minute, since only about 0·5 ml. of fluid can be given by this route, and there were no ill effects. Three patients with infected heart valves were subsequently treated. It seems clear that their infection was not controlled, and that Dawson realized that the available dosage was inadequate. Finally, like Fleming and Paine, he showed that penicillin was effective as a local application in a few cases of eye infections. He presented these findings at the meeting of the Society for Clinical Investigation already mentioned, and his results did not arouse much interest. The progressive muscular weakness of his own disease gradually crippled him, and he died in 1945, unrecognized as the first person to have injected penicillin into a human patient.

On 26 June 1941, Howard and Ethel Florey drove by car, with Heatley, to Bristol. They had with them all the records and data of their work on penicillin and, most important of all, a freeze-dried culture of the strain of *P. notatum* they had used. They spent the night at the Grand Hotel, in Florey's words 'by Mr. Rockefeller's bounty', and at 7 a.m. next day they were driven to the airfield at Whitchurch. Florey and Heatley took off in a blacked-out plane for Lisbon, but the

oute they followed was never revealed to them. All they knew
was that it included a stop at an unidentified airfield, where
hey transferred to a larger but equally blacked-out plane.
Then they landed at Lisbon, where Rockefeller representa-
ives met them and took charge of all formalities. There fol-
owed four days of enforced relaxation at the Hotel Tivoli in
he hot sunshine and in a city of shops and bright lights and
10 shortages. Finally they boarded the clipper flight to New
York, via the Azores and Bermuda, arriving on the afternoon
of 2 July.[6]

Despite the long and tiring flight, and the temperature
92 °F.) of New York, Florey wasted little time when he
rrived. He and Heatley left their bags at their hotel in
Madison Avenue and then went straight to the Rockefeller
Foundation in the R.C.A. Building to see Dr. Warren Weaver
nd to arrange to meet Dr. Alan Gregg, Head of the Natural
Sciences Division, next day. At that meeting, without notes
nd almost without a question or an interruption, Florey gave
 lucid and complete account of all the penicillin work that
he Foundation had supported. It was a performance that
illed Heatley with admiration as he described to Bickel: 'Even
hough I knew the subject well, he showed me new facets,
nd I realized suddenly how great a man he was. None of us
n the penicillin team could have matched him, and he was
o clearly the leader. I count that hour in Gregg's office as
ne of the great experiences of my life.'

On that same afternoon Florey and Heatley went to stay
with the Fultons in their substantial and luxurious New
Haven home, Mill Rock, for the Independence Day vacation.
t was, of course, also the temporary home of Paquita and
Charles. Paquita, now nearly twelve years old, had not
dapted herself very well to American life, and she was de-
ighted to see her father. Charles, five years younger, had
ccepted Lucia Fulton as his second mother. But this visit
was more than a happy family reunion. John Fulton was
nthusiastic about the penicillin project and anxious to help.
He soon began telephoning all over the United States to
nyone who might be both interested and useful. He gave a
arge lunch party on 4 July, to which he had invited several
nfluential people, and also took Florey and Heatley to meet

Dr. Ross Harrison, Chairman of the National Research Council. He, in turn, arranged for Florey to see Dr. Thom, of the Department of Agriculture, the mycologist who had correctly identified Fleming's mould for Raistrick ten years before.

After four days with the Fultons, Florey and Heatley began what was to become a long and arduous exercise in travel and persuasion. Their first and most important interview was with Dr. Thom, at Beltsville, in Maryland. Florey once more produced his data, and he did not minimize the difficulties. He stressed that with their existing knowledge and techniques it needed the material from 2000 litres of mould filtrate to treat a single case of severe sepsis—a daunting prospect for any commercial producer hoping to cover his costs. But Thom, with his unequalled experience of moulds, believed that better ways could be found. The next step must be to enlist the practical collaboration of experts in large-scale fermentation. So he took his visitors to Dr. Wells, head of the Bureau of Agricultural Chemistry, who arranged for them to go to the Agriculture Department's Research Laboratories at Peoria, in Illinois, where a team was working on the production of useful chemicals by fermentation.

At Peoria, Florey told his story once again to the Director of the laboratories, Dr. May, and Dr. Coghill, Head of the Fermentation Division, who were prepared to undertake the project provided Heatley could remain there to get it started. Coghill suggested that deep culture might be more productive than surface culture, once the mould brought from Oxford had become acclimatized to the new conditions, and the assay methods established. So Heatley stayed, and Florey went off on a round of visits to the headquarters of all the likely drug companies throughout the United States.

Heatley fared better than Florey, though his assigned collaborator, Dr. A. J. Moyer, a fervently anti-British isolationist was not the easiest person for an Englishman to work with Only Heatley's quiet good temper and gentle manners made their association, and thus the ultimate success of their work together, possible. For Moyer had a positive flair for making moulds do what was wanted. One of his first innovations wa to try corn-steep liquor—a by-product of the commercia starch-extraction process—as a growth medium instead of yeast extract. The result of this and other alterations in the medium was a tenfold increase in the yield of penicillin which, combined later with the deep culture technique that allowed thousands of litres to be dealt with in each batch, transformed the production process.

Florey had little success with his visits to the drug companies. He disliked his role, in which he felt he was regarded as 'a carpet bag salesman trying to promote a crazy idea for some ulterior motive'. Most of the executives he talked to were indifferent, sceptical, or only mildly interested. But his efforts were not, in the end, useless. Of the firms he approached Merck had already started some experiments, and Pfizer, Squibb, and Lederle were sufficiently interested to keep the idea in mind, and eventually to act on it.

In August, Florey returned to Peoria to learn of the progress there, and to take Heatley on a visit to the Connaught Laboratories in Toronto. There they met Dr. Ronald Hare, who knew the story of Fleming's discovery of penicillin from personal experience in Almroth Wright's laboratory, and who was in 1970 to write the best book about it so far published. But Hare's presence at the Connaught Laboratories did not persuade its Director, Dr. Defries, that penicillin was a worthwhile project. The difficulties of biological production were too great, he decided, and at any moment penicillin might be synthesized.[7] Florey, deeply disappointed, left Toronto and, while Heatley returned to Peoria, sought out his old friend A. N. Richards, with whom he had worked in 1926. Richards, still in Philadelphia, was now a very important man, being Chairman of the Medical Research Committee in the Office of Scientific Research and Development. He had a deep respect for Florey, and for his integrity as a scientist. It was this respect that decided him to accept Florey's judgement of the potential value of penicillin, rather than the somewhat scanty case records he could produce. With the authority of his official status, Richards approached the four drug companies named by Florey as showing interest. He informed them that they would be serving the national interest by undertaking penicillin production, and that there might be financial support from the American Government. All four agreed to send representatives to a meeting to be arranged by Richards and Government officials, but by the time this meeting was held, America was at war, and the matter of penicillin production had become a national priority.

Florey returned to Oxford at the end of September 1941. He wrote to Fulton, thanking him for all his help and hospitality. 'I got back safely, after quite a good journey, except for the last hour when we were bumping along in the birdcage at what seemed to me an extremely low altitude.'[8] Heatley had stayed on in Peoria, and would be there until the winter, when he went to the research and development section of Merck & Co. Inc. at Rahway to help design large-scale production methods. Just before he left Peoria, he went over the details of the paper that Moyer and he had agreed to publish on the revolutionary improvements which had been achieved. Heatley had for some time been uneasily aware that Moyer did not always tell him exactly what was going into the fermentation vessels, but the draft of the paper that he saw seemed complete and explicit. However, the months went by while Heatley was busy at Merck's, and no paper appeared. And when it was finally published over a year later, only Moyer's name appeared on it. The true reason for the delay was more disturbing than Heatley's exclusion from the authorship. In the interval, Dr. Moyer (among others in the United States) had prepared applications for British and American patents covering the essential stages of the improved process. This fact was not generally appreciated until 1945, when British firms discovered that they had to pay royalties on their penicillin production, but Florey had mentioned the matter officially in 1942.[9] Chain's forebodings were therefore justified, as he was not reluctant to point out.

But, in October 1941, these future complications were not a matter of concern. Florey's American visit had undoubtedly stirred official action there, but this was directed to producing penicillin for America, not Britain, a direction determined by the realization that America would soon be at war herself. So Florey, back in Oxford, found himself in as bad a position from which to launch a major clinical trial as he had been in June. Moreover, he had lost Heatley, who was to spend nearly a year in America. But the Dunn School factory was still in operation under Chain's direction, and Gordon Sanders, who had much of Heatley's own genius for improvisation, had constructed what was to be for the next two years the largest penicillin extration plant in Britain. This was based on the principle of large-scale emulsification and was housed in one of Dreyer's lavishly designed amenities, the animal autopsy room, which would have comfortably allowed the dissection of an elephant. Sanders had filled it with ingeniously improvised machinery, including steam-heated dustbins to act as stills, milk-coolers, churns, separators, and other dairy equipment, and a domestic bath as a reservoir for crude mould filtrate (Fig. 2). It was also filled with unavoidable fumes of amyl acetate. Even with the best ventilation, the workers there became affected. Florey introduced a strict rota system by which everyone involved, including himself, took short spells of duty in this noisome atmosphere, and no one, it seems, suffered permanent damage.

The increased production at the Dunn School was supplemented during the next twelve months by material from two commercial firms. The first was the supply of small amounts of penicillin from a pilot plant set up by the Dyestuffs Division of I.C.I. This began to arrive in January 1942 and, added to the material produced in Oxford, built up a stock sufficient to start another clinical trial. The second source was from Kemball, Bishop, and Co., the East London firm originally approached by Wellcome a year before. Now, as a result of persuasion by Sir Robert Robinson, they agreed to produce crude mould filtrate which they would send to Oxford for extraction. After months of experiments in their factory, subjected to repeated air raids and surrounded by bomb damage, they sent the first consignment of 200 gallons to the Dunn School on 11 September 1942, and repeated this at more or less regular weekly intervals and entirely at their own expense. As stocks built up, and production rose, Florey began to plan another clinical trial. He felt that he had to break out of the vicious circle that restricted progress—the dependence of commercial-scale production on clinical proof, and the dependence of clinical proof on increased production. He decided, however, to make a start with what penicillin he had, and to

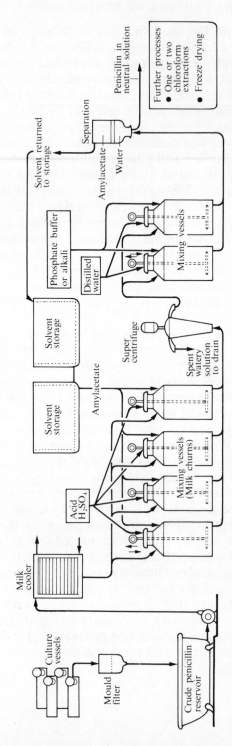

Fig. 2. Line diagram of the penicillin extraction plant in the animal house of the Sir William Dunn School of Pathology, Oxford, 1942. (Reproduced from *Molecules against microbes*, by E. S. Duthie.)

ely on increased supplies as the clinical evidence mounted. Fletcher had by now left Oxford, and another clinician was needed who could devote whole-time attention to the penicillin work. This would involve the selection of patients, the control of treatment, the organization of all tests, and the keeping of notes—a most exacting task that might mean being 'on call' for twenty-four hours a day and seven days a week. Ethel Florey was the person who undertook it, and she did it supremely well. It was a return to clinical medicine that she welcomed, and her determination to combine the best treatment for the patients with the methodical collection of all the data required for the research was admirable. With her children away and her husband seldom at home, she could give all her time and energy to such work.

Her position in the Oxford hospitals was complicated by the fact that she was given no staff appointment—a matter on which Florey felt rather aggrieved. He himself had the title of Consultant Pathologist, and as such he had access to the hospital wards and services, with the agreement of the consultant clinicians who were ultimately responsible for the patients. Ethel was regarded officially as his assistant, though in fact she carried out all the clinical work herself. The patients referred to her by the clinicians for penicillin treatment were not always the most suitable. Florey complained that, in many cases, every other sort of treatment had already been tried in vain until, with the patient at death's door, his doctors turned to penicillin as (in Florey's words) 'a corpse-reviver'. But the fact that penicillin *did* work under these worst of all conditions began to impress even the most sceptical and conservative of the Oxford clinicians, and Ethel Florey found herself more welcome and respected in their company. She lived at the hospital (in the Nurses' Home) during the following year. But The Red House was in good hands. Doris, the Florey children's nanny, had married Jim Kent in 1941, and Ethel rented them rooms for a flat in the basement, which made a convenient home for them for the next twelve years.

There is no need to describe in detail the results of this second clinical trial. It proved that penicillin had a therapeutic power beyond Florey's hopes. In a letter to Sherrington, who had by then retired to Eastbourne, Florey wrote in August 1942:

The penicillin work is moving along and we now have a fairly substantial plant for making it here. It is most tantalising really, as there is, for me, no doubt that we have a most potent weapon against all common sepsis. My wife is doing the clinical work and is getting astonishing results—almost miraculous some of them ... I am afraid the synthesis of the substance is rather distant, but if, say, the price of 2 bombers and some energy was sunk into the project we could really get enough to do a considerable amount. We also have another lot of antibacterial substances from moulds and plants under investigation and apart from the prospect of some immediate use in the war these substances are full of interest and open up quite a vista.[10]

The year 1942 was a time of fulfilment in Florey's scientific career. The enormous effort expended on penicillin—an effort that might have proved a disastrous failure for all involved in it—was achieving a triumphant success. In March, Florey was elected a Fellow of the Royal Society, a proof of his acceptance by the country's best scientists as one of themselves. It was an acknowledgement of the honest, systematic research that he had achieved during the past seventeen years. The penicillin work may have played a part in his election, but a minor one. The selection committees of the Royal Society rely on firmly established evidence in balancing the rival claims of the many candidates, and penicillin was not, at the beginning of 1942, firmly established. A year later, it would have been, when Ethel and Howard Florey published, in *The Lancet* of 27 March 1943, their series of 187 cases of sepsis treated with penicillin. [11] There was then no room for any further doubt that clinical medicine was about to enter a new era.

15

PALMAM QUI MERUIT FERAT

Popular fame is nowadays largely bestowed by those who control the media of mass communication. In the field of inventions and discoveries the public mind is usually conditioned to associate one person with any particular advance. For example, most people, asked to name those deserving credit for the steam engine, the discovery of radioactive elements, or wireless telegraphy, would answer James Watt, Madame Curie, and Marconi. Such a selection would be quite just, since it was the ingenuity and dedication of these pioneers that created the practical achievements. Few people would (or perhaps could) name Savery, Becquerel, and Hertz, the men on whose original discoveries these later achievements were based. Where penicillin is concerned, however, the priorities are reversed. The name that occurs to almost everyone is that of Fleming—the orginal discoverer of an antibacterial strain of mould—while those whose immense labours ten years later made penicillin therapy a reality remain almost unknown. The events that determined this strange inversion are interesting, sad, and rather alarming. They are interesting from the point of view of history. They are sad because people who deserved credit and gratitude from the public got less than they deserved. They are alarming because they reveal the extent to which propaganda can imprint, apparently indelibly, false impressions on the ordinary mind.

On 30 August 1941, the *British Medical Journal* published an annotation on the *Lancet* paper describing the clinical use of penicillin in Oxford. The annotation referred to Fleming's discovery in 1929, and suggested that its therapeutic possibilities had not, at that time, been realized. Fleming immediately took exception to this, writing a letter to the Editor of the *British Medical Journal* on 1 September, in which he complained that insufficient recognition had been given in the annotation to his work. To the impartial observer, it might seem that he had little to complain about. His discovery had been duly acknowledged, and he could hardly expect much more than that, because he really had done almost nothing to promote the therapeutic use of penicillin. His first (1929) paper on penicillin (which had mainly stressed its use in bacteriology) contained the words, 'It may be an efficient antiseptic for application to, or injection into areas infected with penicillin-sensitive microbes.'[1] In 1931, in an article for the *British Dental Journal* he wrote: 'It is quite likely that it [penicillin] or a chemical of a similar nature, will be used in the treatment of septic wounds.'[2] These prophetic remarks, which are clearly confined to the local use of penicillin, seem to be his only public pronouncement on its therapeutic possibilities, and his failure even to attempt animal protection tests must indicate that he had little conception of its potential value. As described in Chapter 8, he took almost no part in the few attempts by others to extract and purify it, and he certainly made no contribution to the Oxford work, though he visited the School of Pathology on 15 September 1940 and again on 18 July 1941.

On 5 August 1942, Fleming did become actively involved. He telephoned to Florey asking for a supply of penicillin to treat a personal friend who had been ill for the past seven weeks at St. Mary's Hospital. The patient had then developed signs of streptococcal meningitis and seemed to be dying. Though Fleming (like the clinicians at the Radcliffe Infirmary) had turned to penicillin only as a last resort, Florey responded at once. He himself took the penicillin—his entire stock at that moment—by the next train to London, and told Fleming exactly how to use it. For six days Fleming gave the recommended doses, but it was clear that the penicillin was not reaching the infection in the brain. So, after consultation with Florey, Fleming injected the penicillin into the spinal fluid, and the streptococcal infection was overcome. Within a few days the patient had completely recovered, and it was agreed between Fleming and Florey that this dramatic result should be included in the current Oxford clinical trial.[3] Penicillin had not before been given by intraspinal injection (except in Florey's animal experiments) and the case was of considerable interest. It was also an episode of considerable

interest and excitement at St. Mary's Hospital—the birth-
place of penicillin—and it was seen there as a clinical triumph
for Fleming, its original discoverer.

News of this 'miracle cure' at St. Mary's Hospital was undoubtedly in the air when *The Times* published a leading
article headed 'Penicillium' on 30 August. It drew attention
to a remarkable new therapeutic substance and, without mentioning names, referred to work in Oxford. Sir Almroth
Wright at once dispelled this anonymity with a letter to the
Editor on 31 August:

Sir,

   In the leading article on penicillin in your issue yesterday you refrained
from putting the laurel wreath for this discovery round anyone's brow. I
would, with your permission, supplement your article by pointing out that,
on the principle *palmam qui meruit ferat* it should be decreed to Professor
Alexander Fleming of this laboratory. For he is the discoverer of penicillin
and was the author also of the original suggestion that this substance might
prove to have important applications in medicine.

On the day of the appearance of Wright's letter, reporters
besieged St. Mary's Hospital and evidently met with little resistance. That night an interview with Fleming appeared in
the *Evening Standard*, and on the following morning the daily
papers published varying accounts of his discovery of penicillin and its astonishing effects. The *News Chronicle* selected
him as 'The Man of the Week' with a four-column article and
his photograph. Meanwhile, Sir Robert Robinson had replied
to Almroth Wright with another letter to *The Times* on 1 September in which he drew attention to the work in Oxford and
wrote that, if Fleming deserved the laurel wreath, 'a bouquet
at least, and a handsome one, should be presented to Professor H. W. Florey'.

Florey's reaction to this concentious publicity was one of
horror. When, following Robinson's letter, the reporters descended on the Dunn School he refused to see them. There
was more behind this reaction than a personal dislike of publicity. He knew that sensational stories about penicillin would
create a demand that could not possibly be satisfied at that
time, and would thus lead to tragic disappointments. He knew
also that the limelight is bad for research workers, being distracting and disruptive, and he did not want his team to be

exposed to it. He remembered, probably, the disastrous publicity that had ruined Dreyer's scientific reputation. Finally, there was at that time a strong ethical disapproval of doctors who advertised themselves or their work by giving personal interviews to the lay press; the General Medical Council had been known to strike such offenders off the Medical Register. Florey would not be liable to such strong measures since he was not in clinical practice, but a disinclination to talk to the press was general throughout the medical profession. So Florey told Mrs. Turner to send the reporters away; and it was not unnatural that they should have gone, somewhat resentfully, to an obviously warmer welcome at St. Mary's Hospital.

During the next two years scores of articles on penicillin and many interviews with Fleming appeared in the public press. Anyone without prior knowledge who read these would be left in no doubt that Fleming was the main—indeed the only—person responsible for the therapeutic use of penicillin. And these versions of the 'penicillin story' are given in such circumstantial detail that even someone familiar with the truth might begin to doubt his own knowledge. This image of Fleming as the sole creator of penicillin therapy was made to seem credible only because certain facts were ignored or distorted. The true facts are: first, that Fleming did nothing obvious to promote penicillin therapy between 1929 and 1941; second, that the Sir William Dunn School of Pathology, Oxford, was the first laboratory to produce penicillin on a large scale and to show that it protected infected animals; and third, that the first effective clinical trial was on patients in an Oxford hospital, in which Fleming took no part. It therefore becomes of interest to note how these various 'awkward' facts are dealt with in what would seem to be a definite press campaign.

The twelve-year period of Fleming's apparent neglect of penicillin was disposed of in various ways. The simplest was the obliteration of ten years. For example: 'It is only three years ago since Fleming described the bacteriostatic action of penicillin ... and subsequently it was shown that it could be employed as a chemotherapeutic agent.' So begins an article published in 1941,[4] the result, perhaps, of a genuine mistake in reporting. But such an error could not be sustained for long, because dates can easily be checked. A more plausible story then emerged and has persisted ever since. This fills the twelve-year gap with a mythical struggle by Fleming to produce penicillin himself with inadequate laboratory equipment and no official encouragement, and frustrated by the indifference or incompetence of the biochemists he tried to persuade to take up the task of purification. After Almroth Wright's death in 1946, this story of a lack of official encouragement at St. Mary's became more specifically an account of the active hostility of Wright to any work on penicillin because it ran counter to his own research plans and theories. These stories are difficult to refute conclusively, and they are featured for example in the biographies of Fleming by Maurois[5] and Hughes[6]. All that can be said here is that there is no evidence whatever to support them.

The fact that it was in Oxford, and only in Oxford, that penicillin was purified, produced in quantity, and first used effectively on patients, is again distorted in various ways. In some versions the penicillin was said to have been produced at St. Mary's Hospital and tried on patients in Oxford under Fleming's direction, or it was said to have been produced under his direction at Oxford and used by him at St. Mary's. In very few accounts is there any mention of Florey. Chain fares a little better, since it was sometimes acknowledged that it was he who had succeeded in purifying penicillin where Fleming's previous biochemical collaborators had failed. Distortion of this sort is epitomized by a book on Fleming published in 1974 and written by a senior bacteriologist who had been a member of the St. Mary's Hospital staff from 1936 to 1970 and had obviously been misled by the publicity campaign. After mentioning Chain's production of a stable preparation of penicillin and Florey's first mouse protection test he writes:

Fleming's reaction to the new situation was at once to place the resources of St. Mary's Hospital at the disposal of the Oxford team. We had a factory on the third floor of the building and in the basement, where vaccines were being produced for the Forces. We had the means of growing larger amounts of mould than Oxford could and our technicians had been making it every week since its discovery. The penicillin in its crude form in broth

was poured off into large churns and put on the passenger trains at Paddington to be collected at Oxford only about an hour later by their technicians.

Fleming also had access to patients in hospital: some of the beds at St.
Mary's were directly under the charge of bacteriologists. This was an advantage that scarcely anyone else had in this country. Florey sent him
samples to try on human cases. The first patient whose life it saved was
a police sergeant. Fleming had the satisfaction of being the first to see the
drug used successfully.[6]

Thus both the 'factory' and the clinical trial are removed
from Oxford to St. Mary's Hospital with such circumstantial
conviction and transparent good faith that it seems necessary
to repeat here that no churns of crude penicillin were ever
received at the School of Pathology in Oxford from Fleming's
department, and that the first patient treated systemically
with penicillin in Britain was the policeman in the Radcliffe
Infirmary who unfortunately died when the supply became
exhausted. How such garbled and distorted accounts came to
be printed, broadcast, and later filmed and televised, can only
be conjectured. Journalists are sometimes more concerned
with the creation of a good story than with the strict accuracy
of their information, and good stories tend to be copied and
embellished. Fleming took an active part in this continued
publicity. Interviews and articles appeared in every sort of
periodical from *Tit-Bits* to *Picture Post* (which named him
'Man of the Year'). He kept cuttings of all of them and seems
to have derived a sort of impish glee from their wilder in-
accuracies. 'All the false stories invented by journalists and
others were in a special file to make up what we called the
"Fleming Myth" ... The more unlikely they were the more
they appealed to him.'[6] Why the 'Fleming Myth' was allowed
to grow unchecked is again mostly a matter of conjecture. It
seems that nothing was done by Fleming himself, nor by
anyone at St. Mary's Hospital, to correct or even to stop the
continued publication of these false stories, which have
become so embedded in tradition that even senior members
of the St. Mary's Hospital staff have clearly taken them to
be true. Fleming, of course, gained from his myth a world-
wide fame and gratitude. Behind Fleming there were two very
strong characters: Almroth Wright, who was intensely ambitious for his department; and Lord Moran, personal physician

to Winston Churchill, and the Dean of St. Mary's Hospital Medical School, whose dearest wish was for its honour, glory, and financial security. And somewhere in this organization one suspects the activities of a most efficient Public Relations Officer.

The reaction in Oxford to this publicity campaign was complex. Initial indifference, based on the belief that 'truth will out', gradually turned to resentment as it became apparent that truth is easily buried by paper. Wilson writes:

Not unnaturally the Oxford team tended to blame him [Fleming] for the more bizarre stories that were printed. They did not know of his detached attitude of almost perverse amusement as he collected cuttings for 'the Fleming Myth'. On the other hand Florey, well known in academic circles as a tough bargainer when it came to fighting for research funds, allowed the credit (which he could justly have claimed for himself and his team) to slip away through his disdainful treatment of the media. The Oxford team can hardly grumble because the press did not print stories that were not made available to them.[7]

The fact that Florey did let the credit slip away, when it was so justly due to his own team, seems strangely out of character, and it both puzzled and disappointed his colleagues. But Florey had not been inactive. On 11 December 1942 he wrote to Sir Henry Dale, then President of the Royal Society:

As you know there has been a lot of most undesirable publicity in the newspapers and press generally about penicillin. I have taken a firm line here and said there was to be nothing whatever done in the matter of interviews with the press or in any other way. Gardner, I know, thinks that I have been rather wrong about this. I had a letter from Fleming in which he assured me he was endeavouring to do the same and I accepted that at its face value and thought that this newspaper publicity would cease. I have now quite good evidence, from the Director-General of the B.B.C. in fact, and also indirectly from some people at Mary's that Fleming is doing his best to see the whole subject is presented as having been foreseen and worked out by Fleming and that we in this department just did a few final flourishes. You can see what I mean in the article published in 'Britain Today', complete with photographs of Fleming and so on. This steady propaganda seems to be having its effect even on scientific people in that several have now said to us, 'But I thought you had done something on penicillin too.'

Florey then went on to ask Dale's advice on the possible publication of an article setting out the true facts—a suggestion urged on him by his Oxford colleagues.[8]

Dale replied promptly and firmly. He asked Florey not to publish anything that might seem to be a refutation of statements attributed to Fleming. He pointed out that Florey was a member of the Council of the Royal Society,* and would thus soon have to act as one of the judges of candidates for election to the Fellowship of the Society. Fleming was a candidate, and any public dispute between Fleming and Florey at this time would prejudice the complete impartiality with which Fleming's claim to be worthy of election must be weighed.[9] Florey accepted this edict from the President of the Society, to which he had himself been elected less than a year before. He remained silent, therefore, in the matter of Fleming's publicity campaign, and he did not divulge the reason for his silence even to his closest colleagues. Fleming was duly elected to the Royal Society in March 1943 and, with his perverse sense of humour, he might have appreciated the supreme irony of a situation that enforced Florey's silence while his own scientific advancement was ensured.

Thereafter the 'Fleming Myth' continued to grow on an international scale, and Florey, in response to continued pressure from his Oxford colleagues, made another attempt to persuade an independent and authoritative body to publish the truth. On 19 June 1944, he sent a letter to the Secretary of the Medical Research Council:[10]

Dear Mellanby,
   I am writing to you to see if it is possible for you to help me out of what has become a somewhat intolerable position...
   It has long been a source of irritation to us all here to witness the unscrupulous campaign carried on from St. Mary's calmly to credit Fleming with all the work done here ... My policy here has been never to interview the Press or allow them to get any information from us even by telephone ... In contrast, Fleming has been interviewed apparently without cease, photographed, etc. ... with the upshot that he is being put over as 'the discoverer of penicillin' (which is true) with the implication that he did all the work leading to the discovery of its chemotherapeutic properties (which is not true) ...
   You of course know how dishonest this is and might reply 'why worry'. This has been our line and would continue to be if it were not that my

---

* Florey had the rare distinction of becoming a member of Council in the year following his election as a Fellow.

colleagues here feel things are going much too far ... I have always asked people who say 'why don't you do something?' what should I do. I will not go in for press publicity and no-one questions the rightness of that. The only suggestion which has been made of which I approve is that the Medical Research Council should issue a statement putting out the relevant facts of how penicillin was really introduced into medicine ... it is the only body likely to have the slightest influence on the Mary's propaganda.

I realise I am in great danger of being accused of trying to 'get something for myself'. This really is not the case and I hope I have always made it clear how much was due to the other workers here. Nor should anyone suppose that we think we have performed any great intellectual feats here. All we did was to do some decent experiments and have the luck to hit on a substance with astonishing properties...

Yours sincerely,

To this appeal Mellanby replied on 20 June, as follows:[11]

Dear Florey,

I was glad to have the conversation with you the other day about the difficult position in which you and your colleagues at Oxford find yourselves, owing to the unusual attitude Fleming has taken up in response to the public acclamation of penicillin discoveries.

I want to assure you in writing, as I did orally, that I think the reticence as regards press interviews and the fairness and even generosity in apportioning credit to Fleming of all in your laboratory have been excellent and above criticism, and whether judged from a short term or long term point of view are, I am convinced, the most desirable. You need have no doubt whatever in your mind that scientific men, in this country at least, and doubtless most of them in other countries, have appraised the situation correctly and know that, from the point of view of scientific merit, your work and that of your colleagues stands on a much higher level than that of Fleming.

I realise how irritating your position must be, if you are at all affected by what appears in the public press, but you can be quite certain that this is an ephemeral reaction which means little or nothing, and the only appreciation which is worth bothering about is that of your scientific peers. In time, even the public will realise that in the development of this story of penicillin, the thing that has mattered most has been the persistent and highly meritorious work of your laboratory. The dish you have turned out is so good that you must swallow the rather nauseating but temporary publicity ingredient with a smile.

Yours sincerely,

This letter could have brought little comfort to Florey or, in its counsel, to his Oxford colleagues who had passively to watch Fleming's meteoric aggrandizement 'with a smile', and to accept the fact that the greater their own success with penicillin the more honours—scientific and academic—would be heaped on Fleming. Mellanby was probably right in some aspects of his letter and certainly wrong in others. He rightly avoided embroiling the Medical Research Council in a wrangle about credit. Such disputes actually tend to bring discredit on all concerned and, in this particular case, might have provided a public counter-charge that the Oxford workers were trying to deprive Fleming of the credit for his genuine discovery. Mellanby was right when he maintained that the merit of the Oxford work was recognized by the leading scientists in the field, and that Florey's avoidance of contentious publicity was commended by those in scientific authority. His statement that 'the only appreciation which is worth bothering about is that of your scientific peers' is an expression of opinion which today might be labelled elitism. Florey might himself also have welcomed some appreciation from outside this small but elevated circle. And Mellanby was patently wrong when he dismissed the Fleming campaign as 'an ephemeral reaction meaning little or nothing', and asserted that in time even the general public would realize that what had mattered most in the development of penicillin was the work in Oxford. Even after forty-five years there is little sign of any such general realization. Mellanby, in fact, underestimated the power of the press and the durability of a myth, and overestimated the influence of leading scientists. He also regarded public opinion as being beneath notice, but Florey himself had, in his letter to his father (see p. 42), recognized it as a personal factor not to be so easily dismissed.

There is no need to dwell much longer on the Fleming Myth. It will probably survive any efforts to establish the truth (such as those in the preceding pages) because a popular myth is one of the most indestructible things on earth. Fleming deserved credit for the discovery of penicillin—it was a stroke of brilliant observation. Many before him had also observed bacterial inhibition by moulds, and it was good luck that Fleming's mould happened to have the properties that were to make it not only the first but also the best antibiotic. Fleming should not be criticized for not having developed his discovery—one might as well blame Johann Strauss for not having composed Brahms's Fourth Symphony. It was in

Fleming's nature to make discoveries (as Strauss made melodies), but not in his nature to develop them by indomitable hard work, ingenuity, and organization. One should not, therefore, begrudge Fleming his world fame: his share of the Nobel Prize, the honours, medals, awards, and degrees conferred on him in almost every civilized country; the hundreds of thousands of pounds that were donated by his admirers to St. Mary's Hospital, nor his burial place in St. Paul's Cathedral beside Nelson, Wellington, and Wren. One can only lament the fact that the public recognition that goes with such honours was not shared fairly with the people in Oxford who deserved it.

Florey often insisted (quite rightly) that the immensely successful work on penicillin in Oxford was the result of a team effort and (quite wrongly) that he himself got more credit than he deserved. What had been done could only have been achieved by a team of experts, but they were also highly gifted individuals who needed to be free to improvise, explore, and invent. Such freedom, unless combined with a common sense of purpose, would soon lead to scientific digression and the loss of co-ordinated progress. There can be no doubt that the penicillin team was Florey's creation: it could not have formed itself into a functioning unit without him. Nor could it have continued to exist as a purposeful group for several years without his leadership. Florey was thus the creator and the leader of the only sort of organization that could have made penicillin a practical proposition at that time.

If, by a brief analogy, Florey is to be regarded as the founder and conductor of an orchestra, then Chain must be accorded the status of a soloist in a concerto of his own choosing. He had the brilliance of the true virtuoso, but it was a brilliance that needed to be directed and co-ordinated with the work as a whole. It seems almost certain that it was Chain's personal discovery of Fleming's paper that fired his enthusiasm for penicillin as a research project, an enthusiasm that he transmitted to Florey. But if it was chance that determined this choice, it was Florey's prepared mind that put the whole operation into effect. For years Florey had been moving in this direction through his work on lysozyme and his search for similar antibacterial substances in the animal body.

Chain's persuasion revealed another approach, and once Florey had recognized this, it was he who supplied the energy and the powers of co-ordination needed to follow the new line. Chain's pioneer biochemical work on penicillin was indispensable to the early progress of the project and his identification, with Abraham, of its molecular structure was a scientific *tour de force*.[12] But the project involved much more than biochemistry. It was an interlocking problem in logistics, like a military campaign in which the organization of the supplies, equipment, personnel, and reconnaissance needed for an advance are all interdependent. Chain depended on the support of such parallel activities for his own specialized work. Before Florey began to organize a full-scale effort, in 1939, Chain had worked for a few months with Epstein on penicillin, and had not made any significant progress. Thereafter, with ever-increasing supplies of crude material, he made rapid strides. But he could not have taken Florey's place, any more than Florey could have taken his. In 1942, while Florey was in America for three months, Chain was in charge of the production team and affairs were soon in sad disarray. By the time that Florey returned in October production had virtually come to a halt, and it needed his firm hand to restart it.[13]

On the production side there was the almost equally indispensable Heatley, on whose genius for invention and improvisation depended the supplies of semi-purified penicillin needed by everyone else in the team. And it was Heatley who was able to convey so much of his technical experience to the American penicillin producers and thus initiate the large-scale manufacture there that finally stimulated British firms to follow suit. Heatley has always preferred to remain in the background in any portrayal of the penicillin work. He is entitled to his dislike of publicity, but this must not be allowed to obscure the historical fact that he was a key figure in the early stages of the Oxford project. Gordon Sanders was an able understudy while Heatley was in America, and one who further developed the role. The work done by the other members of the team, though essential, might perhaps have been done equally well by other appropriate experts. Gardner and Miss Orr-Ewing did all that good bacteriologists could do. The clinical work was carried out with great skill and dedication by Ethel Florey and Charles Fletcher. The seemingly endless biological experiments were done by Florey himself and Dr. Margaret Jennings. Kent was invaluable, not only as the most skilled and efficient animal assistant but as Florey's tireless and devoted laboratory companion, responsive to every mood and anticipating every need. Finally when the Dunn School factory was at the height of its production, there was George Glister and his six 'penicillin girls', working steadily and cheerfully under dangerous and unpleasant conditions. Thus, in this team there were, in the later stages of the work, about fifteen to twenty people at any one time. It is to them, and in particular to their leader, that the world should acknowledge gratitude for the beginning of the antibiotic era.

Three people, however, can be seen as absolutely essential to the development of penicillin, in the sense that, in the absence of any one of them, the project could not have been started, and the antibiotic era would have been indefinitely delayed. These people are, in the chronological order of their contributions, Fleming, Chain, and Florey. The relative value of these contributions may remain a matter of unresolvable dispute, but when the 1945 Nobel Prize for Physiology and Medicine was divided between them, no one could dispute the justice of the choice.

In practical terms, penicillin represents the summit of Florey's scientific career, a summit that towers above the long mountain ranges of medical discovery and endeavour. Before we leave him there, as in this book we must, we should try to see the qualities that made him the most effective medical scientist since Lister. Scientists are, in general, not very different as people from non-scientists; they display the same variety of personalities, emotions, ambitions, and interests that one encounters in most other professions. And not all scientists are dedicated to research: for many, science means routine laboratory work or teaching. But scientists have in common a disciplined scepticism that demands concrete proof for any statement or belief concerning the material world. This is the legacy of the 'scientific method'—a phrase seldom used by scientists themselves, and a method practised by them instinctively rather than consciously. To be a great scientist needs the same sort of refinement and power that raises an artist, an engineer, or a statesman to the highest levels of achievement. Some scientists have become great men in the academic world, in government, or in public life without being great scientists. To be a great scientist one must (among other things) be a pioneer, a discoverer, a dedicated research worker, and one must be successful.

The qualities that made Florey the great scientist he was will have become apparent in the preceding chapters. Predominantly, perhaps, was that most basic one: objective honesty. Self-delusion—the unconscious misreading or selection of results to support some cherished hope or carefully constructed theory and a disregard for misfit facts—this is the greatest enemy of true progress. The literature of science is lumbered with theories—the beloved brainchildren of their creators—that are little more than dogma. Florey was quick to see those that could easily be demolished by experiment, and much of his earlier work consisted of such clearances. He was equally objective about his own researches, distrusting any ideas that could not be put to the test at once, and distrusting laboratory results until they were confirmed again and again. His characteristic attitude of scepticism and laconic understatement, so often and rather dauntingly applied to the work of others, was applied no less rigidly to his own work. But this was the negative side of his approach. On the positive side, he could see more quickly than most the really significant fact in a jumble of observations and pretentious ideas, seize upon it, and develop it.

A most important factor in Florey's practical success as a scientist was his sense of direction in research. His flair for choosing lines that led, not into blind alleys, but into wider and wider fields has been described as 'almost uncanny'.[14] The studies on the micro-circulation, inflammation, vascular pathology, the lymphatic system, the functions of the lymphocyte, mucus secretion, gastro-intestinal function, and lysozyme all preceded his work on penicillin and almost all continued to yield important results throughout the forty years of his active research life. But 'uncanny' is not a word favoured by scientists, and in this connection it conveys the idea of a mysteriously predictive intuition that most would reject.

Chance, of course, plays a very large part in the success or failure of scientific research; and once more it must be stressed that the apparently most fortunate scientist is often, in reality, the most critically observant. Good luck undoubtedly made some contribution to Florey's long list of successful research projects, but it is stretching the bounds of probability too far to suppose that he was merely lucky in happening to choose, among the countless attractive openings that are revealed at every step along any path in research, just those that would lead on and out into wider fields. It is easier to believe that he did possess an instinctive gift, recognizing pointers and indicators not obvious to less gifted people. We admit that other such gifts as artistic inspiration, musical or mathematical genius, or even the homing abilities of birds and animals, are at present inexplicable by the experimental psychologist or physiologist. Perhaps a sense of direction in research is of the same order and neither more nor less mysterious.

The importance of logic in research depends on the sort of work being done. Where the number of variables is small and the basic rules ('laws') are well established (as in some branches of physics or chemistry) the significance of a new observation can be logically deduced with fair confidence. But in biology even the number of variables is uncertain and the rules themselves have often to be discovered as the research proceeds. Deductive logic is thus unreliable and often comes to depend on statistical analysis and unsatisfactory shades of probability. The best biologist is not necessarily, therefore, the best logician, and we cannot entrust his research to a computer. The other aspect of the logical approach—induction— is not a matter of fixed rules and pure reason (despite the canons of John Stuart Mill and the faith of Almroth Wright). Induction requires an effort of the imagination, the creation of a picture of the world that *might* (not must) be true, and which can only be verified by experiment. Florey was essentially an experimentalist. 'If you do the experiment you may not be certain to get an answer,' Florey used to say, 'but if you don't do it you can be certain not to get one.' 'Do the experiment' was his motto, as it had been for John Hunter. From the results of previous work he would think out the experiment most likely to give useful results, and design it with care to combine the greatest economy of time, effort, and materials with the maximum of reliable information. He seldom took more than a single inductive step without verification, and he relied on multiple experiments to offset the inevitable variability of any biological result. But he always tried to achieve simple, clear-cut answers by breaking down a problem into simple clear-cut questions, and he disliked and distrusted experiments so variable that statistics were needed to evaluate their results.

Florey's success was also due to a prodigious industry coupled with supreme technical skill, and a concentration that would allow of no interruption. During the early years it was his habit to do an experiment every day, seven days a week, except for the annual holidays that were often partly spent in foreign laboratories. An experiment for Florey was not a matter of some test-tubes and reagents, or of a few sections and a microscope; it was usually a matter of animal surgery involving hours of operating and the most meticulous and exacting techniques. There are hundreds of such experiments carefully recorded in his notebooks. He acquired, thereby, a mass of experimental data and his published papers were based so soundly that he could be certain of his results and of the validity of conclusions that never went beyond the facts. This style of research is in contrast to the evanescent excursions of workers like Fleming on the one hand, and the profound originality of those like Sherrington on the other. Looking back over Florey's papers one seldom finds a major line of research that began with an original idea of his own. Some of his early work was suggested by Sherrington, and some by his recognition of fallacies in the published papers of other people. The idea of working on mucus secretion, however, prompted by his own gastritis, seems to have occurred to him *de novo*. But if Florey cannot be rated as a brilliant originator of ideas, he was the best and soundest builder of solid factual knowledge in British experimental pathology.

One of Florey's characteristics was his capacity to attract collaborators. He worked with them because of their ability and not necessarily because he liked them. 'I would work with the devil himself—if he were good enough,' he used to say. Though, as his letters have shown, he seems to have thought of himself as somehow isolated from his fellow-men, he never lacked people eager to work with him. Some stayed with him, or in his department, for the greater part of their lives, and many of his postgraduate students came back to him for further working visits whenever they could. At first he worked with one or possibly two collaborators on any one problem; later the idea of the team took shape. Lysozyme occupied five people, including Florey himself, though the association was a loose one. The lymphocyte project, involving eight graduates, was planned in every detail by Florey, and the penicillin work, of course, was a sustained and co-ordinated effort by a still larger group. What was the attraction that brought these people to Florey's laboratory and kept them there to do, in many cases, the best work of their careers with him? It was not the money or the security that he could offer them, since he had little to give of either. It was not the promise of an easy job: Florey bluntly promised them hard work, no certainty of success, and no future in his department if they failed. Nor was it any particular charm of manner: Florey until quite late in life was brusque, sometimes brutally direct, seldom complimentary, and never flattering. Nevertheless, it was Florey's personality that captured so many able young research workers, who had been initially attracted by his scientific reputation. What they found in him was an infectious vitality: great physical energy combined with an independence of mind that seemed to open mental windows and let the fresh air of realism clear away stuffy academic pomposities. Above all they gained something of his own attitude to research, the sense of purpose, the excitement of discovery (which he seldom expressed in words), and the will to work. And he gave them confidence. Despite his show of disparagement and pessimism, they felt that, like a good ship's captain, he was always in control, and knew exactly where he and they were going. Like a good captain, too, he treated his crew fairly and well. He knew (and cared) far more about their personal difficulties and troubles than most of them ever realized, and his offhand manner concealed an unsuspected sensitivity.

The main driving force in Florey's life was his pleasure in experimental research. This, in turn, had two sources. There was, first, the excitement of possible or actual discovery, the lure that is felt by every explorer or experimentalist. Florey, in his letters, expressed the thrill, almost the sense of awe, that he experienced when he realized that he was probably the first person ever to see some natural phenomenon. This involves the sense of priority which is often misunderstood or underestimated by non-explorers. 'Priority is a word that figures prominently in the thoughts and vocabulary of most contemporary scientists,' wrote Florey in 1963. 'Like geographical explorers of old, the scientists likes to be the first to make a discovery, the first to do something.'[15] The second factor in Florey's enjoyment of research was in the design and the technical performance of experiments. He loved overcoming natural difficulties, just as he had enjoyed defeating human adversaries in any form of competition or confrontation. He enjoyed practical planning, but his greatest pleasure was in putting his plans into effect in the laboratory. He preferred research ideas to be simple, but he was quite prepared to pursue them by elaborate and very difficult methods. He was a superb technician and the exercise of his skill gave him the same sort of satisfaction as that experienced by a master craftsman. With the thoroughness that had marked all his work from his schooldays onwards, he had prepared himself to acquire these techniques by studying with their established masters in Europe and America. And once he had perfected a technique he would apply it with tenacity. Florey, as Drury had remarked, was 'a great finisher'. Fleming was like a man who stumbles on a nugget of gold, shows it to a few friends, and then goes off to look for something else. Florey was like a man who goes back to the same spot and creates a gold mine.

Without the evidence of his letters to Ethel, it would be easy to be misled by the picture of his own character that he was at pains to present. Though socially friendly, and certainly amusing in a somewhat astringent way, Florey almost invariably kept his personal relations, even with people he had known and worked with for years, on a superficial level. Except for members of his own family and John Fulton, he did not use Christian names. Since in his professional life he seldom displayed obvious enthusiasm or optimism, the source of his drive and energy was mysterious for most of his colleagues. But we have seen a different character in the writer of the letters, and it is impossible to believe that this character was not still at the heart of his personality, however carefully concealed. The Florey who wrote to Ethel Reed was a sensitive young man, desperately anxious to succeed in research, deeply appreciative of the arts, but so obviously unsure of himself that, even then, one has to perceive his true feelings through a protective screen of understatement. He was lonely and craved the sympathetic companionship that he hoped to attain through marriage. When this failed to materialize he retreated even further into his emotional shell, and his need to justify himself had to be satisfied more and more through his work. Even here he feared disappointments, and he guarded himself against possibly painful experiences by a studied undervaluation of the chances. He kept potential friends at arm's length, where they would be unlikely to disappoint or hurt him, but he had a very real concern for people. He maintained publicly that the idea of 'suffering humanity' was not in his mind when he started work on penicillin, but it had been very much in his mind fifteen years before when he wrote (about a possible treatment for tuberculosis): 'One becomes rather lost in a maze at the thought of stopping the appalling thing of seeing young people maimed and wiped out while one can do nothing.' This emotional reaction surely reveals one of the hidden motives behind Florey's work. If so, few humane intentions have been more abundantly fulfilled.

EPILOGUE

The point at which we ended our story of Florey's emergence as one of the world's great scientists probably marks the summit of his practical achievement. Yet, at the age of forty-four, he was then less than half-way through his career and another twenty-five years of active life and great distinction still lay ahead. They were to be the years of the fruition of researches begun long before; of travel, important posts, and great institutions; of his part in a revolution in medicine and in the relations of government to science. Another book would be needed to describe them. Here, only the barest outline can be sketched to show the dimensions of the subject.

The first part of this sketch should emphasize the tremendous consequences of what is called the 'antibiotic era'. When Florey began his medical career, wards in every general hospital were filled with patients suffering from such infections as pneumonia, and its complications of empyema and lung abscess; children with osteomyelitis, mastoid infection, or meningitis; and patients with every sort of abdominal sepsis and wound infections so virulent that even an accidental needle-prick could mean fatal septicaemia. There were the isolation hospitals for infectious diseases, and the sanatoria for the thousands of tuberculous patients, most of whom would die there. Today, only the doctors and nurses old enough to remember the desperate battles they fought against infection can appreciate the antibiotic revolution. Medical students taking their final examinations within the past twenty-five years may never have seen a case of these previously common and deadly infective diseases—as many an elderly examiner used to discover to his astonishment. Now, of course, the examiners themselves are happily as inexperienced in this field as the candidates.

The effects of this revolution, which, though it began with the sulphonamides, was mainly created by penicillin, are complex. Though whole groups of diseases have virtually disappeared it is difficult to set precise figures on the quantitative effect of antibiotics because other factors, such as improvement in hygiene and preventive medicine, have also been at work. But the reduction in the mortality of previously dangerous infective diseases since 1930 is dramatic, and it can be said with confidence that the introduction of chemotherapy, and particularly the antibiotics, is the major cause. The number of people dying in Britain from the commoner infective diseases was 115,000 in 1930 while the comparable figure for 1960 was 24,000, these being mostly deaths from virus infections that are not sensitive to antibiotics.[1] Figures for the reduction of illness from non-fatal infections are difficult to obtain, but the improvement must be very large. As long ago as 1944, Ethel Florey and R. E. O. Williams found that in a series of thirty-five cases of infected hands treated locally with penicillin 1000 fewer working days were lost as compared with a similar number of cases not treated with penicillin.[2] Now, with the availability of systemic antibiotics, the improvement would be much greater, and prophylactic treatment would prevent such minor sepsis altogether.

The saving of the lives of millions of people has changed world health patterns and created (as major changes always do) a different set of problems. Among those saved, children and young adults predominate, and the number of people now surviving to old age has therefore increased. But in the diseases of old age—for example, coronary thrombosis, strokes, cancer, and arthritis—there has been no great improvement, and the absolute number of such cases has increased almost as dramatically as the number of cases of infection has fallen. Thus the general hospitals tend to become filled by elderly patients, and the sanatoria built to treat tuberculosis now mostly house people with mental illness. Florey, with his prevailing interest in the problem of world population, saw the dangers. Speaking in a recorded interview of the benefits of penicillin, he said, 'But then you've got the reverse side of the medal, because I'm accused of being partly responsible for the population explosion, which is one of the most devastating things that this world has got to face for the rest of this century.'[3]

Let us now return to Florey himself. From 1942, the problems of producing and using penicillin on a large scale occupied much of his energy for several years. Whereas in America commercial production was making rapid progress, in Britain it remained slow and difficult. The early clinical trials by Ethel Florey in several hospitals were limited to the amount of penicillin that could be produced at the Sir William Dunn School of Pathology. And when, in May 1943, Florey and Cairns went to the North African battle zone to carry out a trial on war wounds, supplies were so restricted that the local application of penicillin was the only practical method. Nevertheless, the results were so good, particularly during the invasion of Sicily in July 1943, that the American authorities gave the highest priority to penicillin production and consequently had enough for the Normandy campaign a year later. In December 1943, Florey and Sanders made a dangerous wartime journey to Russia as members of an Anglo-American Mission. Their purpose was to tell our Soviet allies of recent Western medical advances. Penicillin was the most important of these, and the Russians received not only technical and clinical data, but cultures of the mould and samples of the purified product.

After the war Florey and his colleagues continued to follow the lines opened up by penicillin. Despite its almost miraculous success, it was by no means the perfect antibiotic. It did not act against some important pathogenic bacteria, in particular those causing tuberculosis and the enteric fevers, and there was a disturbing tendency for certain initially sensitive organisms to become resistant to penicillin. Also, patients themselves could become allergic to it, and liable to have serious or even fatal reactions. So, in Oxford, the search was on for other natural antibiotics and samples of hundreds of different moulds, fungi, and likely bacteria were cultured and tested. A number of possibly useful antibiotics were discovered, but only two merited intensive study. One of these was micrococcin, which had an activity against the tubercle bacillus *in vitro*. Florey experimented for some years, but without success, on methods designed to overcome the resistance of tubercle bacilli to micrococcin *in vivo*. Meanwhile streptomycin, originally produced by Waksman in America, had been shown to cure tuberculosis when combined with a form of chemotherapy, a regime that has largely banished this terrible scourge from the Western world. The other new antibiotic studied at Oxford proved to be of first-rate importance. A culture of the mould *Cephalosporium* had been sent to Florey from Sardinia in 1948, and five years later Abraham and Newton isolated from it the crystalline substance they called cephalosporin C, which became the basis for later semisynthetic products of great antibiotic value, with a wider range of action than penicillin and without some of its drawbacks.

The vast extent of this antibiotic research in Florey's department is only partially reflected in the two massive volumes of *Antibiotics* which he and six of his colleagues published in 1949. Ethel Florey published the results of her clinical work in two volumes, the first appearing in 1952 and the second in 1957. Florey himself was also the author or co-author of twenty-nine original scientific papers on the subject, and scores of published lectures and review articles. As Mellanby had predicted, the Oxford achievements were recognized and applauded in medical and scientific circles (if not by the general public), and Florey, though he always insisted that the work was that of a team, was rightly honoured as the leader. He was knighted in 1944, and shared with Fleming and Chain the Nobel Prize for Physiology and Medicine in 1945. Chain left Oxford in 1948 to direct chemical microbiological research at the Institute of Health in Rome. His personal relations with Florey had deteriorated. He felt that, in the matters of the patenting of penicillin and the St. Mary's Hospital publicity campaign, Florey had been wrong not to act, and had therefore failed to acquire funds and prestige that would have provided equipment and autonomy on the scale that Chain now wanted, and which Florey could not give him. When it was discovered that British manufacturers had to pay United States royalties on the penicillin produced in Britain, other people too had blamed Florey for 'giving away' penicillin to the Americans. But he had had little choice in the matter. He went to America because no British firm was able or willing to take penicillin seriously in 1942. Patents had not been applied for because patenting was then against ethical medical principles, and it is doubtful if much commercial protection could have been afforded by any patents applicable to the penicillin process as it then existed. The American patents covered, in fact, new inventions in production technique made in America. However, the experience with penicillin changed the climate of official opinion on the patenting of medical discoveries and inventions, and was a factor in the setting up by the Government of the National Research Development Corporation to handle the commercial exploitation of such discoveries, not only in medicine, but in academic science generally. An example of the new attitude was the prompt patenting of the cephalosporin derivatives developed by Newton and Abraham at the School of Pathology, which earned royalties that did something to offset the early losses on penicillin. And on the positive side of the balance, too, is the range of semi-synthetic penicillin derivatives developed in Britain by the Beecham group with the help and advice of Chain.

Florey's scientific standing, which might well have suffered had he embarked on acrimonious public disputes over patents or publicity, grew steadily. He received honours at home and abroad, and many more invitations to lecture or preside at meetings and congresses than he could possibly accept. He was also involved in the frustrating committee work needed to develop and control full-scale antibiotic therapy in Britain. These occupations—the inevitable social and official rewards for any outstanding scientific success—were not much to his taste. He wrote in a letter to Mellanby: 'I could quite readily dispense with all the publicity and "honours" and such-like, in fact I would readily surrender it all for a little sustained peace of mind such as we had when working ... before the war.'[4] Perhaps he had forgotten the material difficulties of which he used to complain in those days, or perhaps he felt that these had been outweighed by the luxury of uninterrupted experimental work.

When, in the mid-1950s, Florey was able to turn back to the research lines that had lapsed for nearly fifteen years, it was probably with some nostalgic feelings. He took up once more the study of mucus secretion, of inflammation, vascular structure, and the function of the lymphocytes. But he could now do so from new technical levels. In the intervening years the electron microscope had been developed, enormously increasing the range of his studies. The introduction of various isotopic or electron-dense markers had made it possible to label cells and follow their migrations and transformations in the body, as the ornithologist studies his ringed birds. Such techniques enabled J. L. Gowans to solve a major haematological mystery, by showing that the supposedly vast rate of production and destruction of the lymphocytes does not exist. Gowans found that while lymphocytes pour continuously into the blood-stream through the main lymphatic vessels (thoracic ducts) they leave the blood-stream as rapidly by passing through the capillary walls into the tissue spaces and thence circle back again into the lymphatic vessels. This was, in its own field, another Harveian revolution. Gowans extended his work to demonstrate the function of the lymphocyte in producing antibodies, thus linking up with Medawar's work on the mechanism of immunity which has made transplant surgery possible. Gowans was elected a Fellow of the Royal Society in 1963, and subsequently became Secretary of the Medical Research Council.

Florey's work on vascular structure soon led him into the field of atherosclerosis, the arterial disease generally blamed for coronary thrombosis and strokes. Besides a good deal of personal experimental work in animals, he helped to found an active international discussion group which included cardiologists, physiologists, chemists, pathologists, epidemiologists, and dieticians. The regular meetings of this group led not only to exchanges of information but planned co-operative research, the sort of approach most likely to make useful advances in a field of almost impenetrable complexity. Florey was the leader in this group, contributing the hard facts of his own observations and his refreshing ability to deflate airy notions, which, in this subject, are particularly prevalent.

Despite his outside commitments, laboratory work remained Florey's greatest pleasure, and he would set aside days for experiments during which he would allow no interruption. Almost to the end of his life he was the most skilful animal surgeon in the department, and he became an adept electron-microscopist. Rhodes Scholars, D.Phil. students and research graduates eagerly sought a place and an unrivalled training in experimental science at the School of Pathology, and their fields of study extended widely in bacteriology, immunology, and the physio-pathology of vascular and gastro-intestinal function. On the walls of the main corridors are the photographs of scores of the graduates who have worked there. Most are shown as young men and women, and many of them have become famous in later life. Not all these pictures are of young people, however. Florey gave laboratory space, and a chance to continue their research and teaching, to mature workers on leave and to two old friends in their retirement, Sir Paul Fildes and Professor Robert Webb, so that both spent many happy years in useful work and congenial company.

Second only to experimental work, Florey's great pleasure was in travel. His international fame gave him opportunities to visit most of the world's interesting places, where he was an avid and highly informed sightseer, usually with some prior knowledge of the local history and customs. Naturally he was particularly appreciated in Australia, receiving inducements to return, including the offer of a State Governorship, which he did not feel able to accept. The Australian project which did occupy him for several years concerned the proposed foundation of a National University at Canberra. This idea, which had been put forward many years before, had not received much support from the State authorities, who had their own established universities to consider. During the war, however, Alfred Conlon (who had great influence both with John Curtin, the Prime Minister, and with General Blamey, the Australian Commander-in-Chief) and R. D. Wright revived the concept as a means of retaining the very able people being trained in the sciences and arts, and of bringing back eminent Australians who had gone abroad. Florey was one of these distinguished expatriates around whom such a centre might be built, and his views on medical research in Australia were sought in 1943. In the following year he visited Australia and discussed the matter with Robert Menzies (in the absence of John Curtin, who was ill) and General Blamey. As a result, Florey was asked to write a report and in this he suggested the foundation of a national school for medical research.

It seems that Florey's report precipitated the decision to found a National University, to include his proposed school. The death of John Curtin delayed the project, but it was adopted by the next Prime Minister, Joseph Chiffley, and the Enabling Act was passed in 1946. In 1948 Florey visited Australia as a member of the Academic Advisory Committee, which included Sir Mark Oliphant, Raymond Firth, and a friend from his early Oxford days, the historian Sir Keith Hancock. He then, with the help of Gordon Sanders, undertook the detailed design of what was to be called the John Curtin School of Medical Research, planned as their ideal laboratory. Much of this planning was done in Oxford, but they went to Canberra several times between 1950 and 1953, and were concerned with the building of the laboratories and of University House, residential quarters with some of the best features of an Oxford college.

From an early stage in this project, Florey's Australian colleagues urged him to accept the Directorship of the new John Curtin School. It is clear that he was strongly tempted. It would have been a triumphant return to the native country for which he felt loyalty and affection. And it would have been an exciting new venture that would give him (he was promised) all the research facilities he needed. But there were opposing factors. The reasons that had first taken him to England still operated to keep him there. In England he was at the centre of his scientific world; in Australia he would be on the periphery. He had, too, a loyalty to his colleagues in Oxford, and the team he had built up there; and changes of plan in Canberra were not to his liking. For the first few years he refused to commit himself on the grounds that the building was unfinished—he would not leave Oxford 'for a hole in the ground and a lot of promises'. Then he began to make conditions. He must, he said, be allowed to bring with him some of his Oxford colleagues, a condition reluctantly accepted, though it would raise difficulties not only in Canberra but in Oxford. Finally he proposed a temporary trial appointment for himself and these colleagues, and when this proved unacceptable he definitely refused the Directorship in 1957. But he continued his connection with the Australian National University. He received an honorary degree there and formally opened the John Curtin School in 1958, and became the University's third Chancellor in 1965.

In November 1960, Florey was elected President of the

Royal Society. The office confers the highest honour that can be awarded to a scientist in Britain, and also great influence not only in the world of science but in the wider relations of science to industry, education, and government. Florey's election surprised some people, and for different reasons. Most were merely surprised because he was the first pathologist and the first Australian to become President. Others, who knew him better, were doubtful on the grounds of personality. The President has high official duties to perform, and great social occasions to command, and Florey had not previously shown much liking for such events; nor, indeed, much of the urbane distinction with which these duties are usually conducted. But in the event, he carried out all these functions (as Medawar has said) superbly well, and in the opinion of the late Sir David Martin (for many years Executive Secretary) he was the most effective President the Royal Society had had.[5]

Florey became President in succession to Sir Cyril Hinshelwood, Professor of Physical Chemistry at Oxford, President of the Classical Association, linguist, scholar, mathematician, artist—in fact a man of almost legendary attainments. And the end of Hinshelwood's term of office was marked by the colourful celebrations of the Society's tercentenary that brought world-wide acclamation. A high standard had therefore been set for the new President, yet from the moment of his inaugural address and the handling of his first council meeting, it was clear that through Florey the Royal Society would gain a new vitality. As he had done at Sheffield and Oxford, he began his practical reforms quietly, and also with a modesty that quite disarmed those who would have resented the 'bush-ranging' methods that used to be attributed to him. Yet soon the strength of his leadership was felt, as Sir Lindor Brown, the Biological Secretary, was later to describe: 'He drove his Officers, persuaded his Council and cajoled his office staff until he had a team that was prepared to work as hard as the President.'[6] He had, in fact, instilled in all of them a common sense of purpose.

Florey's purpose for the Royal Society was a widening of its interests and the metaphorical opening of doors and windows. The passing of three centuries had made the scientific revolution respectable and the Royal Society had acquired the rather vitiated atmosphere of long-established authority. Though as receptive as ever to purely scientific change, as a body it had developed some stiffening of its joints. One of Florey's rejuvenating reforms was to change the attitude to applied science. Hitherto, 'pure' science—the pursuit of knowledge for its own sake—had been regarded as somehow existing on a higher plane than 'applied' science—the process of putting discoveries to practical use. Florey had often stressed that his own aim in research had been new knowledge, but his work had usually resulted in some practical advance, penicillin being the supreme example. Though he does not seem to have said so, he was really a scientific pragmatist, and the gist of the pragmatist philosophy is 'The truth is what works'. He had, therefore, a respect for applied scientists. He persuaded the Royal Society to increase the chances of election for engineers, industrial chemists and physicists, and those who had advanced the practice of medicine, veterinary science, and agriculture. He also introduced new subjects for consideration by the selection committees, including demography, which has a direct bearing on world population problems and their social consequences. To involve the Society more directly in research he raised funds to increase the number of Royal Society professorships (research posts held in University departments) from two to twelve, to set up study groups in various subjects (including population studies), and to institute new lecture subjects, including a series on behavioural science, and on technology, and to endow new awards and medals for distinguished work in applied science.

A matter of wider importance in which Florey was involved as President was the growing interdependence of science and the Government. As its technical complexity increased, scientific research demanded more expensive laboratories and equipment, and the services of highly trained full-time specialists. Private funds could no longer satisfy these demands, and in one way or another government money was being channelled into academic institutions and, of course, into the research establishments that the Government itself was creating. Science, in fact, was becoming increasingly dependent on government patronage; and Government, responsible for national defence, health, and the economy, was increasingly dependent on the practical application of scientific research. But government is essentially the exercise of national control, while science is essentially the international pursuit of natural knowledge and the freedom to publish what is discovered without regard to the consequences. A partnership between such diametrically opposed ideologies is bound to be an uneasy one, and workable only by compromise. Florey recognized that the Royal Society was in a key position to represent the interests of science and scientists in effecting such a compromise, and he played an important part in the setting up of joint councils and advisory committees with the Government during his presidency.

With his usual realism Florey recognized that these expanding interests were physically restricted by the lack of space in the elegant quarters occupied by the Society at Burlington House. Few of the Fellows realized quite how cramped these quarters were. The beautiful Meeting Room, Library, and Council Room were usually adequate for their needs, though on various formal occasions additional space had to be borrowed in the adjoining apartments of the Chemical Society on the one side and the Royal Academy on the other. But the Fellows had no idea of the deplorable conditions under which the executive and office staff of the Society had to work. The question of more space had been considered from time to time for several years, but without a decision, and it was Florey who procured not only the decision but the ways and means to carry it out. As usual he wasted little time in verbal persuasion. For example, at a particular Council meeting he announced that the business would consist of a complete tour of the Society's premises. Every Council member saw for himself basement rooms and attics jammed with desks, records, and office and printing machinery, inadequately lit and ventilated, where the Society's business and publishing work had to be done. As a result, every Council member became an ally in Florey's successful drive to acquire a more spacious home. The actual move to the magnificent building in Carlton House Terrace took place in 1967, after Florey's presidency had ended in 1965, but it was during his term of office that the idea was conceived, the very substantial funds raised, the building acquired, and the plans made. In its new quarters the Royal Society has room to breathe and grow. Its tall windows open on to The Mall and St. James's Park, and its doors are open to eager and active young scientists as they never have been before.

In 1965, Florey became a Life Peer as Baron Florey of Adelaide and Marston, and a member of the Order of Merit. His success as President of the Royal Society was accompanied by an obvious relaxation of his constrained manner and in an increased social confidence. Nevertheless it was another surprise for many of his colleagues when he was offered the Provostship of the Queen's College, Oxford, in 1962, and even more surprising when he accepted it. The post entailed his resignation from the Chair of Pathology, and the virtual end of his collaborative experimental work, though he was given personal laboratory space by Sir Lindor Brown, then recently appointed Waynflete Professor of Physiology. Florey, when asked why he was giving up his department to become head of a college, gave the typically prosaic reason that he could earn a salary for several more years as Provost than he could as a professor. But this was probably only one of his reasons. Another may well have been the challenge that the change of direction entailed. Since the seventeenth century, only three or four scientists (including Dame Janet Vaughan and Sir Henry Tizard) had become heads of Oxford colleges. Florey therefore saw his appointment as something of a test-case for the changing status of science in Oxford.

There are subtle and important differences between an old-established Oxford (or Cambridge) college and a science department or even an equally old-established science institution such as the Royal Society. The discipline of science is a bond between scientists even in quite different branches— they have the same attitude, use basically similar methods, understand each other's problems, and have clear objectives. They are united, in fact, by their work. A college is essentially a community of scholars that still retains something of the academic monasticism of past centuries. It has well-guarded traditions and an ethos that resists change. Yet its Fellows are often strongly individual, jealous of their independence and of their personal reputations as scholars, critics, wits, or eccentrics. They are united, in fact, less by their work than by the college itself. Such a group can be guided, perhaps led, but never driven.

Florey's experience of college life in Cambridge and Oxford had been happy. He had enjoyed his Fellowships at Caius and at Lincoln. They were small, friendly colleges in which scientists were not made to feel that they were beyond the pale of culture. But the Provostship of Queen's brought unfamiliar responsibilities and the highly critical attention of clever men, learned in every sort of academic field except science. If the new Provost had hoped to create some common sense of purpose among his Fellows, he must soon have been disappointed. College politics are essentially devious, and it was suspected that Florey's clear and simple proposals concealed deep and detrimental plans. The contrived meandering of a college meeting is a traditional brake on positive action, and Florey found himself frustrated to an almost maddening degree. But he did, after a year or two, establish good working relations and the sort of practical improvements that he had achieved in every previous appointment. Queen's gained the new Florey Building, with the aid of substantial benefactions that he worked hard to raise, and the creation of a European Visiting Studentships scheme. Florey gained a gracious style of living that he had not before experienced in the beautiful house and garden of the Provost's Lodging. He also gained the respect and affection of many of his Fellows and this was given a practical expression when they invited him to remain as Provost for an extended period beyond the statutory retiring age. And it seems that the results of the 'test-case' of his election were regarded as favourable by other Oxford colleges, since several of them have subsequently elected scientists as their heads.

During his later years, Florey's increasing responsibilities were not backed by a happy and stable domestic life. Ethel's health continued to deteriorate, and it was only her inflexible determination that enabled her to complete her extremely arduous work at four different hospitals on the clinical application of antibiotics in 1957. Her respiratory troubles were complicated by hypertension and incipient heart-failure and the special operation she underwent to relieve her deafness was not successful. In 1957, the Floreys had to leave No. 16 Parks Road, the house they had occupied for twenty-two years, because it was one of several in the area due to be pulled down to make way for new University developments. Florey had anticipated this move by buying a plot of land in the village of Old Marston, near Oxford, and making plans to build a small and attractive house. There was an interval, however, of two or three years before the Marston house was finished, and during this time Florey lived at Lincoln College and Ethel, by now seriously incapacitated, in Edinburgh with her daughter, who had married. She returned to Oxford when the Marston house was ready, and though she left it for the Provost's Lodging when Florey took up his post at Queen's, she did so, apparently, with reluctance. She was, in fact, too ill to carry out the social duties expected of the wife of the Provost, or of the President of the Royal Society, and she soon retired to Marston. Yet, despite a recent coronary heart attack, such was her determination that she insisted on travelling alone to America in 1965 to lecture, and then to Australia. It was a harrowing visit for her friends and relatives in Adelaide, Melbourne, and Canberra, since she could scarcely walk, and yet was determined to see everything connected with penicillin and Howard Florey. In Adelaide, her sister tried to persuade her to stay there to recuperate, but she returned to Oxford. In the spring of 1966 she again insisted on travel, this time to the wedding of her son Charles in America, against all advice and the express wishes of Florey. A few months later she died suddenly, on 10 October, at Marston.

Florey's own health was precarious during the last years of his life. He had made no secret of his chronic digestive troubles and minor ailments, but he told no one except his doctor that he had been suffering from angina since about 1950. It was noticed that he walked more slowly, used lifts rather than stairs, and had given up smoking. He also came to avoid the emotional strain of contentious disputes, with the result that he seemed to have mellowed considerably with advancing years. But he was well aware of the highly uncertain prognosis of his condition, and that he might not be considered for positions of responsibility if his health was known to be at risk. During the years of his Presidency of the Royal Society and his work for the Australian National University his energies and enthusiasm seemed undiminished, but he himself must have suspected that he was living on borrowed time.

In June 1967 Florey married Dr. Margaret Jennings, who had been his colleague and collaborator at the School of Pathology for over thirty years. Their marriage brought great happiness to both of them, but it was to be sadly brief. Florey died suddenly from a heart attack on 21 February 1968 at the Provost's Lodging. He was sixty-nine years of age. The funeral was in the small parish church almost exactly opposite his house at Marston. Later there was a memorial service at Westminster Abbey, a moving and impressive ceremony attended by hundreds of those who remembered him with gratitude. It was a gratitude that millions would have shared, had they known to whom they owed their lives or health.

SOURCES

The main sources of unpublished material are as follows:

1. The Florey archives (designated 98HF) collected and deposited at the Royal Society, Carlton House Terrace, London, by Lady (Margaret) Florey, to which access requires her permission. The material, which is catalogued, consists of 29 volumes of laboratory notes, 27 volumes of papers, and 315 boxes containing files of papers or letters. Each box, file, and item is numbered. Thus, 291, 22, 107 indicates item No. 107 in file No. 22 in box No. 291.
2. The Medical Research Council archives at Park Crescent, London, W.1. The Florey papers are arranged in chronological order in several volumes, each prefixed by the number 1752.
3. A small amount of relevant material has been consulted in the archives of Rhodes House and the Sir William Dunn School of Pathology, Oxford; the universities of Sheffield and Adelaide; and the Australian National Library, Canberra.
4. Extensive personal notes made by Lady (Margaret) Florey; notes and diaries kept by Dr. N. G. Heatley and Dr. Gordon Sanders, and also notes by Dr. Beatrice Pullinger, now living in South Africa.
5. Personal letters to the author from many of Lord Florey's relatives, friends, colleagues, contemporaries, and students.
6. Personal notes of the author's interviews in Oxford, London, Cambridge, Sheffield, Adelaide, Melbourne, and Canberra, with people who had known Lord Florey.
7. Notes made by Professor E. P. Abraham for his own Biographical Memoir of Lord Florey (1971), which he kindly made available.

REFERENCES

References to published material follow the usual form. For scientific papers, the title of the journal, abbreviated in the standard way where possible, is followed by the volume number (bold type), page number, and year of publication.

Chapter 1

1. Blackett, P. M. S. and Coombs, H. C. Memoiral to the Rt. Hon. Lord Florey, O.M., F.R.S. Pamphlet issued by the Florey Memorial Appeal Committee of the Australian National University, Canberra, and the Royal Society, London 14 February 1969.
2. McKie, D. Organ of the new philosophy. In *The Royal Society Tercentenary*. p. 22. The Times Publishing Co., London (1961).
3. Dobson, Jesse. *John Hunter*. p. 152. E. & S. Livingstone, Edinburgh and London (1969).
4. Scott, E. L. 'Edward Jenner, F.R.S., and the cuckoo.' *Notes Rec. R. Soc. Lond.* **28,** 235 (1974).
5. Jenner, Edward. *The origin of the vaccine inoculation.* D. N. Shury, London (1801).
6. Crookshank, E. M. *The history and pathology of vaccination.* H. K. Lewis, London (1889).
7. Gladstone, G. P. and Abraham, E. P. Acquired immunity. In *Lectures on general pathology* (ed. H. W. Florey.) p. 409. Lloyd-Luke, London (1954).
8. Chick, H., Hume, M., and Macfarlane, M. *War on disease.* p. 20. Andre Deutsch, London (1971).
9. Duncan, W. G. K. and Leonard, R. A. *The University of Adelaide, 1874–1974.* Rigby, Adelaide (1975).
10. Dubois, R. J. *Louis Pasteur.* Gollancz, London (1951).
11. W. W. C. 'Lord Lister.' Obituary notice of Fellows Deceased. *Proc. R. Soc.* **36,** 1 (1913).
12. Godlee, R. J. *Lord Lister.* Macmillan, London (1917).
13. (*a*) Fraser-Moodie, W. *Proc. R. Soc. Med.* **64,** 87 (1971); (*b*) Florey, H. W. *Canad. Med. J.* **71,** 417 (1954); (*c*) Guthrie, D. J. *Lord Lister: his life and Doctrine,* E. & S. Livingstone, Edinburgh (1949).
14. Sherrington, C. S. First use of diphtheria antitoxin made in England. *Notes Rec. R. Soc. Lond.,* **5,** 156 (1948).
15. Hill, C. *Reformation to Industrial Revolution.* Penguin Books, Harmondsworth. (1959).
16. Atlay, J. B. *Sir Henry Wentworth Acland.* Smith, Elder, London (1903).
17. Robb-Smith, A. H. T. *A short history of the Radcliffe Infirmary.* Church Army Press, Oxford (1970).

18. Vernon, H. M. and Vernon, K. D. *A history of the Oxford Museum.* Clarendon Press, Oxford (1909).

19. Darwin, F. *Life of Charles Darwin.* p. 237. John Murray, London (1902).

20. Burdon-Sanderson, Lady, Haldane, J. S., and Haldane, E. S. *A memoir of John Burdon-Sanderson.* Clarendon Press, Oxford (1911).

21. Franklin, K. J. *Ann. Sci.* **1,** 431 (1936).

22. Sinclair, H. M. In *Oxford medicine* (ed. K. E. Dewhurst). p. 1. Sandford Publications, Oxford (1970).

Chapter 2

1. Goadby, Mrs. Joan. Personal communication, Standlake, Oxfordshire (1977).

2. Gardner, Dr. Joan. Personal communication, Melbourne (1975).

3. Oborn, P. and Chinner, C. *Mitcham Village Sketchbook.* Rigby, Adelaide (1974).

4. Bowen, Mrs. Mollie. Personal communication, Adelaide (1975).

5. Florey, Mr. James. Personal communication, Standlake (1977).

6. See Chap. 1, Note 9. Note: some of the references to Professor Watson are derived from the oral tradition still current in Adelaide.

7. Florey, H. W. Interview recorded by Mrs. de Berg for the National Library of Australia, Canberra, on 5 April 1967.

8. Florey family papers in the possession of Dr. Joan Gardner, Melbourne.

9. University of Adelaide archives.

10. Florey, H. W. Unpublished letters to Mary Ethel Hayter Reed. Royal Society archives, 98HF, 287, 1, 1–55 (1922); 2, 1–13 (1923); 3, 1–31 (1924); 4, 1–26 (1925); 5, 1–27 (1926).

11. Wylie, Francis. In *The First fifty years of the Rhodes Trust and the Rhodes Scholarships.* Blackwell, Oxford (1955).

Chapter 3

1. Florey, Anne. Letter to her mother, 25 January 1922. Florey papers in the possession of Dr. Joan Gardner, Melbourne.

2. See Chap. 2, Note 11.

3. Liddle, E. G. T. 'Charles Scott Sherrington.' *Obit. Not. Fell. R. Soc. Lond.* **8,** 241 (1952–3).

4. Sherrington, C. S. *The assaying of Brabantius and other verse.* Oxford University Press (1925).

5. See Chap. 1, Note 22.

6. Clark, R. W. *Tizard.* p. 10. Methuen, London (1965).

7. Sherrington, C. S. Letter to H. W. Florey, 12 July 1923. Royal Society archives, 98HF, 290, 2, 1.

8. Merton College records, Oxford.

9. Abraham, E. P. Personal papers.

Chapter 4

1. Binney, George. In *Spitsbergen papers.* Vol. 2, p. 23. Oxford University Press (1929).

2. Elton, Charles. Letter to Professor E. P. Abraham, 30 September (1968).
3. Young, F. G. *The rise of biochemistry in the nineteenth century.* Vol. 20. Publications of the Wellcome Institute of the History of Medicine, London (1971).
4. Florey, H. W. *J. Physiol.* **59,** 83P (1925).
5. Florey, H. W. Letter to his mother, 6 April 1925. Florey papers in the possession of Dr. Joan Gardner, Melbourne.
6. Florey, H. W. *Brain,* **48,** 43 (1925).
7. Florey, H. W. *J. Path. Bact.* **28,** 645 (1925).
8. Stephens, J. G. and Florey, H. W. *Br. J. exp. Path.* **6,** 269 (1925).

Chapter 5

1. Florey, H. W. Letter to Francis Wylie, 8 November 1925. Rhodes House archives, Florey file.
2. See Chap. 4, Note 8.
3. Florey, H. W. and Carleton, H. M. *J. Path. Bact.* **29,** 97 (1926).
4. Reed, Ethel. Letters to H. W. Florey, 1925–6. Royal Society archives, 98HF, 288, 2, 1–34.
5. Lamphee, Dr. Alan. Personal communication, Adelaide (1975).
6. Testimonial from a Nursing Sister (for Dr. Ethel Reed). Royal Society archives, 98HF, 288, 1, 16.
7. Heller, A. *Centralblt. Med. Wiss.* p. 545 (1869).

Chapter 6

1. Bulloch, W. and Fildes, P. G. Haemophilia. In *Treasury of human inheritance.* Dulau, London (1911).
2. Gladstone, G. P., Knight, B. C. J. G., and Wilson, G. S. 'Paul Gordon Fildes.' *Biogr. Mem. Fellows R. Soc.* **19,** 317 (1973).
3. Clark-Kennedy, A. E. '*The London.*' Pitman, London, Vol. 1 (1962), Vol. 2 (1963).
4. Florey, H. W. and Witts, L. J. *Lancet,* **1,** 1323 (1928).
5. Carleton, Dr. Alice. Personal communication, Oxford (1976).
6. Burrin, James. Letters to Anne Florey, 1902–4. In the possession of Dr. Joan Gardner, Melbourne.
7. Carleton, Dr. Alice. Personal letter to the author. 20 August (1976).
8. Florey, H. W. Memorandum to Ethel Florey (undated, but probably September 1934). Royal Society archives, 98HF, 288, 3, 1.

Chapter 7

1. Florey, H. W. and Carleton, H. M. *Proc. R. Soc.* **B.100,** 23 (1926).
2. Florey, H. W. *Proc. R. Soc.* **B.100,** 269 (1926).
3. Florey, H. W. and Carleton, H. M. *J. Path. Bact.* **29,** 97 (1926).
4. Florey, H. W. *J. Physiol.* **62,** 267 (1927).
5. Witts, Prof. L. J. Personal communication, Oxford (1976).
6. Drury, Sir Alan. Personal communication, Cambridge (1976).
7. Florey, H. W. Laboratory notebooks. Royal Society archives, 98HF 1–7.
8. Florey, H. W. *J. Physiol.* **63,** 1 (1927).

9. Florey, H. W. and Fildes, P. *Brit. J. exp. Path.* **8**, 393 (1927).
10. Florey, H. W. The physiology and pathology of the circulation of the blood and lymph. Thesis submitted for the degree of Ph.D., Cambridge (1927).
11. Abraham, E. P. 'Howard Walter Florey.' *Biogr. Mem. Fellows R. Soc.* **17**, 255 (1971).
12. Kent, Mr. James. Personal communication, Oxford (1976).
13. Webb, Prof. R. A. Personal communication, Oxford (1976).
14. Florey, H. W. Letter to Francis Wylie, 30 September 1928. Rhodes House, Oxford, archives, Florey file.
15. Wilkinson, L. P., *Cambridge University Reporter*, **1961–2**, p. 1960.
16. (a) Florey, H. W. and Marvin, H. M., *J. Physiol.* **64**, 318 (1928); (b) Florey, H. W., Marvin, H. M., and Drury, A. N., *J. Physiol.* **65**, 204 (1928).
17. Barcroft, J. and Florey, H. W. *J. Physiol.* **66**, 231 (1928).
18. Florey, H. W. and Witts, L. J. *Lancet*, **1**, 1323 (1928).
19. Drury, A. N., Florey, H. W., and Florey, M. E. *J. Physiol.* **68**, 173 (1929).
20. Barcroft, J. and Florey, H. W. *J. Physiol.* **68**, 181 (1929).
21. Florey, H. W. and Carleton, H. M. *J. Obstet. Gyn.* **38**, 550, 558 (1931).
22. Florey, H. W., Szent-Györgyi, A., and Florey, M. E. *J. Physiol.* **67**, 343 (1929).
23. Williamson, R. Early history of the Department of Pathology at Cambridge. In *Cambridge and its contributions to medicine*. Publications of the Wellcome Institute of the History of Medicine, Vol. 20 (1971).
24. See Note 12.
25. Ayer, A. J. *Metaphysics and common sense.* p. 172. Macmillan, London (1969).
26. Britton, Karl. Portrait of a philosopher. *The Listener*, **53**, 1072 (1955). Quoted by G. Pitcher, *The philosophy of Wittgenstein.* p. 12. Prentice-Hall, London (1964).

Chapter 8

1. Florey, H. W. Laboratory notebook. Royal Society archives, 98HF, 2.
2. Goldsworthy, N. E. and Florey, H. W. *Brit. J. exp. Path.* **11**, 192 (1930).
3. Colebrook, L. 'Almroth Edward Wright.' *Obit. Not. Fell. R. Soc. Lond.* **6**, 297 (1948–9).
4. Maurois, André. *Fleming: The man who cured millions.* Methuen, London (1961).
5. Hare, Ronald. *The birth of penicillin.* p. 57. Allen and Unwin, London (1970). Quotations occurring later in this chapter will be found on pages 62, 34, 35, 64, 98–9, 147, and 140, in that order.
6. Colebrook, L. 'Alexander Fleming.' *Biogr. Mem. Fellows R. Soc.* **2**, 117 (1956).
7. Wright, A. E. *Alethetropic logic.* Heinemann, London (1953).
8. Fleming, A. *Brit. J. exp. Path.* **10**, 226 (1929).
9. Fleming, A. *Proc. R. Soc. Med.* **34**, 342 (1941).

10. Fleming, A. *J. R. Inst. Pub. Health and Hyg.* **8,** 36, 63, 93 (1945).
11. Jennings, M. A. in *Antibiotics* p. 1152. Oxford University Press (1949) (see Chap. 12, Note 16).
12. Birkinshaw, J. H. Harold Raistrick. *Biogr. Mem. Fellows R. Soc.* **18,** 489 (1972).
13. Clutterbuck, P. W., Lovell, R., and Raistrick, H. *Biochem. J.* **26,** 1907 (1932).
14. Reid, R. *J. Bact.* **29,** 215 (1935).

Chapter 9

1. Florey, H. W. *Brit. J. exp. Path.* **11,** 251 (1930).
2. Miles, Sir Ashley. Personal communication, London (1975).
3. Sherrington, C. S. Letter to H. W. Florey, 12 August 1928. Royal Society archives, 98HF, 290, 2, 15.
4. Anderson, H. K. Letter to H. W. Florey, 8 October 1928. Royal Society archives, 98HF, 290, 6.
5. Harding, Dr. H. E. Personal communication, Whiteparish, Hants (1976).
6. Chapman, A. W. *The story of a modern university.* p. 364. Oxford University Press (1955).
7. Peters, R. A. 'J. B. Leathes.' *Biogr. Mem. Fellows R. Soc.* **4,** 185 (1958).
8. Dale, H. H. 'Edward Mellanby.' *Biogr. Mem. Fellows R. Soc.* **1,** 193 (1955).
9. See Chap. 2, Note 7.
10. Pullinger, Dr. Beatrice. Personal letter to the author, 8 August 1976.
11. Kent, Mr. James. Personal communication, Oxford (1977).
12. Florey, H. W. Application to the M.R.C., 4 February 1932. Medical Research Council archives, London, file No. 1752, Vol. 1.
13. Poulton, E. P. Letter to H. W. Florey, 27 February 1933 (and subsequent letters on the same subject). Royal Society archives, 98HF, 6, 1–22.
14. Pullinger, Dr. Beatrice. Personal letter to the author, 6 April 1976.
15. Hall, A. J. Testimonial for H. W. Florey, December 1934. Royal Society archives, 98HF, 291, 5, 6.
16. Peters, Sir Rudolph. Personal letter to the author, 5 July 1976.
17. Florey, H. W. *Brit. J. exp. Path.* **13,** 349 (1932).
18. Florey, H. W. Letter to W. M. Fletcher, 30 May 1932. Application to M.R.C., 23 June 1932. Letter to Fletcher, 2 December 1932, and Fletcher's reply, 5 December 1932. Medical Research Council archives, London, file No. 1752, Vol. 1.
19. Fleming, A. *Proc. R. Soc. Med.* **24,** 808 (1931).
20. Paine, C. G. In *Antibiotics.* Oxford University Press (1949). p. 634. See Chap. 12, Note 16.
21. Harding, Dr. H. E. Personal letter to the author, 6 March 1976.
22. Dale, H. H. Letter to H. W. Florey, 25 April 1934. Royal Society archives, 98HF, 4.
23. Florey, H. W., Harding, H. E., and Fildes, P. *Lancet,* **2,** 1036 (1934).

Chapter 10

1. See Chap. 2, Note 7.
2. Harding, Dr. H. E. Letter to the author, 6 March 1976.
3. Pullinger, Dr. Beatrice. Letter to the author, 20 May 1976.
4. Douglas, S. R. 'Georges Dreyer.' *Obit. Not. Fell. R. Soc.* **1**, 569 (1932–5).
5. Gardner, A. D. and Stewart, M. J. 'Ernest William Ainley-Walker.' *J. Path. Bact.* **71**, 293 (1956).
6. See Chap. 1, Note 22.
7. Obituary notice. 'Georges Dreyer.' *J. Path. Bact.* **39**, 707 (1934).
8. Sherrington, C. S. Letters to H. W. Florey, 3 and 26 October 1934. Royal Society archives, 98HF, 290, 2, 24–5.
9. Pullinger, Dr. Beatrice. Personal notes (1976).
10. Testimonials for H. W. Florey, December 1934. Royal Society archives, 98HF, 291, 5, 1–8.
11. Florey, H. W. Letter to the Warden of Rhodes House, 25 January 1935. Rhodes House archives, Florey file.
12. Florey, M. E. Letter to Valetta Florey, 22 January 1935. In the possession of Dr. Joan Gardner, Melbourne.
13. Harding, Dr. H. E. Personal communication, Whiteparish (1976).
14. See Chap. 7, Note 11.
15. Gardner, A. D. Letter to the author, 29 May 1976.
16. Gardner, A. D. Unpublished memoirs, 'Some recollections'. The Library, University College, Oxford.
17. Florey, H. W. Letter to John Fulton, 15 October 1937. Royal Society archives, 98HF, 289, 3, 2.
18. Cairns, Hugh. Letter to H. W. Florey, 22 November 1935. Royal Society archives, 98HF, 291, 8, 17.
19. Obituary notice. 'Sir Hugh Cairns.' *Br. med. J.*, 26 July 1952, p. 233.
20 (a) Cooke, A. M. *Sir E. Farquhar Buzzard.* p. 28. Church Army Press, Oxford (1975); (b) Robb-Smith, A. H. T. *A short history of the Radcliffe Infirmary.* p. 157. Church Army Press, Oxford (1970).
21. Florey, H. W. Letter to Sir Jeffrey Jefferson, 1 July 1952. Royal Society archives, 98HF, 141, 5, 10. (Also much correspondence on the Nuffield scheme filed in Box 141.)
22. Buzzard, E. F. *Br. med. J.* **2**, 163 (1936).

Chapter 11

1. Florey, H. W. and Harding, H. E. (a) *J. Path. Bact.* **40**, 211 (1935); (b) *Proc. R. Soc.* **B.117**, 68 (1935); (c) *Q. J. exp. Physiol.* **25**, 329 (1935).
2. See Chap. 7, Note 11.
3. Gardner, A. D. Letter to the author, 30 May 1976.
4. Gowland Hopkins, F. Letter to H. W. Florey, 28 May 1935. Royal Society archives, 98HF, 291, 8, 15.
5. Chain, E. B. *J. R. Coll. Physicians, Lond.* **6**, 103 (1972).
6. Kent, Mr. James. Personal communication, Oxford (1975).
7. Pullinger, Dr. Beatrice. Letter to the author, 8 August 1976.
8. Author's personal recollection.

9. Medawar, Sir Peter. Letter to the author, 20 August 1976.
10. Gowans, J. L., *Br. J. exp. Path.* **38,** 67 (1957).
11. Sanders, Dr. A. G. Letter to the author, 1 October 1976.
12. Ebert, R. H., Florey, H. W., and Pullinger, B. D. *J. Path. Bact.* **48,** 79 (1939).
13. Gardner, Dr. Joan. Personal communication, Melbourne (1975).
14. McMichael, Mrs. Paquita. Letter to the author, 10 August 1978.
15. Wright, Prof. R. D. Letter to the author, 1 November 1976.
16. Florey, H. W., Jennings, M. A., Jennings, D. A., and Castro-O'Connor, R., *J. Path. Bact.* **49,** 105 (1939).

Chapter 12

1. See Chap. 10, Note 20 (a).
2. (a) Florey, H. W. and Barry, H. C. *Lancet,* **2,** 728 (1936); (b) Pullinger, B. D. and Florey, H. W. *J. Path, Bact.* **45,** 157 (1937).
3. Sanders, A. G., Florey, H. W., and Barnes, J. M. *Br. J. exp. Path.* **21,** 254 (1940).
4. See Chap. 10, Note 3.
5. Baker, Prof. J. R. Letter to the author, 30 June 1976.
6. Abraham, E. P. and Robinson, R. *Nature, Lond.* **140,** 24 (1937).
7. See Chap. 11, Note 15.
8. Gibson, W. C. Letter to Lady (Margaret) Florey, 27 October 1970
9. Chain, E. B. and Duthie, E. S. *Br. J. exp. Path.* **21,** 324 (1940).
10. Epstein, L. A. and Chain, E. B. *Br. J. exp. Path.* **21,** 339 (1940).
11. Chain, E. B. *J. R. Coll. Physicians, Lond.* **6,** 103 (1972).
12. Florey, H. W. *Mem. Proc. Manch. lit. phil. Soc.* **86,** 191 (1945).
13. Papacostas, G. and Gaté, J. *Les Associations Microbiennes.* Doin et Cie, Paris (1928).
14. Paine, Dr. C. G. Letter to Dr. N. G. Heatley, 30 June 1977.
15. See Chap. 2, Note 7.
16. Florey H. W., Chain, E. B., Heatley, N. G., Jennings, M. A., Sanders, A. G., Abraham, E. P., and Florey, M. E. *Antibiotics. A survey of penicillin, streptomycin, and other antimicrobial substances from fungi, actinomycetes, bacteria, and plants.* Oxford University Press (1949).
17. Pasteur, L. and Joubert, J. F. *C.R. Acad. Sci., Paris,* **85,** 101 (1877).
18. See Chap. 1, Note 13 (c).
19. Florey, H. W. *Canad. Med. Assoc. J.* **71,** 417 (1954).
20. Meneces, A. N. T. Letter to *The Times,* London, 21 July 1977.
21. Florey, H. W. Application to the M.R.C., 27 January 1939. Medical Research Council archives, London, file No. 1752, Vol. 1.
22. Heatley, Dr. N. G. Personal communication, Oxford (1977).
23. Falk, L. A. (formerly Epstein). *J.A.M.A.* **124,** 1219 (1944).
24. Hare, Ronald. *The birth of penicillin.* p. 103. Allen and Unwin, London (1970).
25. Bickel, Lennard. *Rise up to life.* p. 65. Angus and Robertson, London (1972).
26. Masters, David. *The miracle drug.* p. 73. Eyre and Spottiswoode, London (1946).

27. The author's personal recollection.
28. Thomson, Mrs. J. R. Personal communication, Adelaide (1975).
29. Brebner, Mrs. E. Personal communication, Adelaide (1975).

Chapter 13

1. Florey, H. W. and others, *Q. J. exp. Physiol.* **27,** 381 (1938); **28,** 115 (1938); **28,** 207 (1938); **29,** 303 (1939). *J. Path Bact.,* **48,** 79 (1939); **49,** 105 (1939). *J. Physiol.* **96,** 13P (1939). *Br. J. exp. Path.* **20,** 342 (1939).
2. Florey, H. W. and Mellanby, E. Correspondence, June–September 1939. Medical Research Council archives, London, file No. 1752, Vol. 1.
3. Mellanby, E. and Florey, H. W. Correspondence, July 1939. Royal Society archives, 98HF, 36, 6, 3–4.
4. Mellanby, E. Letter to H. W. Florey, 22 October 1937. Royal Society archives, 98HF, 291, 8, 18.
5. Dubos, R. J. *J. exp. Med.* **70,** 1, 11 (1939).
6. Chain, E. B. Oliver Sharpey Lecture. In *Symposium on advanced medicine* (ed. N. Compston). Royal College of Physicians, London (1964); Pitman Medical Publications, London (1965).
7. Sir William Dunn School of Pathology, Oxford, archives.
8. Florey, H. W. The Dunham Lectures, Boston (1965). Manuscript in the possession of the Dean of Harvard Medical School.
9. Heatley, Dr. N. G. Personal communication, Oxford (1975).
10. See Chap. 12, Note 24.
11. Bickel, Lennard. *Rise up to life.* p. 91. Angus and Robertson, London (1972).
12. Chain, Sir Ernst. Personal communication, London (1978).
13. Burn, Prof. J. H. Letter to the author, 2 July, 1976.
14. Gardner, A. D. *Nature, Lond.* **146,** 837 (1940).
15. Abraham, E. P. and Chain, E. B. *Nature, Lond.,* **146,** 837 (1940).
16. Heatley, N. G. Personal notes, Oxford.
17. Florey, Lady (Margaret). Note on the author's manuscript (1977).
18. Clarke, H. T., Johnson, J. R., and Robinson, R. *The chemistry of penicillin.* Pergamon Press, New Jersey (1949).
19. Bickel, Lennard. *Rise up to life.* p. 115. Angus and Robertson, London (1972). But Mrs. Paquita McMichael in a letter to the author, 10 August 1978, does not agree that she used the quoted wording.
20. Cutting from a Canadian newspaper. Royal Society archives, 98HF, 291, 2, 4.
21. Fulton, John. Letter to H. W. Florey, 24 July 1940. Royal Society archives, 98HF, 291, 2, 1.
22. Heatley, Dr. N. G. Personal communication, Oxford (1978).
23. Chain, E. B., Florey, H. W., Gardner, A. D., Heatley, N. G., Jennings, M. A., Orr-Ewing, J., and Sanders, A. G. *Lancet,* **2,** 226 (1940).
24. Landsborough-Thomson, A. *Fifty years of medical research.* Vol. 2, p. 53. H.M.S.O., London (1973).

Chapter 14

1. Abraham, E. P., Chain, E. B., Fletcher, C. M., Gardner, A. D., Heatley, N. G., Jennings, M. A., and Florey, H. W. *Lancet*, **2**, 177 (1941).
2. Bickel, Lennard. *Rise up to life.* p. 121 (Oxford trial); p. 124 (New York trial). Angus and Robertson, London (1972).
3. Fletcher, Dr. C. M. Letter to the author, 4. November 1976.
4. Florey, H. W. and Mellanby, E. Correspondence, April 1941. Medical Research Council archives, London, file No. 1752, Vol. 1.
5. Chain, Sir Ernst. Personal communication, London (1978).
6. Heatley, Dr. N. G. Personal notes, Oxford.
7. Hare, Ronald. *The birth of penicillin.* p. 173. Allen and Unwin, London (1970).
8. Florey, H. W. Letter to John Fulton, October 1941. Royal Society archives, 98HF, 289, 3, 3.
9. Wilson, David. *Penicillin in perspective.* p. 222. Faber and Faber, London (1976).
10. Florey, H. W. Letter to Sir Charles Sherrington, 2 August 1942. Sherrington Collection, University of British Columbia.
11. Florey, M. E. and Florey, H. W. *Lancet*, **1**, 387 (1943).

Chapter 15

1. Fleming, A. *Br. J. exp. Path.* **10**, 226 (1929).
2. Fleming, A. *Br. dent. J.* **52**, ii, 105 (1931).
3. Florey, M. E. and Florey, H. W. *Lancet*, **1**, 387 (1943).
4. Unsigned article in *The Retail Chemist*, September 1941.
5. Maurois, André. *The life of Alexander Fleming.* Jonathan Cape, London (1959).
6. Hughes, W. H. *Alexander Fleming and penicillin.* Priory Press, London (1974).
7. Wilson, David. *Penicillin in perspective.* p. 235. Faber and Faber, London (1976).
8. Florey, H. W. Letter to Sir Henry Dale, P.R.S., 11 December 1942. Royal Society archives, 98HF, 38, 3, 1.
9. Dale, H. H. Letter to H. W. Florey, 17 December 1942. Royal Society archives, 98HF, 38, 3, 2.
10. Florey, H. W. Letter to E. Mellanby, 19 June 1944. Royal Society archives, 98HF, 36, 4, 107.
11. Mellanby, E. Letter to H. W. Florey, 30 June 1944. Royal Society archives, 98HF, 36, 4, 108.
12. See Chap. 13, Note 18.
13. Florey, H. W. Letter to N. G. Heatley, 9 October 1941. Royal Society archives, 98HF, 46, 1, 4.
14. Obituary notice. 'Lord Florey.' *The Times*, London, 23 February 1968.
15. Florey, H. W. *Nature, Lond.* **200**, 397 (1963).

Epilogue

1. Chain, E. B. *Nature, Lond.*, **200**, 441 (1963). Note: the numbers of deaths quoted in the text of the Epilogue have been calculated from the figures in Table 1 of this paper.
2. Florey, M. E. and Williams, R. E. O. *Lancet*, **1**, 73 (1944).
3. Florey, H. W. Tape recording. See Chap. 2, Note 7.
4. Florey, H. W. Letter to E. Mellanby. See Chap. 15, Note 10.
5. Martin, Sir David. Personal communication, London (1975).

INDEX

1 (*a*) Coreega. The Florey family home from 1906 to 1920. (From a sketch made *c.* 1970 by Pamela Oborn. Reproduced from *Mitcham village sketchbook* by P. Oborn and C. Chinner, by kind permission of Messrs. Rigby Ltd., Adelaide.)

1 (*b*) The Florey family outside their country house, Nunkerri (in Belair). Joseph at the wheel of his Darracq, with Anne, Valetta, Howard, and Hilda in the back. Bertha and Charlotte on the veranda. *c.* 1907.

2 (a) Howard Florey aged 7.

2 (b) Howard Florey, medical officer aboard the S.S. *Otira*, December 1921.

2 (c) A family group at Coreega: Joseph and Bertha Florey with their daughters.

3 Ethel Hayter Reed *c.* 1920.

4 (a) Howard Florey, medical officer aboard the S.S. *Polar Bjorn*, August 1924.

4 (b) Ethel Florey with Charles, and Phyllis the dog, *c*. 1935.

5 Howard Florey with Charles and Paquita, *c.* 1935.

6 Dr. Margaret Jennings (later Lady Florey).

7 (*a*) Dr. Norman Heatley.

7 (*b*) Dr. Gordon Sanders.

8(a) Professor A. D. Gardner.

8(b) Sir Ernest Chain.

physiology = science of functioning
of living organisms.

virus = disease producing organism.

bacteria = microscopic organism : some
producing disease.

obstetrics = of childbirth etc

pathology = science of diseases.